To Mark,

You're not only a good scientist, but a good guy as well.

Enjoy this

Warmest Regards

Einul

MOLECULAR
SPECTROSCOPY
WORKBENCH

MOLECULAR SPECTROSCOPY WORKBENCH

Advances, Applications, and Practical Advice on Modern Spectroscopic Analysis

Emil W. Ciurczak

Contributing Editor
Spectroscopy Magazine

A WILEY-INTERSCIENCE PUBLICATION

JOHN WILEY & SONS, INC.

New York / Chichester / Weinheim / Brisbane / Singapore / Toronto

This book is printed on acid-free paper. ∞

Copyright © 1998 by John Wiley & Sons, Inc.

All rights reserved. Published simultaneously in Canada.

Library of Congress Cataloging-in-Publication Data:
Ciurczak, Emil W., 1945–
    Molecular spectroscopy workbench : advances, applications, and practical advice on modern spectroscopic analysis / Emil W. Ciurczak.
        p.   cm.
    "A Wiley-Interscience publication."
    Includes bibliographical references (p.   —) and index.
    ISBN 0-471-18081-5 (alk. paper)
    1. Spectrum analysis. I. Title.
QD95.C56   1998
543—dc21   97-24060
                                                                              CIP

Printed in the United States of America.

10 9 8 7 6 5 4 3 2 1

To Diana, Alex, Adam, Alyssa, and Button and Buster (the Wonder Dogs)

CONTENTS

III. General Troubleshooting

PREFACE

When I first agreed to take over the "Molecular Spectroscopy Workbench" from John Coates, I wasn't sure of the direction or tone that the articles should take. Indeed, to be quite candid, I had no direction for the first few issues. The compass that I came to follow was pointing in the direction I have always taken as a teacher or group leader in industry: explain the technique as if the person listening knew little about it. This works, simply, because a person who has heard it before nods in agreement (feeling comfortable in a review) while a novice appreciates the simple explanation. As the column continued, I made a point of cross-pollination of techniques, attempting to dispense with the esoteric and arcane portions that keep newcomers at bay. I write, in short, for new scientists, students, and practitioners who are interested in hearing about what other spectroscopists, NOT in their particular discipline are doing.

The positive response to "Molecular Spectroscopy Workbench" during the eleven years it has appeared in *Spectroscopy* magazine encouraged me to gather a number of the more interesting articles in one volume. I imagined that the articles might tell an interesting story if put in some logical order. The eight years of columns for which I am responsible have been categorized and arranged in a logical manner (or as logical as the eclectic nature of the column allows) as an anthology. Updates are included whenever information has been "corrected," amended, or I have seen mistakes. Interspersed throughout the anthology are several research papers (either by me or other authors) that also appeared in *Spectroscopy* that fit into this collection. These are complementary to the column articles, as they demonstrate unique applications of vibrational spectroscopy and/or diffuse reflectance behavior.

Since not all possible topics in spectroscopy have been covered in the column in the past seven years, the scope of this anthology does not (and cannot) cover all issues of the art. Rather, a chronology of development may be seen. In fact, several of the chapters are as timely today as they were at printing, several are hopelessly outdated, and most are, to quote Goldilocks, "Just right!" I, myself, was surprised how timely and informative a five- or six-year-old article could still be. Apparently, I chose subjects and guest authors well. The instrument reviews from the various conferences might be seen as a year-by-year introduction of "state-of-the-art" equipment and when it was first introduced and serve as an historical footnote.

My whole approach to writing the column has always been one of wonder: "I wonder what that word means?" For instance, "What does 'CCD' stand for?" My search of this particular question led to the Hubble Space Telescope, which uses several Charge-Coupled Devices to create images. I had good fortune asking "experts" in the fields where I was a little undereducated. They seemed to like to show off to a humble novice, such as yours truly.

I have tried to mix up my subjects, but, I confess, that I keep coming back to near-infrared spectroscopy (NIRS). I have been playing with NIRS for over 15 years and am still amazed at what it can do. I am equally amazed by the versatility of most instruments in the hands of experienced practitioners. If I seem to have more NIR chapters than any other it may be an inverse rule: There are so many other texts on the other spectroscopic techniques and only three to date on NIRS (1–3).

Perhaps the fact that I am genuinely delighted to learn from the people who help me with columns comes through in the writing. I like to think that my "Oooh-Ahhh!" attitude carries across the page to the reader. I purposely did not date the columns in the body of work, just to show how some have kept their value after half a decade. You may, of course, look at the publication dates in the index, but play along and read the column first. You may be surprised to discover it came out here first!

Since some of the authors have changed jobs, some topics have changed dramatically, or I found "mis-statements" in the text, or there are some small grammatical corrections from the original text. Where a comment will suffice, the italicized words are my editorial comments. Where major rewrites were needed, the new information is likely incorporated into the text. Updates may appear at the end of a particular column . . . what happened subsequent to appearing in the column.

In any case, "Enjoy!" and, hopefully, learn (as did I) something from these pages.

EMIL W. CIURCZAK

Calverton, MD
January, 1997

REFERENCES

1. Williams, P., and Norris, K., *Near-Infrared Technology in the Agricultural and Food Industries*, American Association of Cereal Chemists, St. Paul, MN, 1987.
2. Osborne, Fearn, & Hindle, *Practical NIR Spectroscopy with Applications in Food & Beverage Analysis 2nd Ed.*, Langman Science & Technology with J. Wiley & Sons, NY, 1993.
3. Burns, D. A., and Ciurczak, E. W., *Handbook of Near-Infrared Analysis*, Marcel-Dekker, New York, 1992.

ACKNOWLEDGMENTS

First and foremost, I acknowledge the authors of the columns and research papers that I have included: All have taught me something about spectroscopy. To the authors whose names appear on the chapters: Thank you for writing such good articles. To the many people I bothered for information, I also say, "Thank you." You are legion in number, and this book is already long enough, so a general "Thanks" will have to do.

I would like to single out Mike MacRae, once and future editor of *Spectroscopy*. I started with Mike, worked with several other (quite capable) editors, and am back with Mike. He left for a short time and rejoined the *Spectroscopy* family again. He is largely responsible for helping me put this together . . . which is why I won't give out his home address!

My wife, Diana, is the world's greatest proofreader (in addition to being a Ph.D. chemist) . . . she is why MacRae can print my columns without much editing. She usually has helpful hints, such as "Wasn't your column due last week?" and "Didn't you write about that last year?" and the ever-popular, "That doesn't make any sense!" Such are comments that spur me on to greatness, and I love her for it.

My (unpaid) consultants are Howard Mark and Jerome Workman, both also contributing editors for *Spectroscopy*. I have had the pleasure of teaching the ACS Short Course on NIR with both and learned as much as any paying customer. Howie and Jerry keep my observations closer to the "Learning Channel" and away from "X-Files."

And, of course, all the people who read my doggerels and do not demand its removal from *Spectroscopy*. I am always amazed when someone comes up to me at a meeting and admits to being a follower of my scribblings. Thank you, kind people whom I may never meet, save through the printed word. I hope I have done some good and helped with a problem or two. (A smile from one of my bad jokes is appreciated, as well.)

MOLECULAR
SPECTROSCOPY
WORKBENCH

CHAPTER I.A.1

ECHOES OF FUNDAMENTALS PAST OR I DON'T THINK WE'RE ALONE IN THIS SPECTRUM

This column is an answer to all those criticisms about near-infrared that begin, "But it's only overtones and combinations ..." I wrote this in attempt to show those "real" spectroscopists that they have been using overtones and combination bands for years without thinking. Not only that, but chose to open the anthology with one of my favorites.

My subject this month is one near and dear to my heart: overtones and combinations in vibrational spectroscopy. While these are the heart of near-infrared (near-IR), they are quite common in midrange IR as well. The example I am using to illustrate overtones and combinations is the simple gas SO_2. (1)

From the common, everyday formula for vibrational modes (number of modes $= 3N - 6$), it is expected that gaseous SO_2 would have three bands in the IR region. The first vibration (*I used calculations based on the vibrational "Hooke's Law" approach, which matches quite well with higher-level quantum-mechanical calculations, well beyond my simple calculator.*) is ν_1, the *symmetric stretch* (Fig. 1a), and it gives a peak at 1151 cm^{-1}. (This can be calculated by group theory, or you can take my word for it!) A second stretch ν_3 is called the *asymmetric stretch* (Fig. 1b) and has a frequency of 1361 cm^{-1}, or slightly more energy than the first. A third absorbance at 519 cm^{-1}, ν_2, comes from the O–S–O bending and is called (cleverly enough) the *bending mode* (Fig. 1c). (Note: The numbering of these bands is consistent with conventions that cannot be explained in a column of this length.)

The above assignment of vibrational modes would be a perfect example of

1

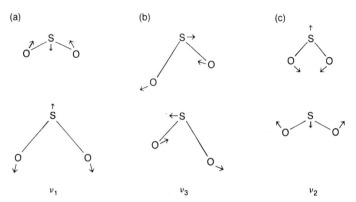

Figure 1. The vibrational modes from gaseous SO_2: (a) the symmetric stretch, peak 1151 cm^{-1}; (b) the asymmetric stretch, peak 1361 cm^{-1}; and (c) the bending mode, peak 519 cm^{-1}.

quantum theory at work were it not for the inconvenience of bands appearing at 606, 1871, 2305, and 2499 cm^{-1}. The three bands mentioned above are the *allowed* bands, according to quantum theory. They are based on the assumption that the motion of any two atoms may be described using a harmonic oscillator as a model. The assumption holds reasonably well for the first stretching mode. For the molecule to acquire enough energy to be promoted from the ground state to a second level (first overtone) or higher is impossible for an (*ideal*) harmonic oscillator.

For an ideal oscillator, the representation of the energy of the bond versus the distance between the atoms would resemble Fig. 2a. In this representation, the bond strength is assumed not to diminish with distance; also, there is no repulsion between approaching nuclei. In real molecular bonds, the energy graph looks more like Fig. 2b. As the atoms diverge, the energy curve slopes off until the atoms actually dissociate. When the atoms are pushed together, the energy increases so rapidly as to become a barrier to further compression.

In the former model, each level is exactly the same energy above the previous level. Thus overtones are now allowed. In the real world, the changing shape of the energy well allows the energies between levels to vary. This "anharmonicity" allows overtones to exist. However, because these levels are strictly forbidden, the probability that they will be occupied will be diminished. A good rule of thumb is that each succeeding overtone is approximately 10 times lower in intensity than the previous one. If the fundamentals are given the arbitrary value of one (1.0), the first overtone has a value of one-tenth (0.1), and so on.

Armed with this knowledge, let us look anew at the "extra" bands we observed earlier. The band at 2305 cm^{-1} would suggest that it is double the frequency of a fundamental. If the ν_1 is doubled, the value 2302 cm^{-1} is obtained. The difference of a mere 3 cm^{-1} is easily accounted for by the anharmonicity that must exist for the band to be seen. The *[origins of]* other "extras" are not so obvious. This introduces the idea of combination bands.

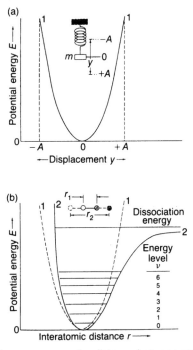

Figure 2. Graph of bond energy versus the distance between atoms: (a) representational and (b) real.

Instead of doubling or tripling a single fundamental, two or more different fundamentals may be combined in a constructive or destructive mode. That is, we may see $\nu_a - \nu_b$ or $\nu_a + \nu_b$ as a new band (minus the energy of anharmonicity, of course.) With this additional information, we can now arithmetically assign the remaining bands (see Table I.)

In like manner, any unknown bands may be deduced from first principles. When the bands in question are C–H, N–H, and O–H (4000–2500 cm^{-1}), the

Table 1. The combination of different fundamentals leads to remaining band assignments.

$\nu(\text{cm}^{-1})$	Assignment
519	ν^2
606	$\nu_1 - \nu_2$
1151	ν_1
1361	ν_3
1871	$\nu_2 + \nu_3$
2305	$2\nu_1$
2499	$\nu_1 + \nu_3$

overtones and combinations make up the backbone of what is called the "near-infrared." Thus the so-called indecipherable region of the spectrum is not nearly so macabre as was once implied by spectroscopists. We simply did not have the computer power to deal with an entire region that was considered forbidden. Just as in early maps of the world, where knowledge was missing, one could find the inscription "Here There Be Dragons and Sea Monsters," so did early spectroscopists warn of falling off the end of the "known spectrum." However dull reality may sound, the region is just as bound and predicted by the *[selection]* rules of spectroscopy as any other "normal" region.

So, sail on, brave analysts. Let the monsters beware.

REFERENCE

1. Russell S. Drago, *Physical Methods in Chemistry* (W. B. Saunders Co., Philadelphia, 1977).

CHAPTER I.A.2

FOURIER TRANSFORM SPECTROSCOPY: A DISPERSIVE MODEL

Ron Williams

Professor Ron Williams from Clemson has written an interesting story about how FT-IR can be explained in dispersive terms. Personally, I didn't think that this was possible, but he has done a fine job. I'm sure you will enjoy this unusual and original tutorial as much as I did.

In the late nineteenth century, Albert Michelson invented the interferometer for an experiment he was conducting to test for the presence of "ether" in the universe. For this, he received a Nobel prize in physics (1). He also realized that the interferometer could be used to make spectral measurements (2), but, unfortunately for him, computational equipment to do the necessary calculations did not become readily available until the latter part of the twentieth century. In the interim, less computationally intensive dispersive spectroscopy developed rapidly and grew to dominance. Because of this chronology, all students of interferometry started as dispersive spectroscopists, and interferometry seemed difficult and arcane.

There is nothing inherently difficult about interferometry, now more commonly called Fourier transform spectroscopy. It is simply a different approach to making measurements. Rather than producing the spectrum directly, interferometry requires two steps: collecting the signal and mathematically processing it to produce the spectrum. Why bother with this two-step process when the spectrum is readily available dispersively? There are several significant advantages offered by the interferometer that, in certain applications, offer significant improvements in the collected spectra.

In this column, I will present a model of interferometry using equipment familiar to dispersive spectroscopists: prisms, choppers, amplifiers, and the like. This "working model" of the interferometer is presented as a transitional step between interferometry and dispersive spectroscopy. Although crude, it explains many of the advantages and disadvantages of Fourier transform spectroscopy.

THE INTERFEROMETER

The interferometer, as illustrated in Fig. 1, is a mechanically simple device consisting of only three pieces: two mirrors and a beam splitter. The latter device, the most mysterious part, is simply a "half-mirror"; that is, it reflects half the light that strikes it and transmits the other half. (Actually, no beam splitter operates in exactly this way. A portion of the beam is also absorbed by the beam splitter—it is more common for 40% of the light to be reflected, 40% transmitted, and 20% absorbed.)

Light enters from the source, strikes the beam splitter, and is divided into two beams that are sent down two separate "arms" of the interferometer to the mirrors. All that happens at the mirrors, of course, is that the light is reflected back to the beam splitter where the same transmission–reflection occurs again. The net result is that half of the light from the source returns to the source, and the other half goes to the detector.

Things are quite simple from the detector's point of view. It looks at the two images of the source simultaneously, one from each mirror. The signal recorded by the detector is the apparent brightness of the source measured as one of the mirrors moves. This intensity will increase and decrease as the moving mirror sweeps out its path, generating what is called an interference pattern—hence the name *interferometer*. (The size of this signal is greatest at the point where

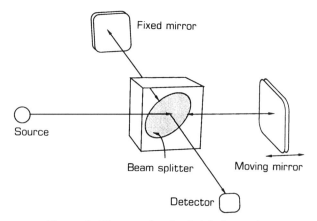

Figure 1. Diagram of a simple interferometer.

both mirrors are the same distance from the beam splitter. This is referred to as zero path difference, or ZPD.)

In most explanations, the interference of the photons is introduced at this point to explain why the intensity varies with mirror movement (3). However, the operation of the interferometer can also be explained using more familiar concepts.

A DISPERSIVE MODEL

The essence of interferometry is expressed in Fig. 2. For simplicity, we assume a source that emits only two wavelengths having the intensities shown in Fig. 2(b). These impinge on a prism, something very familiar to dispersive spectroscopists, and each is bent at a different angle and then sent through two separate choppers that are operating at different frequencies. The blue light has been bent more by the prism than the red, and its chopper operates at a correspondingly higher rotation rate than that of the red. The detector sees both of these signals simultaneously because the second prism is adjusted to make the light collinear again. (Note the similarity to the roles of the beam splitter: One prism divides the light, and the second recombines it.)

The signal observed by the detector in this example is the sum of the two square waves caused by the rotation of the choppers, shown in Fig. 2(b). The blue light is approximately twice as intense as the red light. In this example, its square wave occurs at a frequency of 6 Hz while the frequency of the red light is 2 Hz. (In the jargon of Fourier transform spectroscopy, we say that the source has been modulated by the interferometer.)

The major advantage of interferometry over dispersive spectroscopy is also illustrated in this figure. The sum of the two signals is recorded by a single

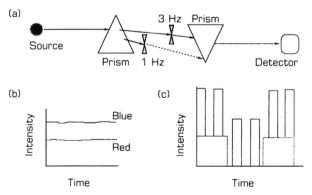

Figure 2. (a) Diagram of a dispersive interferometer. Assuming a two-wavelength source (b), the signal observed by the detector (c) is the sum of the two square waves caused by the rotation of the choppers.

detector. This is the origin of what is known as the multiplex advantage. The detector records both signals simultaneously; thus detector noises are spread over both signals equally—that is, each square wave ends up with one-half of the total detector noise. A dispersive spectroscopist would leave out the second prism and record each wavelength separately. If we spend one second recording the signal shown in Fig. 2, the interferometer will have recorded one second's worth of noise to be split between two signals. A dispersive measurement would require that we measure first the blue light for one second and then the red. Each signal would get a full measure of one second of detector noise. Notice that the more wavelengths of light that are present in the spectrum, the bigger this effect.

In summary, each wavelength of light incident on the interferometer is converted to a chopped frequency that is proportional to its energy; that is, shorter-wavelength light shows up at higher frequencies. The signal is monitored as a function of time, and this is why the interferogram is referred to as a time-domain signal. (All frequencies are measured with a time abscissa.)

EXTRACTING THE SPECTRUM

But how do we get the spectrum from this information? The frequency is directly related to the wavelength, as we have seen; however, we need some method for quantitation. One approach is to listen to the signal (because your ears are frequency processors.) A monochromatic signal would sound like a single note from a piano. As more frequencies are added to the spectrum, the sound would become more complex, forming complicated chords.

This is clearly unsatisfactory for analytical spectroscopy. In order to extract useful information we must have the spectrum—that is, intensity as a function of wavelength (or some related parameter such as wavenumber.) The first approach taken by a dispersive spectroscopist would be to use a bank of tuned amplifiers to produce the spectrum. We would have to supply one amplifier tuned to the frequency of each chopper. The output of the amplifier (peak-to-peak voltage) would correspond to the intensity of the spectrum at that frequency, and because each frequency is proportional to the wavelength of light producing it, we could plot those numbers and produce the spectrum.

This approach, however, would be far too expensive—and there is a better way. By allowing a computer to store the signal from the detector, we can use a mathematical approach to extract the amounts of each frequency. The counterpart of the tuned-amplifier approach uses digital filtering to process the data over and over sequentially, extracting each piece of frequency information. This also is extremely inefficient. The most efficient approach discovered so far is to use a mathematical technique known as the Fourier transform to extract this information. In essence, it performs the same function as the bank of tuned amplifiers: It extracts the frequency identities and intensities in parallel.

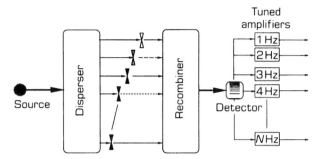

Figure 3. A complete spectrometer.

PHASE CORRECTION

Let us return to our dispersive model in Fig. 3. An even more sensitive processing scheme would be to replace the tuned amplifiers with lock-in amplifiers. In this case, each amplifier would process the signal from the detector using a reference signal generated from its chopper, which would allow discrimination using frequency and phase.

The role of phase in our dispersive interferometer comes into play because each chopper may not have started spinning at the same time. In interferometry, this can be caused by imperfections in the beam splitter and electronics, or it can arise from the sample. The effect that this phase shift (that is, phase error) has on the recorded data is illustrated in Fig. 4, which shows what happens

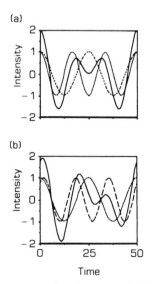

Figure 4. Effect of phase errors on an interferogram: (a) in phase—both choppers start at the same time, and (b) out of phase—the choppers start at different times.

when two frequencies are recorded slightly out of phase. If both choppers in our example had started at the same time, the frequencies would be in phase and the sum seen by the detector would have symmetry, as shown in (a). However, if the two choppers had started at slightly different times, the signal would be slightly out of phase, as shown in (b). The summation recorded by the detector loses its symmetry as a result. This loss of symmetry is the indication that phase errors are present in the interferogram.

In the tuned-amplifier example, this error would have no measurable effect on the resulting spectrum. But because the Fourier transform processes all the frequencies as a group, the slight shift shows up as "derivativelike" features in the spectrum (4). In order to avoid this, several mathematical procedures, termed *phase correction*, are used to fix these slight phase shifts either before or after transformation.

FREE SPECTRAL RANGE AND ALIASING

One problem that will result from our model is that if we have a broadline source, literally millions of different wavelengths will be produced. How can we produce enough choppers to encode each at a different frequency? The answer is that we can't. The number of choppers (unique frequencies) available from the interferometer is controlled by the total distance that we allow the mirror to travel. If it travels 2 cm, we will have twice as many available choppers than if it only travels 1 cm. In either case, all the wavelengths in the spectrum will be distributed equally across the available frequencies. In other words, the resolution (total number of frequencies produced in the spectrum) of the interferometers is controlled by the distance that the mirror is driven. In theory, we can continue increasing the number of choppers *ad infinitum* to achieve any resolution. In practice, one of the advantages of interferometry is its ability to provide extremely high resolution (5) in relatively compact instrumentation.

We still have one subtle point left to make about frequency encoding. The frequencies produced by the interferometer begin at 0 Hz and increase to some maximum value. How do we know what this maximum value is? It would seem that as we increase the distance that the mirror travels we would increase this upper limit, but this is not the true. The choppers will run from 0 Hz up to a maximum that is determined by the shortest wavelength present in the spectrum regardless of mirror travel. The interferometer is capable of producing a unique frequency for every frequency in the spectrum, no matter how short the wavelengths are. However, this signal must be digitized by a computer before it can be processed. This introduces a general data-acquisition problem called *aliasing*, and it is related spectroscopic concept of free spectral range (6).

Aliasing, as the name implies, describes the process in which a frequency shows up in the digitized interferogram with the wrong identity. The frequency at which aliasing begins is the upper limit of the free spectral range, and any wavelengths of light producing frequencies above this limit will "fold" back into the spectrum like an accordion. This process is controlled by the rate at which the computer records data from the interferometer and has nothing to do with how far the mirror travels. (In fact, the exact frequency of the chopper in our example is controlled by the speed at which the mirror is driven back and forth. If the mirror velocity is doubled, then the frequencies of all the choppers is also doubled. This is why it is so important that the mirror be driven at a constant velocity in an interferometer: All velocity errors show up as errors in the spectrum.)

If the computer reads data at a rate of 1000 Hz, then the highest frequency that will be recorded properly is 500 Hz. (This is known as Nyquist's theorem.) A frequency of 600 Hz will not be recorded properly and, in fact, will show up in the digitized spectrum as 400 Hz; that is, it will alias back in the spectrum by an amount equal to how far it is located above the Nyquist frequency. The free spectral range of the interferometer in this case would be 500 Hz. All frequencies above this value will be recorded with the wrong identity, unless they are removed by electronic filtering prior to digitization. Aliasing is not necessarily bad because there is a relatively simple relationship between the frequency and its alias. But it is imperative that you know that aliasing has occurred before you can use it. In our example, the presence of a frequency at 400 Hz means that either the spectrum has 400 or 600 Hz in it. If we happen to know that 400 Hz is not produced by the source or that the detector cannot respond to light from that frequency, then the presence of 400 Hz indicates that a 600 Hz signal is present. You can think of this as a kind of data compression.

LIGHT INTO SOUND

To conclude, we will present one final analogy of interferometry. The net effect of the interferometer is that light with oscillations in the 10^{15} range are converted into oscillations in the audio-frequency range (10^3). Light has been converted into sound. Remember, it is sometimes useful to think with your ears rather than with your eyes when thinking about the interferometer.

The applications of Fourier transform spectroscopy span the range of far-IR to ultraviolet using absorbance, emission, and fluorescence techniques. It excels in those applications in which broad spectral measurement is required in short time frames and high resolutions. In cases in which only a single wavelength region is of interest, dispersive spectrometers still dominate.

However, perhaps the greatest contribution of interferometry to dispersive spectroscopy is in the introduction of computing and automation to spectroscopy in general. Fourier transform spectroscopy has been and continues to be a driving force in the advancement of spectroscopic computation.

ACKNOWLEDGMENT

This paper was presented in part at the FACSS XVII Meeting in Cleveland, Ohio, October, 1990.

REFERENCES

1. A. A. Michelson, *Philos. Mag.* **31**, 338 (1891).
2. P. Connes, *Infrared Phys.* **24**, 69 (1984).
3. G. Horlick, *Appl. Spectrosc.* **22**, 617 (1968).
4. P. R Griffiths and J. A. deHaseth, *Fourier Transform Infrared Spectroscopy* (John Wiley & Sons, New York, 1986), p. 25.
5. A. Thorn, *Anal. Proceed.* **22**, 63 (1985).
6. G. Horlick and W. K. Yuen, *Anal. Chem.* **47**, 775A (1975)

CHAPTER I.B.1

NEAR-INFRARED SPECTROSCOPY: WHAT IT IS AND WHAT IT IS NOT

So many people were interested in NIR, but were also so confused, I had to try to set the record straight about this growing technique. By letting people understand what a technique cannot *do, you define the strengths of the technique. If the explanation seems oversimplified, remember, I only have two printed pages to write my stuff.*

One of the rules of writing is, simply put, "write about what you know best." So I'm going to dedicate this first technical column to near-infrared spectroscopy (NIRS). I have found in my travels that NIRS is underestimated, overestimated, or just misunderstood. Before you infer that I will be advocating NIRS for every problem, allow me to assure you that I have had my fill of "ultimate techniques."

I was once told that GC was the ultimate methodology, then LC, then MS–MS, and so on. It should now be obvious that there are specific areas for which each approach excels. In other words, some techniques were "made for" certain types of analyses, while they fail miserably for others: e.g., who would use HPLC to determine trace water in solvents? Perhaps, in my NIRS tale, several niches for it to fill will become apparent.

Historically, the discovery of the near-infrared region is attributed to Herschel at the turn of the nineteenth century. Nothing substantial (in terms of instrumental work) was accomplished for over a century. In the 1950s instruments such as the Cary Model 12, Beckman's DK, and Perkin–Elmer's Spectracord 400A were introduced in the United States. In England, the Mervyn NPL, the Optica CF4DR, and the Unicam SP700 appeared. All these grating instruments made the NIR region of the spectrum available to spectroscopists. Virtually all work, however, was performed on pure liquids or solutions in the transmission mode. Robert Goddu (1) wrote a classic chapter on the early uses and instrumentation.

Because the spectrum generated in the near-IR region (800–2500 nm) consists entirely of overtones and combinations of primary bands within the mid-IR region (4000–17,000 nm), interpretation was often difficult at best. All species, solvent and solute alike, absorb in the near-IR, overlapping at most wavelengths. This fact alone makes single-band spectroscopy and band assignment nearly impossible. Add to this the fact that extinction coefficients in the near-IR are 10–1000 times less than those in the mid-IR. Thus useful spectroscopic applications on anything resembling a routine scale seem out of the question, right? Wrong!

INSTRUMENTATION

Modern instruments (as supplied by manufacturers such as Bran + Leubbe, Dickey-john, LT Industries, Perten, and NIR Systems, to name a few of the larger suppliers) make use of the combination of high-energy sources and low-noise detectors to generate exceptional spectra. The quartz–halogen (now tungsten, as well) lamps used are usually on the order of (*now*) 20 watts and controlled by high-quality constant-voltage sources. This high-energy output allows high-resolution holographic gratings to be used for wavelength selection, and environmentally insensitive quartz optics replace the "salt" windows needed in mid-IR work. For detectors , PbS crystals are used as resistance detectors in a high-voltage Wheatstone bridge. This arrangement routinely allows noise levels in the microabsorbance unit range. Signal-to-noise ratios are usually in the 1000–100,000 range. This sensitivity and wavelength reproducibility allow for exceptionally repeatable spectra.

Luckily for analytical chemists, the equipment being built has been conducive to more than grain analyses. Because most of the samples to be analyzed were not easily made into dilute solutions, diffuse reflectance instruments were designed and built. The two major approaches involved detectors placed at 45° off the vertical or similar detectors built into an integrating sphere. The light collected in this fashion was called "reflectance." Thus the negative log of the reflectance was thereafter referred to as "absorbance."

DATA HANDLING

The dawn of modern near-IR came with Karl Norris of the U.S. Department of Agriculture (2–5). He generated spectra of mixtures containing all levels of analytes to be determined (usually agricultural products such as wheat and soymeal.) Then, using multiple linear regression routines on the log 1/reflectance spectra, he generated equations that basically resembled the following:

$$\% \text{ analyte} = b_o + b_1 A_1 + \cdots + b_n A_n$$

where the bs represent the constants generated by the software (roughly proportional to the inverse of the extinction coefficients), and the As are the absorbance values (log $1/R$) values at each wavelength.

Every manufacturer has commercial software that chooses the wavelengths for the analyst based on statistical parameters. In short, there are "step-up" and "combination" algorithms in common use. The step-up program looks at each wavelength and the values attached to the spectrum by the user. It constructs a Beer's law plot for each wavelength, then uses a linear regression program to select the most linear wavelength. This wavelength is then used in combination with a second wavelength, and so on until n wavelengths are chosen to fulfill the equation. As long as certain statistical parameters are met, an equation can be generated in minutes (*now seconds!*). The combination approach is slightly different, but, in essence, looks for the best four or five combinations of three, four, five, or however many wavelengths are presented to the computer.

But some still viewed NIRS as a kind of "voodoo." Because the technique was used initially by working farmers—not chemists—there was no pressing need to determine which bend, stretch, or rotation (or combination thereof) was chosen by the algorithm. While the instrumentation and software became rather sophisticated, the terminology used in trade journals became peppered with terms like "the protein peak." And, of course, the technique was used on uncontrolled natural products, not "real" chemicals and products. To pure spectroscopists, this made NIRS a pseudoscience and not worth investigating.

APPLICATIONS

Fortunately, analytical chemists are not pure, but simply wish an accurate answer in the shortest possible time. Near-IR was looked at more closely and eventually found useful for polymers (6–10), textiles (11–16), and pharmaceuticals (17–21), as well as for the foods and grains for which it was developed.

With polymers, NIRS is used for such obvious applications as blend composition, copolymerization ratio, hydroxyl number, acid number, and end-group analysis. In addition, because of the strong influence of hydrogen bonding, such parameters as degree of crystallinity and orientation are also measured.

In textiles, such physical–chemical parameters as degree of heat cure can be measured, so that, for example, nylon fiber can yield information as to how long it was cured and at what temperature. Qualities such as weave and thickness are easily monitored via NIRS.

And lately, in pharmaceutical manufacturing, NIRS is being used to measure particle size, composition, and polymorphism of pure materials and dosage forms. The diffuse reflectance nature of the technique lends itself to solid and semisolid dosage forms.

SUMMARY

Although NIRS is not the be-all and end-all of analysis, there are a number of difficult and interesting applications for which it may be perfect. Because I work in the field, doubtless you will be seeing some of my ideas in future installments of this column.

REFERENCES

1. R. F. Goddu, *Adv. Anal. Chem. Instrum.* **1**, 347–424 (1960).
2. K. H. Norris, *Principles and Methods of Measuring Moisture in Liquids and Solids* (Vol. 4, Reinhold, New York, 1965), pp. 19–25.
3. I. Ben-Gura and K. H. Norris, *Israel J. Agri. Res.* **18**, 125 (1968).
4. I. Ben-Gura, *J. Food Sci.* **33**, 64 (1968).
5. K. H. Norris and P. C. Williams, *Cereal Chem.* **61**, 158 (1984).
6. R. J. W. LeFevre *et al.*, *Aust. J. Chem.* **13**, 169 (1960).
7. H. Dannenberg, *SPE Trans.* pp. 78–88 (January, 1963).
8. W. H. Grieve and D. D. Doepken, *Polym. Eng. Sci.* **112**, 303 (1968).
9. A. Giammarise, *Ana. Lett.* **2**(3), 117 (1969).
10. E. W. Crandell and A. N. Jagtap, *J. Appl. Polym. Sci.* **21**, 449 (1977).
11. C. M. Ashworth *et al.*, *Analyst* **108L**, 1481 (1983).
12. S. Ghosh, *Proc. 8th Internat'l Symp. on NIRA* (Technicon, Tarrytown, New York, 1985), pp. 33–35.
13. J. Connel and G. Brown, *J. Text. Inst.* **69**, 357 (1978).
14. T. Slack-Smith *et al.*, *J. Text. Inst.* **70**, 33 (1979).
15. J. Rodgers, *Proc. 7th Internat'l Symp. on NIRA* (Technicon, Tarrytown, New York, 1984), pp. 16–18.
16. H. Starkweather and R. Moynihan, *J. Polym. Sci.* **22**, 366 (1956).
17. R. G. Whitfield, *Pharm. Manuf.* p. 31 (April, 1986).
18. B. Buchanan, E. W. Ciurczak, A. Q. Grunke, and D. E. Honigs, *Spectrosc.* **3**(5), 54 (1988).
19. E. W. Ciurczak and R. P. Torlini, *Spectrosc.* **2**(3), 41 (1987).
20. J. J. Rose, Y. Prusik, and J. Mardelcian, *J. Parentr. Sci. Technol.* **36**(2), 71 (1982).
21. E. W. Ciurczak and T. A. Maldacker, *Spectrosc.* **1**(1), 36 (1986).

CHAPTER I.B.2

A CLOSER LOOK AT THE
WHOLE INFRARED REGION

[Since this article was written to compare the IR and NIR equipment, FT-NIR instruments have had a marvelous renaissance. The noise level of the systems has plummeted by two orders of magnitude, and the software needed is now provided. The tone of this article is assuredly dated. I humbly acknowledge that FT instruments, in many cases, are equal to and often superior to dispersive types.]

I wish to thank John E. (Jack) Carroll, (*former*) marketing manager for NIRSystems, Inc., for much of the inspiration and information for this column. Jack has spent so many years in spectroscopy that some of his earlier notebooks have dates in Latin! Those of you wondering what Jack looks like can find his visage in an ad for "another" company, in which he is holding a small UV/Vis spectrophotometer ... or remember Uncle Fester (without a light bulb.)

I grew up thinking that the infrared (IR) region of the spectrum stretched from about 4000 to 400 cm^{-1} (or 2500 to 25,000 nm.) Lately, the near-IR has been "discovered" by any number of analysts (like Columbus discovered the New World with a million or two people already on board.) This "new" region is roughly from 750 to 2500 nm and isn't so new, having been discovered by the English astronomer Herschel in 1800. Because near-IR consists of overtones and combinations of fundamental bands originating in the mid-IR, it is, in truth, part of the infrared.

Differences between the mid- and near-IR come into play in the instruments used for work in these regions and the sample types amenable to analysis by them. All IR instruments have certain main components in common:

- a source of radiant energy
- a spectral sorter (grating, interferometer, absorption or interference filters)

- a sample holder
- a detector
- a spectral display (recorder, screen, computer)

SOURCES

Because IR radiation is essentially heat, hot wires, light bulbs, or glowing ceramics are used as sources. The energy distribution from these blackbody sources tends to peak at about 1000–2000 nm and then tail off into the mid-IR. As the temperature is increased to boost output, the peak intensity shifts to lower wavelengths (near-IR/vis; red-hot to white-hot), and the researcher must seek better throughput to overcome this low-intensity light (*in the mid-IR*). A larger slit allows more light but shoots the heck out of resolution. An example of low energy would be a double-beam mid-IR spectrophotometer operating at 8 cm^{-1} resolution. The detector sees only about 0.2% of the total radiation available from the source. The source in mid-IR usually operates between 1200 and 1700 K, while a typical near-IR source operates between 2500 and 3000 K. This results in approximately 10 times more radiation impinging on the near-IR sample.

Raising the energy (and, thus, the temperature) of the light that reaches the sample would only deteriorate the sample (*due to its higher absorbencies*). As a consequence an interferometer is a better choice for increasing throughput. (The throughput advantage is known as Jacuinot's advantage and is limited to a maximum factor of about 300 times.)

SPECTRAL SORTER

The search for a throughput advantage leads to the interferometer. The Michelson is the most common, but other types, such as the Steele, are also in use. The general idea is to split the incident beam and recombine it in an interference pattern [see Chapter I.A.2]. The whole of this recombined light beam is shone on the sample (*and, thus, through to the detector*), greatly increasing the signal-to-noise (S/N) ratio. Another advantage of this sorter is Fellget's, or multiplex, advantage. Because an interferometer measures all IR frequencies simultaneously, it yields a spectrum with resolution comparable (or better) than a grating, but in a far shorter time span. The Fellget's advantage for mid-IR can be as much as 250 times that of the near-IR.

Near-IR spectrophotometers use, as a general rule, holographic gratings to "sort" wavelengths. (*Since this article was printed, at least five companies introduced FT-NIR equipment of the same high quality as the FT-IR instruments already on the market: Perkin–Elmer, Bruker, Buhler, Bomem, and Nicolet. More will have been introduced before this goes to press, no doubt.*) Because the sources are generally more efficient in this range, much more of the radiant energy reaches the sample.

SAMPLE HOLDER

The sample holder becomes important when considering the different materials that are analyzed by each instrument. In the mid-IR, the extinction coefficients are from 10 to 1000 times greater than in the near-IR. Because of this, the mid-IR is quite useful for analyzing dilute or microscopic samples. In instances where sufficient amounts of sample are available, the sample must be diluted before it can be used in this region. Dilution can take the form of KBr pellets, thin films, mulls, or solutions. In the near-IR, samples are mostly run "neat" as powders, slurries, or solutions, with no dilution. In cases of pure material in solid form, diffuse reflectance is the mode of choice, while liquid or semisolid samples may be run in transmission.

The pathlengths of samples in mid-IR are limited to between 15 μm and 1 mm because of the weaker signal and stronger absorptivities in this range. In near-IR, the pathlengths may routinely be as long as 100 mm. In mid-IR, there is always a concern that a representative sample may not have been taken, since the amount is so small, especially in heterogeneous mixtures. The near-IR, using undiluted, larger samples and longer pathlengths, allows for more representative sampling. The water absorptions that are nearly fatal in mid-IR not only do not bother near-IR, but water is commonly used as a solvent for analyses.

DETECTORS

Detectors for each case are essentially heat detectors. One of the more sensitive detectors in the mid-IR is deuterated triglycine sulfate (DTGS), a pyroelectric bolometer. Its D^* (detectivity, a measure of a detector's electrical output versus photonic input) is about 2×10^9. The lead sulfide (PbS) detector most commonly used in near-IR has a D^* of 6×10^{10}—an advantage of roughly 60 times for the near-IR.

The operating noise range for the mid-IR is typically in the milliabsorbance range (*note: the newest, circa 1996 models are much, much quieter*), while the near-IR detectors operate with microabsorbance noise levels.

SPECTRAL DISPLAY

The displays or computers used in modern instrumentation are indispensable. Both FT-IR and near-IR instruments use the computer to control the scan as well as "crunch" the data. No advantage on this count: Data reduction programs abound for both types of equipment.

CONCLUSIONS

Both regions of the IR spectrum are valuable for almost any kind of work. The classic use of mid-IR has been in the elucidation of the structure of molecules. It is still the best as far as molecular spectroscopy goes. Newer attachments (microscopes, CIRCLE cells, ATRs, and so forth) allow mid-IR to be adapted to quantitative work. Near-IR has traditionally been used for quantitative work because of its nonsample prep rep(utation). (Sorry, I can be serious for just so long, then I get a brain cramp.) Because of qualitative algorithms introduced over the past decade, near-IR is fast becoming a method for rapid identification of materials, as opposed to merely identifying previously identified samples.

For structure elucidation, or for identification of limited amounts of pure materials, I would put my money on the mid-IR. For forensics, mid-IR will always be valuable because of its sensitivity. Organic chemists will always depend on the mid-IR for instant confirmation of the success or failure of their reactions.

For large numbers of samples or production process streams, the near-IR appears to have an edge. The nonintrusive fiber-optic attachments and the on-line systems are suited to the production environment. Of course, there is considerable overlap between the two portions of the IR—and that is good.

In short, there is only one IR region, but it has two important parts: the near- and the mid-range. Which is the best? Both!

CHAPTER I.C.1

CHIROPTIC PHENOMENA, PART I: CIRCULAR BIREFRINGENCE—OPTICAL ROTATION AND OPTICAL ROTATORY DISPERSION

Harry G. Brittain

[The guest author of this chapter is Harry G. Brittain (formerly) *from Bristol–Myers Squibb Pharmaceutical Research Institute. Harry taught at Seton Hall University before his present position. Why should I mention this? Because there will be a number of SHU papers in this volume. When surrounded by professors performing cutting-edge research, lean on them for contributions (I always say).* (Actually, my working "life thesis" is simply, "A friend is not someone to use and discard—a friend is someone to be used over and over again!)*

Chiroptic phenomena are often used, and even more often misunderstood, by analysts. I am pleased that Dr. Brittain consented to set me (and through me, you) straight as to what is happening in optical rotational instrumentation.

Next month will see my third Pittsburgh Conference wrapup column on products for molecular spectroscopy. I've wrestled, played line in football, and argued with my mother. Nothing compares with the agony of trying to cover all the new things at PittCon, but, Lord knows, I'll try yet again (it's really no worse than rooting for Rutgers year after year after year...)]

When a compound's mirror images cannot be superimposed on each other, that compound is said to be *chiral*, and the mirror images are denoted as *enantiomers*. The phenomenon of enantiomorphism in optically active compounds has become especially important in pharmaceutics as both industry and regulatory bodies discover the importance of drug chirality (1). The enantiomers of a compound may or may not exhibit equivalent pharmacokinetic or pharmacologic properties, and this situation must be determined during a new drug's development (2).

The interaction of polarized light with chiral compounds is of great interest because chiroptical techniques are extremely useful as methods of characterization. It is equally true that although most scientists are aware that enantiomerically rich solutions will rotate the plane of light, the origins of this effect are not as simple as might be imagined. In Part I of this column, the phenomena of polarimetry and optical rotatory dispersion will be discussed. Part II will concern the related phenomenon of circular dichroism.

E. T. Malus discovered that images of objects viewed through calcite crystals appeared to double (3). This phenomenon of double refraction was ultimately ascribed to linear birefringence, where the indices of refraction along the axes of the anisotropic crystal are unequal. A beam of collimated light passes through a calcite crystal and emerges as two beams, the electric vectors of which are shown to be orthogonal. The blockage of one of these beams yields an optical element known as a linear polarizer. It was demonstrated that light polarized in this manner would not pass through a second linear polarizer whose crystal axis was rotated by 90° with respect to the original plane of polarization.

J. B. Biot and others found that when certain materials were placed between these crossed polarizers, light could be transmitted by the second optical element. This medium could be crystals cut along certain directions (for example, slabs of quartz cut perpendicular to the hexagonal axis) or solutions of some organic compounds (such as solutions of camphor or sugars). It was quickly realized that the initial plane of polarization appeared to be rotated upon passing through the "optically active" medium between the polarizers. This phenomenon led to the development of the polarimeter, with which the amount of optical rotation induced by a given solution was quantified. Biot developed the convention that positive optical activity corresponded to a clockwise rotation of the linearly polarized light. A block diagram describing the action of a polarimeter is shown in Fig. 1.

An explanation for this effect was only possible once Fresnel applied the

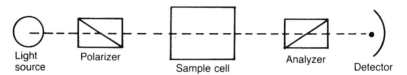

Light source   Polarizer   Sample cell   Analyzer   Detector

Figure 1. Block diagram of a simple polarimeter. Monochromatic light from the source is linearly polarized by the initial polarizer and then allowed to pass through the sample medium. The angle of polarization associated with the light leaving the medium is determined by rotating the analyzer polarizer to the new null position. A variety of photoelectric methods are available that can be used to automatically determine the observed angle of rotation.

new transverse wave theory of light to the phenomenon of optical activity. He differentiated between linearly polarized light (light whose electric vector lies along a single axis as it propagates through space) and circularly polarized light (light whose electric vector rotates about the axis perpendicular to its direction of propagation through space). Right circularly polarized (RCP) light represents a clockwise rotation of the electric vector, while left circularly polarized (LCP) light represents a counterclockwise rotation. These relations are illustrated in Fig. 2. Fresnel proved that linearly polarized light is actually the result of two coherent circular components that have opposite rotations.

The phenomenon of optical activity is determined by the relative indices of refraction for LCP and RCP light in the medium under consideration. In an optically inactive medium, the refractive indices of LCP and RCP light are equal. Because the two polarization senses would remain in phase at all times during passage through the medium, the resultant vector leaving the medium would be unchanged with respect to the vector that entered the medium [see Fig. 3(a)]. In an optically active medium, the refractive indices for LCP and RCP light are no longer equal. The components are then no longer coherent when they leave the medium. When viewed directly along the direction of the oncoming beam, the resultant vector appears as a rotation of the initial plane of polarization [Fig. 3(b)]. Optical activity is, therefore, a manifestation of circular birefringence.

In units of degrees, the observed optical rotation (α) is given by:

$$\alpha = [1800b(n_L - n_R)]/\lambda_o \tag{1}$$

where b is the pathlength in decimeters, λ_o is the wavelength of light used in

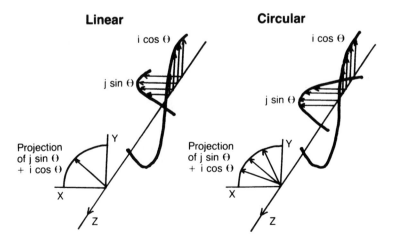

Figure 2. Relationships of the electric vectors associated with linearly and circularly polarized light as these are propagated.

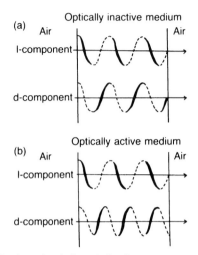

Figure 3. Behavior of the two circularly polarized components of linearly polarized light as they pass through (a) an optically inactive medium, and (b) an optically active medium.

the determination, and n_L and n_R are the refractive indices for LCP and RCP light. The specific rotation of a material dissolved in a fluid solution is given by:

$$[\alpha] = 100\alpha/bc \qquad (2)$$

if c is the solute concentration of grams/100 mL of solution. The molar rotation of a solution is defined as:

$$[M] = FW\alpha/bc \qquad (3)$$
$$= FW[\alpha]/100 \qquad (4)$$

where FW is the formula weight of the optically active solute. When the solute concentration is given in terms of molarity (M), eq. (3) becomes:

$$[M] = 10\alpha/bM \qquad (5)$$

Biot observed that the optical rotation of tartaric acid solutions was a function of the wavelength used for the determination. The wavelength dependence of optical rotation is termed *optically rotatory dispersion*, or ORD. When measured outside absorption bands, an ORD spectrum [as illustrated in Fig. 4(a)] will consist either of a plain positive or plain negative dispersion curve. When measured inside an absorption band, the ORD will exhibit anomalous dispersion, which is referred to as the Cotton effect. A positive Cotton effect consists

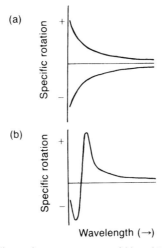

Figure 4. Optical rotatory dispersion curves as would be obtained in (a) optically inactive media (both plain positive and negative curves are illustrated), and (b) optically active media (a positive Cotton effect has been shown).

of positive ORD at long wavelengths, and negative ORD at shorter wavelengths [see Fig. 4(b)]. In the simplest measurement of ORD, the fixed-wavelength source in Fig. 1 is replaced by a tunable source (such as a xenon arc combined with a monochromator).

One of the most intensive studies involving ORD concerns the $n \rightarrow \pi*$ transition in the 300 nm spectral region of ketones, where very strong Cotton effects can be observed. The effects were particularly useful in the characterization of steroids and molecules containing smaller ring systems. Djerassi, Klyne, and others established that the optical activity observable within this band was determined by the location of the ketone group in the molecular ring structure, the absolute configuration of neighboring ring junctions, the nature and configuration of adjacent substituents, and the conformation of the rings that make up the molecule (4). These observations led to the development of the "octant rule," which represents an attempt to generalize the ORD trends in light of stereochemical structure (5). In this system, the three nodal planes of the n and $\pi*$ orbitals of the carbonyl group are considered to divide the molecular environment into four front and four back octants.

A group or atom situated in the upper left or lower right rear octant (relative to an observer looking at the molecule parallel to the $C=O$ axis) induces a positive Cotton effect in the $n \rightarrow \pi*$ band. A negative Cotton effect would be produced by substitution within the upper right or lower left back octants. Although exceptions to the octant rule have been demonstrated, its utility in predicting molecular stereochemistry on the basis of observed ORD has been extremely valuable to synthetic organic chemists.

REFERENCES

1. W. H. DeCamp, *Chirality* **1**, 2 (1989).
2. F. Jamali, R. Mehvar, and F. M. Pasutto, *J. Pharm. Sci.* **78**, 695 (1989).
3. For an excellent review of the history and principles associated with optical rotation and optical rotatory dispersion, see T. M. Lowry, *Optical Rotatory Power* (Dover Publications, New York, 1964), pp. 89–148.
4. Much of this work is summarized in C. Djerassi, *Optical Rotatory Dispersion: Applications to Organic Chemistry* (McGraw–Hill, New York, 1960).
5. W. Moffitt, R. B. Woodward, A. Moscowitz, W. Klyne, and C. Djerassi, *J. Am. Chem. Soc.* **83**, 4013 (1961).

CHAPTER I.C.2

CHIROPTIC PHENOMENA, PART II: CIRCULAR DICHROISM

Harry G. Brittain

[This month's offering is the second part of the series from Harry G. Brittain. He explains chiroptical phenomena, specifically, the ins and outs of circular dichroism—and not in a circuitous manner, either! Of course, he gives examples as well as describing the technique. I, for one, have learned quite a bit from Harry's writing, and I thank him for that. Chapter I.C.3 describes how these techniques are applied to DNA research.]

In the preceding installment in this series (1, Chapter I.C.1), the phenomenon of circular birefringence was discussed. This effect is manifested as optical rotation and optical rotatory dispersion (ORD), and is determined by the unequal indices of refraction for left circularly polarized (LCP) light and right circularly polarized (RCP) light in a chiral medium. Within an absorption band, the ORD spectrum exhibits anomalous dispersion, which is referred to as the Cotton effect. Full understanding of ORD and anomalous dispersion requires a more detailed examination of the properties associated with refractive indices.

The refractive index of a material is actually the sum of a real and imaginary part:

$$n = n_o + ik \qquad (1)$$

where n is the observed refractive index at a given wavelength; n_o, the refractive index at infinite wavelength; and k, the absorption coefficient of the substance. By definition, k will equal zero outside an absorption band. For an achiral compound, no difference in refractive index exists for LCP light as opposed to RCP light. For a chiral material, the refractive indices for LCP and RCP light are no longer equal, and can be expressed as

$$n_L = n_{o(L)} + ik_L \qquad (2)$$
$$n_R = n_{o(R)} + ik_R \qquad (3)$$

The criterion for a substance to exhibit either optical rotation or ORD is that $(n_L - n_R)$ is not equal to zero. Within an absorption band it can be concluded that if $(n_L - n_R)$ does not equal zero, then $(k_L - k_R)$ will not equal zero either. The quantity $(k_L - k_R)$ is defined as the circular dichroism (CD).

CD therefore represents the differential absorption of LCP and RCP light. The effect of this differential absorption is that when the electric vector projections associated with the LCP and RCP light are recombined after leaving the chiral medium, they describe an ellipse whose major axis lies along the new angle of rotation. The measure of eccentricity of this ellipse is defined as the ellipticity, or Ψ. These relations are illustrated in Fig. 1. It is not difficult to show that (2)

$$\Psi = [\pi z (k_L - k_R)]/\lambda \qquad (4)$$

where z is the pathlength in centimeters, and λ is the particular wavelength at which Ψ is measured. If C is the concentration of absorbing chiral solute in moles per liter, then mean molar absorptivity, a, is derived from the absorption

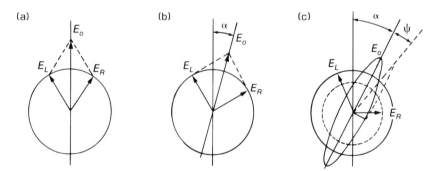

Figure 1. Phase relations associated with the passage of circularly polarized light through various media. (a) For optically inactive media, the recombination of left (E_L) and right (E_R) circularly polarized light yields linearly polarized light whose electric vector (E_o) is unchanged with respect to the incident axis. (b) For optically active media outside of an absorption band, recombination of E_L and E_R yields a resultant E_o vector rotated from the incident axis by the angle α. (c) For optically active media inside an absorption band, recombination after differential absorption of E_L and E_R yields a resultant E_o vector rotated from the incident axis by the angle α, and describing an ellipse of ellipticity ψ.

index by

$$a = (4\pi k)/[(2.303)\lambda C] \tag{5}$$

In that case, the ellipticity (still in units of radians) becomes

$$\Psi = [(2.303)Cz(a_L - a_R)]/4 \tag{6}$$

The expression of ellipticity in radians is cumbersome, and, consequently, this quantity is converted into degrees by the equation

$$\Theta = \Psi(360/2\pi) \tag{7}$$

and then

$$\Theta = (a_L - a_R)Cz(32.90) \tag{8}$$

The molar ellipticity is an intrinsic quantity, and is calculated from

$$[\Theta] = [\Theta(FW)]/[Lc'(100)] \tag{9}$$

where FW is the formula weight of the solute in question; L, the medium path in decimeters; and c', the solute concentration in units of grams per milliliter. The molar ellipticity is related to the differential absorption by

$$(a_L - a_R) = [\Theta]/3298 \tag{10}$$

Most instrumentation suitable for measuring CD is based on the design of Grosjean and Legrand (3), and a block diagram of their basic design is shown in Fig. 2. Linearly polarized light is passed through a dynamic quarter-wave plate, which modulates it alternatively into LCP and RCP light. The quarter-wave plate is a piece of isotropic material, which is rendered anisotropic through the external application of stress. The device can be a Pockels cell (in which stress is created in a crystal of ammonium dideuterium phosphate through the application of alternating current [ac] high voltage), or a photoelastic modulator (in which the stress is induced by the piezoelectric effect [*note: this is the effect described by M. Curie, husband of Mme. Curie*]). The light leaving the cell is detected by a photomultiplier tube, whose current output is converted to voltage then split. One signal consists of an alternating signal proportional to the CD, and is due to the differential absorption of one circularly polarized component over the other. This signal is amplified by means of phase-sensitive detection. The other signal is averaged and is referred to as the *mean light absorption*. The ratio of these signals varies linearly as a function of the CD amplitude, and is the recorded signal of interest.

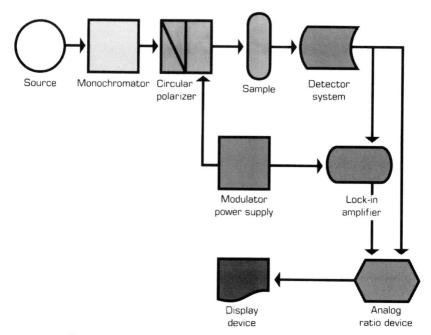

Figure 2. Block diagram of a circular dichroism spectrometer.

As an example of the pharmaceutical utility of CD spectroscopy, we will consider the example of captopril. Captopril, 1-[(2S)-3-mercapto-2-methylpropionyl]-(S)-proline:

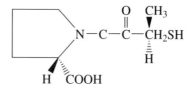

is a potent inhibitor of the angiotensin-converting enzyme and is used in the treatment of hypertension and congestive heart failure. The compound contains two centers of dissymmetry, one associated with the (S)-proline portion and the other associated with the 3-mercapto-2-methylpropionic acid side chain. The compound is normally administered as enantiomerically pure, with both centers of dissymmetry being completely resolved.

The CD spectra of captopril and its enantiomer, 1-[(2R)-3-mercapto-2-methylpropionyl]-(R)-proline, are shown in Fig. 3. As would be anticipated for a true enantiomeric pair, the CD spectra of these two compounds were found to be exact mirror images. Values calculated for the CD and molar ellipticity have been found to be equal, but opposite in sign (4). The chiroptical data also indi-

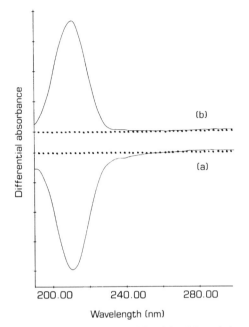

Figure 3. Circular dichroism spectra obtained for (a) 14.3-μg/mL solution of captopril, 1-[(2S)-3-mercapto-2-methylpropionyl]-(S)-proline, and (b) 13.4-μg/mL solution of 1-[(2R)-3-mercapto-2-methylpropionyl]-(R)-proline.

cate that the spectroscopic properties of the individual components of captopril are significantly altered when these are combined. The 193-nm negative CD band of (S)-proline appears to undergo a red shift to 210 nm in captopril, with the molar ellipticity being greatly increased by the process. At the same time, the positive 210-nm CD band of proline is essentially canceled by the negative CD band of the (2S)-3-mercapto-2-methylpropionic acid side chain (4).

The CD spectra of the two mixed diastereomers of captopril, 1-[(2S)-3-mercapto-2-methylpropionyl]-(R)-proline and 1-[(2R)-3-mercapto-2-methyl-propionyl]-(S)-proline, are found in Fig. 4. While the CD spectra of these compounds are mirror images of each other (as would be expected for an enantiomeric pair), they are distinctly different from captopril itself. The low-wavelength proline CD band is evident as a CD peak at 195 nm, although greatly intensified. In the mixed diastereomers, the second proline band now has the same sign as that of the chiral side chain. As a result, a strong CD peak is now observed at 218 nm (also intensified with respect to the component bands.)

CD spectroscopy has played an important role in the characterization of optically active molecules, and will continue to do so as long as chemists are interested in small molecules. High-performance liquid chromatography (HPLC) detectors that make use of the CD effect also show great potential as the chiroptical detectors of choice for work with dissymmetric compounds (5).

CD spectroscopy is not limited to the study of small molecules and has

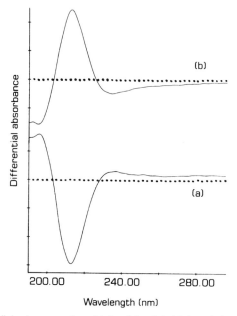

Figure 4. Circular dichroism spectra obtained for (a) 13.4-μg/mL solution of 1-[(2S)-3-mercapto-2-methylpropionyl]-(R)-proline, and (b) 14.1-μg/mL (0.0649-mM) solution of 1-[(2R)-3-mercapto-2-methylpropionyl]-(S)-proline.

become extremely important in the characterization of biomolecules. The secondary structure of proteins can be characterized through studies of the CD associated with amide chromophores. Using a combination of models and calibration spectra, it is possible to deduce the relative contributions to the overall secondary structure made by α-helix, β-sheet, β-turn, and random coil portions of the polypeptide (6). With the increasing use of such agents in the pharmaceutical industry, it is not difficult to envision the role that chiroptical spectroscopy will assume in the future.

REFERENCES

1. H. G. Brittain, *Spectroscopy* **6**(3), 10–13 (1991).
2. E. Charney, *The Molecular Basis of Optical Activity* (John Wiley & Sons, New York, 1979).
3. L. Velluz, M. Legrand, and M. Grosjean, *Optical Circular Dichroism: Principles, Measurements, and Applications* (Verlag Chemie, Weinheim, Federal Republic of Germany, 1965).
4. H. G. Brittain and H. Kadin, *Pharm. Res.* **7**, 1082 (1990).
5. N. Purdie, *Prog. Analyt. Spect.* **10**, 345 (1987).
6. W. C. Johnson, *Ann. Rev. Biophys. Chem.* **17**, 145 (1988).

CHAPTER I.C.3

MONITORING CONFORMATIONAL TRANSITIONS IN SYNTHETIC DNA OLIGOMERS USING CIRCULAR DICHROISM

The topic this month is, in reality, the same as for the last few months: chiroptics. In this, the last of the "trilogy," Dr. Richard Sheardy (Department of Chemistry, Seton Hall University, South Orange, New Jersey 07079), an organic chemist whose research interests are in nucleic acid DNA chemistry, explains how circular dichroism (CD) spectropolarimetry is a powerful tool for determining DNA conformations and for monitoring conformational changes. His report describes some CD techniques applied to short synthetic DNA oligomers capable of undergoing salt and thermally induced conformational transitions. The results of these studies can be used to evaluate certain thermodynamic properties of particular DNA conformations and their allowed transitions. We are fortunate to be the beneficiaries of such up-to-date data.

In the early 1950s, Watson and Crick proposed a right-handed double helical structure for DNA (1). It is now recognized, however, that deviations from the Watson–Crick structure are not only possible, but also biologically interesting (2). As an example, the sequence-specific recognition of a particular protein for a particular segment of DNA must arise from this intrinsic polymorphism of DNA [for example, the lac repressor–lac operator system (3)]. Of course, each conformation of DNA available has its own set of physical properties as well.

Much effort has been devoted to examining the relationships between structure and dynamics for a wide variety of nucleic acid conformations.

The techniques available to this end are numerous. Dynamical properties can be evaluated through ultraviolet/visible (UV/vis) absorption spectroscopy, CD spectropolarimetry, and various nuclear magnetic resonance (NMR) techniques. Efforts at developing a high-resolution structure of DNA include techniques such as infrared (IR) and Raman absorption spectroscopies, NMR spectroscopy, and X-ray diffraction (XRD). These techniques are limited, though, because of the complexity of polymeric DNA and the variety of structures that may be present. Thus the use of short, specifically designed DNA oligomers as models for the study of structure and dynamics has come to the fore.

CD STUDIES OF DNA OLIGOMERS

Our interest lies in the polymorphism of the DNA oligomers shown in Fig. 1 (4–7). Duplex **I** is an alternating purine–pyrimidine octomer arising from self-complexation of (d5meC–dG)$_4$ (abbreviations given in Ref. 8); **II** is also an octomer of random sequence arising from the duplexation of two different strands; **III** is a 16-mer generated by essentially joining **I** and **II** together; **IV** is a variant of **III**. All the components of these molecules were synthesized on a model 389B DNA synthesizer (Applied Biosystems, Foster City, CA) and purified using standard techniques (4). Duplexes **II, III,** and **IV** were generated by mixing equimolar amounts of the individual strands. CD spectra were determined with a model 60DS spectropolarimeter (AVIV Associates, Inc., Lakewood, NJ) equipped with a temperature-controlled cuvette holder.

As noted earlier, many allowed conformations are available to DNA. Furthermore, left-handed conformations are tenable to certain DNA oligomers and polymers. At least two conditions need to be met before a DNA molecule can assume a left-handed conformation. First, the sequence of the DNA must be one of alternating purine–pyrimidine (9). Second, the presence of a left-handed con-

I. 5'--C*-G--C*-G--C*-G--C*-G--3'
    3'--G--C*-G--C*-G--C*-G--C*-5'

II. 5'--A--C--T--G--A--C--T--G--3'
     3'--T--G--A--C--T--G--A--C--5'

III. 5'--C*-G--C*-G--C*-G--C*-G--A--C--T--G--A--C--T--G--3'
      3'--G--C*-G--C*-G--C*-G--C*-T--G--A--C--T--G--A--C--5'

IV. 5'--C*-G--C*-G--C*-G--C*-G--A--T--C--G--A--C--T--G--3'
     3'--G--C*-G--C*-G--C*-G--C*-T--A--G--C--T--G--A--C--3'

Figure 1. The synthetic DNA oligomers discussed in this report. C* is 5-methyl-deoxycytidine.

formational inducing agent must be present in high concentration. Such agents may be cations or dehydrating solvents (10). The concentration of cation necessary to induce the left-handed conformation of an alternating purine–pyrimidine sequence depends on the charges of the cation. Thus the rank of effectiveness follows the order $Co^{3+} > Mg^{2+} > Na^+$. Finally, the presence of modified bases in the alternating purine–pyrimidine sequence usually enhances the ability of the polymer to assume the left-handed conformation (10, 11). For example, substituting 5medC for dC in poly(dG–dC) greatly reduces the amount of NaCl required to induce the left-handed conformation. The [NaCl] required to obtain the midpoint of the right- to left-handed transition for poly(dG–dC) is 2500 mM, while that required for poly(dG–5medC) is only 700 mM (11).

Salt-Induced Transitions

The CD spectral characteristics of a DNA oligomer under a certain set of conditions can be used to determine its conformation. The CD spectrum of **I** under low-salt conditions (for example, 10 mM phosphate buffer, pH 7.0, 100 mM NaCl) is shown in Fig. 2a. This spectrum is characterized by a trough at 255 nm and a peak at 280 nm and, hence, is consistent with that of a right-handed double helical conformation (4–7), designated as *B*-DNA. When this oligomer

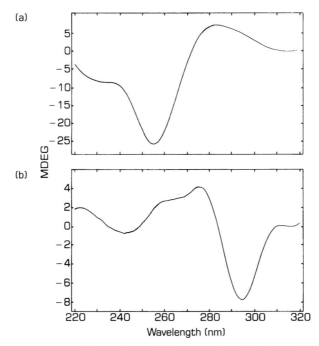

Figure 2. The CD spectra of **I** in buffer at 25 °C at (a) 100 mM NaCl; and (b) 2.5 M NaCl. For all CD experiments, the buffer used is 10 mM phosphate, 1 mM EDTA, pH 7.0.

is prepared in phosphate buffer with 2.5 M NaCl, the CD spectrum of Fig. 2b is obtained. This spectrum is characterized by a trough at 295 nm and a peak at 277 nm. This "inverted" spectrum is typical for DNA polymers and oligomers in left-handed configurations (11), designated as Z-DNA. Thus *B*-DNA is characterized by a trough at 255 nm, while Z-DNA is characterized by a trough at 295 nm. The low-salt spectrum of **II** resembles that of the low-salt spectrum of **I**. However, there is little difference between the low-salt spectrum of **II** and its high-salt spectrum (data not shown). Obviously, **II** retains a right-handed conformation under both salt conditions.

DNA oligomer **III** is thus composed of two segments: one capable of undergoing the *B*-to-*Z* conformational transition, and another segment incapable of such a transition. We were interested in determining if the potential *Z*-forming segment would undergo the transition and, if so, what would be the structure of the interface between the left- and right-handed conformations. Figure 3a shows the CD spectrum of **III** in phosphate buffer at 100 mM NaCl. Again the spectrum indicates that the oligomer assumes a fully right-handed conformation at this low-salt condition. As the [NaCl] is increased, changes in the CD spectra are gradually observed until the [NaCl] is 4.5 M (that is, no further spectral changes are observed at [NaCl] greater than 4.5 M). The resultant

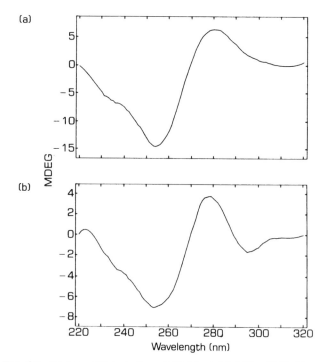

Figure 3. The CD spectra of **III** in buffer at 25 °C at (a) 100 mM NaCl; and (b) 5.0 M NaCl.

high-salt spectrum is shown in Fig. 3b. This spectrum has characteristics of both
"pure *B*" (that is, trough at 255 nm) and "pure *Z*" (trough at 295 nm) spectra.
Thus **III** has both left- and right-handed conformations present at high salt *concentrations*. This molecule therefore contains a *B–Z* conformational junction at
high salt *concentrations*. The presence of this junction in **III** has further been
verified through ^{31}P and ^{1}H NMR studies. These studies have shown that the
first six base pairs (reading from left to right) are in the *Z* conformation, and
the last seven base pairs are in the *B* conformation (5).

DNA oligomer **IV** was designed to determine the effect of sequence near
the *B–Z* junction-forming region on the ease of junction formation. (**III** has a
single AT base pair abutted to the Z-forming block.) Both the low- and high-salt
spectra of **IV** are identical to the low- and high-salt spectra, respectively, of **III**
(see Fig. 5a). Thus **IV** is fully right handed at low salt, but also possesses a
B–Z junction at high salt *concentration*. However, the concentration of NaCl
required to effect the transition of **IV** is less than that for **III**.

The [NaCl] at the midpoint of the *B*-to-*Z* conformational transition for each
oligomer can be used to determine the ease of the transition (6). This midpoint
is obtained from the following equation:

$$\% \text{ transition} = \frac{\delta\Theta_{295}(\text{NaCl})}{\delta\Theta_{295}(\text{total})}$$

where $\delta\Theta_{295}(295) = \Theta_{295}(0 \text{ M NaCl}) - \Theta_{295}([\text{NaCl}])$ for a particular [NaCl] and
$\delta\Theta_{295}(\text{total}) = \Theta_{295}(0 \text{ M NaCl}) - \Theta_{295}(5.0 \text{ M NaCl})$. The change of ellipticity at
295 nm is chosen because that is the wavelength for which Z-DNA has a trough.

Plots of percent transition versus [NaCl] for **I**, **III**, and **IV** are shown in
Fig. 4. Note that the plots for the junction-forming molecules **III** and **IV**

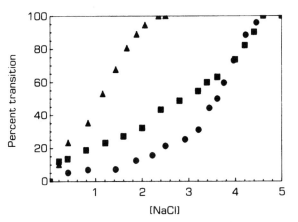

Figure 4. Plots of the percent transition as a function of [NaCl] as determined by CD for
DNA duplexes **I** (▲), **III** (●), and **IV** (■).

are shifted to higher [NaCl] than for **I**. Furthermore, the plots for **III** and **IV** are also different from each other. Midpoints for transitions extrapolated from these plots are at 1.0, 3.55, and 2.95 M, respectively, for **I**, **III**, and **IV**. From these data, we were able to calculate the ΔG_J (free energy of junction formation) to be 4.7 kcal mol^{-1} and 3.9 kcal mol^{-1} for **III** and **IV**, respectively (6).

Thermally Induced Transitions

Finally, thermally induced denaturations can also be monitored using CD spectropolarimetry. Figure 5a shows the CD spectra of **IV** in low and high salt *concentrations* at 25°C, while Fig. 5b shows the low- and high-salt spectra at 90°C. The spectral differences observed for **IV** at low and high salt at 25°C are similar to those observed for **III**. What is interesting is that even the 90°C spectra are not the same, suggesting different conformations for the single strands in low and high salt at this temperature. For comparison, the temperatures of the overlays of the low-salt spectra (at 25°C and 90°C) and of the high-salt spectra are shown in Figs. 6a and 6b, respectively.

In a manner similar to monitoring the salt-induced transitions, plots of percent transition versus temperature can be constructed to determine the midpoints for the thermally induced transitions. This midpoint, referred to as T_m

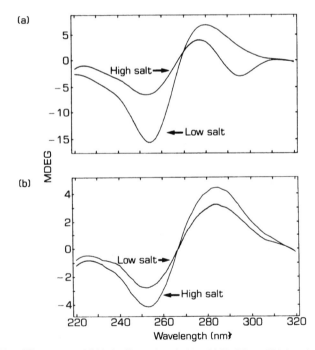

Figure 5. The CD spectra of **IV** in buffer at low (100 mM NaCl) and high salt (5.0 M NaCl) at (a) 25°C; and (b) 90°C.

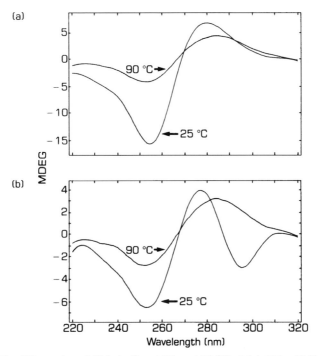

Figure 6. The CD spectra of **IV** in buffer at 20 and 90 °C at (a) 100 mM NaCl; and (b) 5.0 M NaCl.

(melting temperature), can also be determined using UV/vis absorption spectroscopy. Figure 7a shows the plot of percent transition (as determined by the percent change in ellipticity at 255 nm) versus temperature for **IV** in low salt. The midpoint of this transition occurs at ~73°C, in excellent agreement with the UV/vis-determined T_m of 73.6°C at the same salt and DNA concentration. For the high-salt spectra, transitions at both 255 and 295 nm can be monitored. These plots, shown in Fig. 7b, indicate slightly different T_ms for the B end of the molecule (that is, 63°C) and for the Z end of the molecule (that is, 70°C). The UV/vis-determined transition is very broad, with a midpoint at about 70.3°C, in close agreement with the 295 nm CD transition. In this particular case, where different conformations can exist within the same molecule and when these transformations have their own particular CD signal, determination of the T_m using CD spectroscopy may result in better resolution.

CONCLUSIONS

CD spectroscopy can be a useful tool to the structural biophysicist for determining DNA conformations and for monitoring induced conformational changes. However, it should be pointed out that CD results cannot stand on their own.

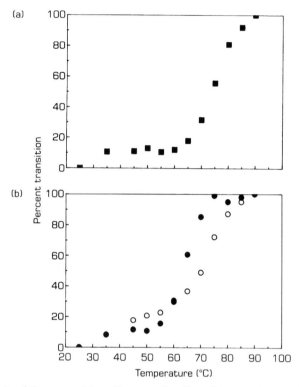

Figure 7. Plots of the percent transition as a function of temperature as determined by CD for **IV** at (a) 100 mM NaCl at 255 nm (■); and (b) at 5.0 M NaCl at 255 nm (●) and 295 nm (○).

Additional proof of these conformations and conformational transitions must also accompany the CD results. This proof can be obtained using a combination of UV/vis and NMR spectroscopies and other physical methods.

ACKNOWLEDGMENTS

The author is grateful to the National Science Foundation for its generous support of this work (DMB-8996232) and to Professor Neville Kallenbach for the use of the CD spectropolarimeter.

REFERENCES

1. J. C. Watson and F. H. C. Crick, *Nature* **171**, 737–38 (1953).
2. R. D. Wells and S. C. Harvey, eds., *Unusual DNA Structures* (Springer-Verlag, New York, 1987).

3. R. Boelens, R. M. Scheek, J. H. van Boom, and R. Kaptein, *J. Mol. Biol.* **193**, 213–16 (1987).

4. R. D. Sheardy, *Nucleic Acids Res.* **16**, 1153–67 (1988).

5. R. D. Sheardy and S. A. Winkle, *Biochemistry* **28**, 720–25 (1989).

6. M. J. Doktycz, A. S. Benight, and R. D. Sheardy, *J. Mol. Biol.* **212**, 3–6 (1990).

7. S. A. Winkle and R. D. Sheardy, *Biochemistry* **29**, 6514–21 (1990).

8. Abbreviations used are dG, 2'-deoxyguanosine; dA, 2'-deoxyadenosine; dC, 2'-deoxycytidine; and T, thymidine. The purines are dA and dG, while the pyrimidines are dC and T. DNA sequences are written from 5' to 3' direction. In the duplexes discussed in this report, the cytidine base has been modified using methylation at the 5-position of the ring for those cytidines in the Z-forming segment of the oligomers (designated as C*).

9. A. Rich, A. Nordheim, and F. Azorin, *J. Biomed. Struct. Dyn.* **1**, 1–19 (1983).

10. T. M. Jovin, L. P. McIntosh, D. J. Zarling, M. R. Robert-Nicoud, J. H. van de Dande, K. F. Jorgenson, and F. Eckstein, *J. Biomol. Struct. Dyn.* **1**, 21–57 (1983).

11. M. Behe and G. Felsenfeld, *Proc. Natl. Acad. Sci. USA* **78**, 1619–23 (1981).

CHAPTER II.A.1

NEAR-INFRARED ANALYSIS AT DUPONT: AN HISTORICAL PERSPECTIVE

Eugene S. Taylor

I am truly a soft touch for a historical anecdote. At the 1989 Pittsburgh Conference in Atlanta, Gene Taylor told me about the pioneering work in near-infrared (NIR) spectroscopy performed at DuPont in the early 1950s. I found his discussion interesting because I had been taught to believe that, in the beginning, Karl Norris raised his hands and said, "Let there be almost light!" and near-IR spectroscopy was born. Having been "discovered" in 1800, the NIR spectral region was bound to be put to practical use by someone. Indeed, the mid-IR–near-IR connection was demonstrated by the clever engineering work done by the folks from Delaware. Taylor's account of these endeavors can help us remember that instruments do not emerge full blown from instrument manufacturers. Rather, instrument design is an evolutionary process that requires the time and effort of many clever people. This column chronicles the efforts of such people, and I hope you enjoy it as much as I did.

The column is a brief history of the use of near-IR spectroscopy in one particular company, namely, the one in whose engineering department I have worked since 1951: the DuPont Company. Like most young engineers when they are first exposed to something new and exciting, I set forth with a lot of "wows" and "gee whizzes" when self-filtering IR analyzers were first explained to me. These devices were developed in Germany during World War II to measure CO and CO_2. The instruments consisted of two beams: one that had a filter cell filled with the gas to be measured and a second, positioned alongside the first, without a filter cell. The device also used a sample cell common to both

43

beams: two detectors, of course, which were thermopiles, bolometers, or some sort of pneumatic device; and finally, a chopper to allow radiation to strike one detector and then the other. How simple and elegant!

Because the source of energy in these analyzers was generally something hot, the wavelength range was not well defined, but it surely covered primarily the mid-IR or fingerprint region. These units, of course, were only good for the analysis of gases, but in those days nobody was analyzing anything. We had to begin somewhere, and gases were pretty easy to work with. Calibration didn't require too elaborate a setup, and partial pressure and similar parameters were all we needed to know. If our gas mixture had an interferant, then a compensation cell exposed to both beams was also provided; this cell was filled with the interferant and served to remove the common wavelengths from both beams! I cut my teeth applying devices of this sort made by Leeds & Northrup—and they worked! This all happened to me in 1952.

IN THE BEGINNING

In the beginning (1952), only gases were considered for on-line analysis, but in the same year, Baird came out with a unit for liquids. The cells had to be fairly thin, and I don't think we made great use of this analyzer. But it was a beginning. Then the Liston–Becker self-filtering unit appeared on the scene, and it showed a good deal of promise for gases, using a pneumatic detector. That detector provided me with lessons from the school of hard knocks, and I learned some things that are just as true today as they were then. For a hypothetical example, let's take one of those pneumatic detectors (without the diaphragm) and hook it up to a good vacuum (essentially a block of aluminum). While connected to this vacuum, we heated the detector to ~300°C for about three days and cooled it to room temperature. Then we charged the cavity with one atmosphere of CO_2, sealed the block, and put it on a shelf. After two weeks, we measured the pressure in the block—one atmosphere, right? Wrong! It's a vacuum. The CO_2 disappeared into interstitial crevices of the metal, which were cleared out in the vacuum-heat treatment. H_2O and other small molecules behave the same way. Interesting!

THE ADVENT OF PbS DETECTORS

The period around 1952 was interesting for another reason, although the significance was not immediately evident—the lead sulfide (PbS) photoconductive detector was invented. It had sensitivity roughly between 1 and 2.6 μm; this region of the spectrum is known as the near-IR and comprises overtone and combination bands of fundamental bands that have their source in the mid-IR from 2.5 to 25 μm. The law of absorption is known as the Beer–Lambert law, where $A = abc$. A is the absorbance and is $\log_{10}(1/T)$, where T is the fraction of light transmitted at a particular wavelength; a is the molar absorptivity in liters per mole centimeter; b is the optical pathlength in centimeters; and c is the concentration in moles per liter. The molar absorptivities of gases in the mid-IR are quite significant, but are one to two orders of magnitude less in the near-IR. The thinking up to that time had been focused on gases, but what about liquids? As previously mentioned, to analyze liquids in the mid-IR required a cell thickness of only a few thousandths of an inch, but if one were able to identify a peak in the near-IR, then a cell thickness of reasonable magnitude would do quite nicely.

EMPHASIS ON INTERNAL INNOVATION

Our philosophy at DuPont was, when possible, to buy commercial equipment for use in the plant, and otherwise develop it ourselves. This approach, along with the realization that there was a nice liquid water band at about 1.9 μm, got some people thinking. Water is, without a doubt, the most sought-after material to measure in a world of different materials—it certainly is important in just about everything DuPont makes. By 1954, a group in the engineering department had developed what we called the model 2 near-IR process analyzer for internal use. It used the PbS detector and an Ebert monochromator for dispersion. The wavelengths were selected by employing two exit slits—one was used for the analytical wavelength, strongly absorbed by the constituent to be measured, and for the other for a reference wavelength that was not, resulting in $\log_{10}(A/R)$. Quite a few of these were built. In this same time frame, Perkin–Elmer's engineers were also busy and came out with their bichromator: a PbS detector with a dispersion means.

An Alternative to Dispersion Devices

A few years later, I'll say 1958 or 1959 because I'm not sure, an equally significant invention came to our attention—the optical interference filter. Basically, it was made of a substrate transparent to the wavelength region of interest, and vacuum deposited on the surface were 1/4-wavelength thicknesses of dielectric materials. This development had the potential to free us from the expensive dispersive devices and to allow us to come up with a reasonably priced analyzer. Again the same laboratory in the engineering department came up with such a device for internal work in 1961, and it became the workhorse for near-IR process measurement, handling from one to four analytes in a single instrument; we called it the model 800 near-IR analyzer.

RAPID GROWTH IN THE EARLY 1970S

With the growing interest in the field of near-IR, developments started to pile up. The PbSe (lead selenide) detector expanded the wavelength region to about

4.5 μm and combined with thermoelectric cooling (TEC) to improve further the performance of detectors. In general, analyses could now be performed in the plant on most gas applications. More and more liquid analyses could be handled, but still only a comparatively few solid analyses were feasible. The field of reflectance drew our attention, and in 1972 a reflectance version of the model 800 was developed. This model, the 804, used an integrating sphere. It was soon followed in 1975 by a second variation that used focused optics to overcome a drawback of the integrating sphere, which required a sample to be in physical contact with the sphere exit. Commercial units were starting to come along as well, such as the Anacon and, later, the Infrared Engineering and Moisture Systems units.

One other development of significance for us at this time period occurred in 1971, when we were unable to find an adequate spectrophotometer with reflectance attachments; so we built our own using an integrating sphere, an Ebert monochromator driven by a stepping motor, and a PbS detector. We used phase-lock loop electronics all driven by a PDP 8 computer. It worked just fine.

The work of near-IR pioneer Karl Norris came to our attention around 1975 through contact with the Neotec division of Pacific Scientific (now NIR-Systems), and I can remember visiting both Norris' laboratory at the U.S. Department of Agriculture and Neotec's facilities. Norris' work on agricultural products for the USDA was based on an integrating sphere and tilting interference filters; he showed us spectra looking through apples and tomatoes! [*Editor's note: At about the same time Dickey-john also manufactured and sold a general-purpose near-IR spectrometer, principally for agricultural purposes.*] Things really started to rush by after this, and it is difficult for me to recall the time sequence of so many events. Trying not to be judgmental, I will list a few of the most significant. Neotec was about the first to capitalize on Norris' work, followed by Technicon (*now Bran + Leubbe*), Then came the development of spectrometers and the introduction of chemometrics [*Editor's note: I was adjuncting at Stevens Institute of Technology in 1979 when Ed Malinowski showed me what he termed "chemometrics." He developed it for his HPLC work, to quantify overlapping peaks. I teased him, saying "Maybe better chromatography or newer columns would be a better idea!" Little did I know how much I would be using his algorithms later in life—I started NIR in 1982, just three years later!*], in which there are now at least four major players: NIRSystems, Bran + Leubbe, LT Industries, and Guided Wave (*now UOP*). (With the introduction of FT instruments, this list, as of late 1996, includes Perkin–Elmer, Buhler, Bruker, Bomem, Nicolet, and virtually any company making FT-IR instruments!)

WHERE ARE WE NOW?

The current status of all this is that there are now capabilities out there to make most measurements, whether the samples are gases, liquids, or solids. In addition, of course, other measurement techniques including the chromatographic family, for example, have undergone similar growth in capabilities.

A tussle continues as to whether it is better to use a chemometric spectrometer on line or develop the model in the laboratory and use a filter device for process analysis. Actually, the right technique depends on the application, and we are pursuing both options and looking for the most cost-effective approach. However, the feeling is strong today that whatever device is used in the field, it should be smart, communicate with the host computer, include a built-in expert system, and be able to withstand an industrial environment.

CHAPTER II.A.2

INFRARED SAMPLE PREPARATION, OR, IT'S NOT THE SAME OLD GRIND

[In an effort to prove that there is little new under the sun, I used some material from a text, now out of print, to help demonstrate the reoccurring nature of mistakes. University courses seldom dwell on the mundane, choosing to discuss the "sexy" theory, ignoring the sample preparation necessary to produce the spectacular spectra seen in texts.]

I recently received a bundle in the mail from the editor of this fine journal that contained many fine comment cards from readers. I was quite pleased to discover that I seemed to have touched a positive nerve with my "primer" articles that have appeared during the past six months or so. My goal has always been to offset other journals' attitude that "if you have to ask how a technique works, you can't possibly understand it." I had grown tired of articles that assumed I knew all about a technique and all prior work done with it. Perhaps you, gentle readers, don't mind such statements as "Building on the work of Smith and Wesson (220–34), we proceeded to . . .," but I really don't have time to look up endless references while reading an article for general knowledge.

That brings us to the topic of this month's column. When I was teaching in college, I observed an organic chemistry teacher showing students how to make a mull. She placed a few milligrams of material on a salt plate, covered it with a few drops of mineral oil, then placed a second plate on top. She proceeded to grind the two plates together. This, she claimed, was a proper mull. Fanning and smelling salts brought me around, but I promised that, someday, I would set this "difficulty" to rights.

I am lucky enough to have taken a Coblentz Society course on sample preparation at PittCon in the early 1970s. I would like to share what I learned there and in the lab during the intervening 20 years.

IR SAMPLE PREPARATION

Infrared (IR) is the most versatile spectrometric method for structure elucidation—if used correctly. If we remember that the energies are low and the extinction coefficients relatively high, much of what I am discussing will make sense. The majority of sample preparation techniques in IR are designed to dilute the sample; for example, a mull, KBr pellet, thin film (plates or ATR), or a dilution in a liquid cell.

Thin Films

To make a thin film, a drop or two of the liquid sample is placed on an IR tranparent plate like KBr or NaCl. A second plate of the same material is placed over the liquid and slight pressure applied. No special clamp or attachment is necessary, because done properly, the film has enough surface tension to hold the plates together. Various attachments from reputable companies (see annual product guides for examples) allow thin films to be introduced into the light beam.

Caution should be used to prevent air from getting back into the sample after it has been compressed. These bubbles cause edge effects and possible fringe effects along the baseline. Care should be taken to avoid evaporation of volatile samples. The cooling effect of evaporation may cause condensation and create water peaks in the spectrum or cloud the salt plates. Evaporation could also cause a disproportionation of materials in a mixture of materials by concentrating the less volatile ones.

Attenuated Total Reflectance (ATR)

This is a simple and elegant method to create a short pathlength without destroying the sample. The cell usually consists of a trapezoidal prism of IR-transparent material. The sample (liquid or solid) is placed on one or both parallel faces. The light beam is introduced so that it "grazes" the sample at a critical angle, creating a sawtooth pattern through the cell. The beam impinges on the sample once with a shorter cell or many times with a longer cell (a frustrated multiple internal reflection optical flat). Because the IR beam doesn't penetrate very deeply into the sample, dilutions are not needed.

One consequence of this is that the optical flat may be the side or bottom of a chemical reactor. This allows a reaction to be followed without withdrawing samples. Another use is to analyze coatings on paper, metal, or glass without the interfering spectrum of the matrix on which the sample is coated.

Some precautions are in order. First and foremost, the depth of the penetration will vary with wavelength. This could cause problems if you are using a set of standards developed on pure samples through another technique. A computer search usually consists of peak intensity ratios. In ATR, the ratios differ from transmission spectra. This causes even worse problems in quantitative *[in the original article, "qualitative" was mistakenly inserted here; since no one seemed to notice, I am a little disappointed in how closely the article was read!]* work (for which near-IR is the preferred mode of analysis), where ratios could be truly skewed if wavelengths used for quantification are relatively far apart in the spectrum. However, if precautions are taken, this is a great technique.

Mineral Oil Mulls

A relatively simple sampling method for softer organic samples is a mull. The proper approach is to use an agate mortar and pestle (small, of course.) Place a few milligrams of the sample into the mortar and grind it until it looks like a thin film. At this point, add a drop of mineral oil and continue to grind. The particle size must be reduced before the sample can be lubricated. It is virtually impossible to grind a sample while it is sliding all over the mortar.

When the paste is placed on the salt plates, use as little force or lateral motion as possible. Even in its finely ground state, the sample is merely a suspension of particles. This mix makes a fine abrasive. Unless you have an unlimited supply of salt plates (in which case, send me some, please), care must be taken at this step. *[It seems that I suggest care at **every** step. Well, so I do ... perhaps there are no unimportant steps.]* Clearly, this technique causes less chance of evaporation of the solvent, but loss of film contact with both plates should be prevented as well (edge effects and all that, you know).

Of course, you might guess that the mineral has a considerable spectrum of its own, being a hydrocarbon of high molecular weight. The C–H peaks are off the chart. A second spectrum of the unknown must be taken using a solvent with no C–H peaks to speak of, such as CCl_4. The portions of the two spectra that are "solventless" are then combined. This little fact is sometimes lost in the heat of discovery.

Potassium Bromide Pellets

My favorite sample preparation technique is the KBr pellet. In this method, a 0.3–0.5% solid solution of the organic sample in dried potassium bromide is prepared. The two materials are ground to a fine powder in a mortar and pestle (agate, again) or in a small stainless steel or plastic container using hard balls of the same material and a reciprocating motion (a vibrating mixer, a.k.a. Wig-L-Bug, seems to be the method of choice).

The mixture is then poured into a pellet press consisting of two "pistons" in a smooth, cylindrical chamber. Pressures of up to 25,000 psi are then applied (under a vacuum, whenever possible, to remove solvents) for varying amounts of time. The pistons are removed, and the pellet is popped out and placed in a holder (some hand-tightened dies allow analysts to place the pellet in the sample beam while it is still in the die). In most cases, a clear pellet is produced. However, several major faults may appear (1). I'll treat them by their appearance.

The first problem appears as a white, translucent pellet. You may have used too much sample. If the pellet is more than, say, 1–2 mm thick, you probably should regrind and remake it with less material. For "quick and dirty" scans, you could probably get away with this first pellet, but you may well lose fine details because of the lack of transmitted light.

The second problem I call the *half-moon*. The pellet displays a half-moon shape (or "Cheshire-cat" smile). This is simply due to the fact that you poured the powder down one side of the die. (You can assure yourself of this by measuring the thickness of both the clear and opaque sides.) Again, regrind and remake.

A third problem is the freckled look. Tiny but distinct spots are apparent throughout the pellet. This arises from insufficient grinding. The sample may be used, but a sloping baseline will be most obvious between 4000 and 3000 cm⁻¹. (This phenomenon also appears in mulls—a good diagnostic tool showing that more grinding is required.) Once more, regrind and remake.

Another common problem is one I call *chocolate chip cookies*. The spots in this case are larger than in the previous case, more like chips in a cookie. They are caused by lack of or insufficient vacuum during pressing and/or poorly dried KBr; that is, the sample is wet. The extreme form of this is seen when a perfectly thin pellet comes out of the press cloudy. The water has dispersed throughout the entire sample. You know the routine by now.

General *Caveats*

Beware of cellulose peaks in an ATR of a polymer on paper. The light may penetrate just deep enough to "see" the matrix. Another common problem is cross-contamination from mortars. I didn't specify *agate* mortars and pestles because I own stock in them: I mentioned them because they aren't porous. Cheaper porcelain models can easily cross-contaminate your samples. They may be easier to use, but "cheap is cheap" (2). If you *must* use porcelain, clean them with salt grindings between samples ... *NEVER* wash them out; you will always have water peaks in your samples!

Another source of ghost peaks is the use of polypropylene balls to grind the KBr and sample. While disposable, they obviate the need for cleaning, but severely limit the grinding time to between "poorly ground" and "where the heck did these peaks come from?" [*A measure of time smaller than that needed for a New York taxi driver to demonstrate the State Bird of NY.*] The stainless steel or agate balls should be used for sensitive work.

Sinusoidal waves may appear and coincide with the spectrum. These are most often caused by internal reflectance between two new or replaced windows. I used to eliminate the waves by breathing on the windows to cause micropitting (garlic acts as a catalyst for this reaction; so don't perform this feat immediately after consuming shrimp scampi!) The sine waves may be used to measure actual pathlength ... the technique is covered in most good instrumental texts.

Basically, analysts must remember that the spectrum will only be as good as the effort put into the sample preparation. "No matter how hurried you are, you will always have time to do it right the second time" (3).

REFERENCES

1. E. W. Ciurczak, *Instrument Troubleshooting: A Basic Guide, 2nd Edition* (Stevens Institute of Technology Press, Hoboken, NJ, 1984), pp. 55–59
2. Emil Ciurczak (the Elder) in a lecture to his stubborn son, Emil W. (the Lesser), Northern New Jersey, ca. 1961.
3. *Ibid.*, later in the same lecture, regarding painting the house, if I recall correctly.

CHAPTER II.A.3

CULINARY SPECTROSCOPY— FT-IR ANALYSIS OF COOKING AND CUISINE

Gerald J. DeMenna

Gerry DeMenna, affectionately known as Dr. Demento, is an interesting character. I asked him to talk about food ... as if he ever stops! He is a gourmet cook, and we trade recipes from time to time. I fear he throws out the ones I give him, however.

It almost appears as if I haven't left New Jersey recently. This month's column is by another Garden Stater (Gerald DeMenna), who was employed as a regional applications scientist at Laser Precision, Piscataway, NJ (formerly known as Analect Instruments), and is now president of Chem-Chek Consulting, P.O. Box 8307, Piscataway, New Jersey 08854. *The truth is: I'm a sucker for Italian cooking! The work presented herein meets many of the criteria that I believe make it a worthwhile column: It is a practical application; the matrix is complex (read "nasty"); and modern molecular spectroscopy is involved. Yes, you can really sink your teeth into the meat of this article. In all seriousness, as entertaining as the writing may be, the methodology is quite clever and may be applicable to similar matrices. Gerry received his B.S. and M.S. degrees from Rutgers University* (Who didn't?), *where he is completing his Ph.D. thesis. When not analyzing samples, he plays the role of gourmet chef at home and reportedly* never *refers to a cookbook. Enjoy!*

Mangiare! The word is the one you hear most growing up in an Italian family. Thirty years ago, it was commonplace to go to the market and buy fatty pork sausages and pepperoni, salty provolone and Parmesan cheeses, rich and creamy cannolli and tortes, and oil-soaked garlic bread. Thirty years ago, it was not common knowledge that salt and fat and sugar were *not* the primary food groups to ingest, nor were things like fiber and monounsaturates and cholesterol widely mentioned in nutritional circles. Today, the more open attitude about unusual foods and the trendiness of "nouvelle cuisine" have markedly changed the dietary styles of many people, though not always for the better. The U.S. Surgeon General says we should cut down on red meat and dairy fat, increase fiber consumption, and reduce alcohol intake, as these factors will reduce (possibly) the chances of heart disease and cancer. The once unpopular "health food" stores and restaurants are now as common as "normal" shops and eateries. As a scientist, one must evaluate *all* data before redefining a nation's eating habits.

My ancestral background shows a predominant age of 85 years before mortality, despite a diet consisting mostly of sausages and pepperoni, cheeses and pastries, and oily garlic bread for most of those 85 years. Why? How come they can do it and we can't? There is some old, yet interesting census data from the United Nations World Health Organization. WHO completed a 10-year study in 1975 that revealed the incidence of arteriosclerosis in northern Italian provinces to be approximately 25% *higher* than that of the central and southern provinces. Why? This is the question that begins our investigation.

"THAT'S THE WAY IT WAS (AND IS!) ..."

If one is involved in the food industry today, one can still see the proliferation of classical manual analytical techniques developed nearly 75 years ago—not only for physical properties, but for chemical composition as well. Some simple, yet reliable, instrumentation is used in refractometry, polarimetry, viscosimetry, and colorimetry to monitor color and pour characteristics, but most chemical analyses use complex extractions or other separations followed by assorted titrimetric or gravimetric techniques. Unsaturation is measured by titration with iodine to give an iodine number. Free acid moieties are titrated with bases to give a saponification value. Spot tests with permanganate are used to identify adultrants and with dichromates to measure antioxidants. Thiosulfate titrations are used to quantify peroxide formation. Karl Fisher devices are employed routinely for water-content determinations. And a virtual plethora of other reagents exists for the multitude of tests required under government and industry regulations.

These procedures are relatively low cost, though labor intensive. They suffer from some interferences, but most can be corrected with blanks and extracted samples. Just within the past decades have commercial analytical instruments become popular additions to classical food laboratories. Near-IR has been a tremendous boon to analysts, allowing rapid and quantitative determination of moisture, proteins, and some carbohydrates. But such measurements represent a bulk determination; so you may have 5% fat, but not know what kind of fat. And today's consumer wants to know if it's good fat or bad fat. Chromatographic techniques are now commonplace, having superb sensitivity and moderate selectivity. A major disadvantage lies in the fact that they are secondary techniques and must be referenced to a primary analytical standard (if one exists). UV/vis spectrophotometric methods are prevalent in most labs, showing good sensitivity and fair selectivity (1–6).

FT-IR Spectroscopy

One spectroscopic technique that has been a standard tool in the chemical, petroleum, pharmaceutical, and polymer fields, but not in the food products environment, is Fourier transform infrared (FT-IR) spectroscopy. It combines the qualities of resolution and selectivity from dispersive-style IR with the speed and sensitivity of Fourier digitization. It allows bulk samples to be analyzed without prior separation. It can elucidate the structures of the fats, lipids, fiber, or vitamins in a sample without having to run a primary standard, since you are getting primary spectroscopic data. It allows samples with water and samples with high-molecular-weight species, as well as heterogeneous samples, to be evaluated simply and rapidly. Ideally, one could buy a "meal" from fast-food restaurant A and another from B, homogenize them in an Osterizer or Cuisinart, smear the mass on a crystal, and get comparative spectra of both within minutes (then you can decide where to risk eating). Or one can look at the ingredients thrown into the ever-popular crock pot and see the compositional changes that occur after the cooking process. It sounds easy enough for a technician to run, is relatively inexpensive, and can be automated (if Cuisinart develops a robot) for additional speed. The following examples are the first documented applications of *culinary spectroscopy*.

MAMA ... MAMA MIA!

Being Italian, living in New Jersey, growing up with a melting pot of friends, and having more skill at gravies than on the gridiron led to my observation, evaluation, and modification of my family's staple recipes. The big favorite is the tomato-based sauce with meat and vegetables that is poured over pasta, known to Neopolitans as gravy. My paternal grandmother and my mother taught me the secrets of good gravy. The only one I can reveal is as follows:

You have a large saucepot with several gallons of fresh or canned tomatoes simmering. In a large frying pan, fry the sausages, stew beef, and pork chops that will go into the gravy. When done, strain the meat out of the melted fat (cholesterol, diglycerides) in the pan, and put the meat into the saucepot, continuing to simmer. Into the melted fat in the pan, add several cans of tomato paste (this is predominantly dehydrated tomatoes and contains about 5% total free carboxylic acids—malic, citric, oxalic, ascorbic, and tartaric) and fry the paste at high heat, stirring constantly, until the fat disappears. Add the fried tomato paste to the saucepot and continue to simmer (7).

The meat fat in the pan (before treatment) gels to a white waxy solid on cooling. After treatment, there is a light oil that floats on the surface of the fried tomato paste but does not solidify. Infrared spectra reveal the formation of esters after the treatment (as shown by bands appearing at 1710 and 1230 cm^{-1}), where little or none existed before. One hypothesis—an esterification occurs in the heat of the frying pan, accompanied by the evolution of lots of steam as the byproduct of the waxy meat fat, with the tomato acids generating a liquid ester or polyester. This material is not metabolized by the body to form arterial plaque. The consumption of beef, pork, and veal throughout Italy is fairly homogeneous. But the central and southern areas tend to use tomato-based dishes, while relatives in the north concentrate on cream and cheese-based sauces; therefore, the trend mentioned earlier from the WHO census seems to have some dietary explanation. Theory: Esterify your meat with tomato first, then eat! It's apparently worked for family and friends for many generations.

The instrument used for this and the following studies was an Analect RFX-40 spectrometer with an FX-70 operating system data station (Laser Precision, Irvine, CA). A horizontal attenuated total reflectance (ATR) sampling accessory with zinc selenide and KRS-5 plates was used. *Note:* If you try this at home, please do not lick the sample off the plate after the analysis! Both of these materials are extremely toxic! The actual spectra of the gravy sample are presented in Fig. 1.

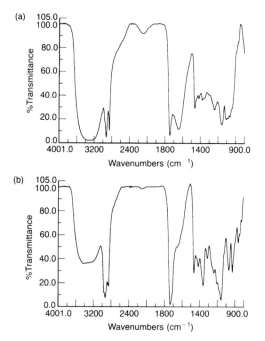

Figure 1. FT-IR spectra of tomato-based "gravy" with meat (a) before cooking and (b) after cooking for 36 hours.

"IS THE SEAFOOD FRESH? (IT WAS WHEN I OPENED THE CAN)"

Another classic sauce for pasta (particularly the flat linguini noodles) is white clam sauce, consisting very simply of clams, clam juice, garlic, oil, herbs, and a little black pepper. If the clams aren't fresh or the canned variety haven't been thoroughly cleaned, there will be a noticeable "off odor" and taste to the sauce. This is due to the oxidation of the various amines that comprise marine animal flesh. This "off taste" is partially corrected by utilizing olive oil rather than corn or vegetable oils because large amounts of unsaturated fatty acid esters (oleic, linoleic, and linolenic) tend to stabilize the amines under conditions of heat and air (this is monitored by the reduction of the $-C{=}C-$ bands at 740 and 970 cm^{-1}, and the $={=}C-H$ band at 3079 cm^{-1}). Another simple chemical fact—most ionized materials (salts) are not volatile and don't smell. Use that fact and *neutralize* the free amines with an appropriate acid, like that of lemon juice (citric and ascorbic). Toward the end of the simmering part of the cooking process, add increments of fresh lemon juice, stir, and sniff until the "fishy" smell is reduced or eliminated (spectra show a shift in two bands, 1235 and 952 cm^{-1}, as the amine nitrogen gets protonated). The analysis here was performed by thin-layer chromatography and FT-IR using a Chromalect accessory from Laser Precision and the RFX-40 spectrometer. Elutions were done on silica and THF and AcCN (Fig. 2). Simple chemistry, and a simply delicious meal, too (8)!

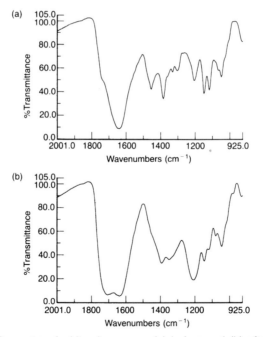

Figure 2. TLC-IR spectra of white clam sauce (a) before and (b) after the addition of lemon juice.

"PLEASE PASS THE SALT AND PIPER NIGRUM L."

In our home, we use more pepper than salt as a table condiment because there probably is more salt put in our foods for us than we really need. There are black peppers (piper nigrum L.), and there are red peppers (capsicum frutescens L.), and although both are hot on the tongue, they act drastically different in the pot. Black pepper has active components that are alkaline in nature (piperine and piperidine) and should be used only in neutral or basic dishes (like New England–style clam chowder.) If used in acidic dishes, they will form salts that will not have nice aromatic fragrances or tastes (that's why we use black pepper in our aforementioned clam sauce). Conversely, red pepper (cayennes, paprikas, chilis) derive their flavors from unsaturated, aromatic monoamides (capsaicin and analogues) that are acidic in nature, stemming from the vanillin moiety that makes up half the capsaicin molecule. The unique absorption band at 3278 cm^{-1} is easily visible, and can be monitored during cooking (especially in classic chili con carne; remember, meat fats are alkaline). Examples of both black and red peppers added to a cream-based sauce (pH = 8.1) are shown in Fig. 3.

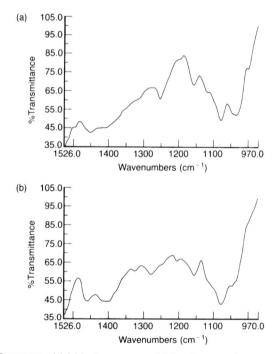

Figure 3. FT-IR spectra of (a) black pepper and (b) red pepper in creamy New England-style clam chowder.

"NOTHING LIKE A NICE STEAK ... UNLESS YOU'RE A VEGETARIAN"

For those of us who enjoy barbecuing meat on the grill, there are many ways of telling when the meat is done. Some of us cut a piece off the original mass and check the color. Others look for a certain degree of surface charring. And a few even time the cooking to the exact minute for the given weight of the meat. The procedure outlined here will provide a guaranteed reproducible technique for preparing meat on the grill, using the common household FT-IR and classic vertical ATR attachment. Ideally, a piece of meat should have the open ends of the cut muscle fibers sealed by searing the surface with short exposure to high heat, then the rest of the mass should be slowly cooked by inductive radiation of heat. This will minimize the overtly destructive pyrolysis of the meat (with its subsequent loss of moisture and protein) and maximize the slow denaturing of the muscle proteins within the meat. The final product will be uniformly tender and juicy, with less potentially harmful pyrolysis byproducts on the surface. Front-surface ATR of the meat surface on a KRS-5 crystal shows less intense polyamide band structure between 1450 and 1650 cm^{-1}, due to pyrolysis and hydrolysis of protein to individual amino acids and finally carbonaceous char. ATR of a cooked inner surface shows intact protein structure within 5 mm of the surface in a properly cooked steak. Examples are shown in Fig. 4.

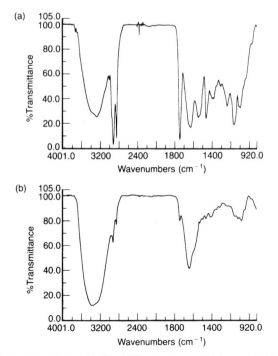

Figure 4. FT-IR spectra, taken with the ATR accessory, of (a) oven-broiled and (b) charbroiled club steaks.

A BUTTER WAY TO BEAT BAD BREATH?

I am not a lobbyist for the American Dairy Association, nor do I have any affiliation with the state of Wisconsin, but I'd like to make my stand for butter. Garlic and onions are important ingredients of many dishes. In years past, a pat or two of butter was used to sauté these members of genus allium L., which have in common the chemical diallyl disulfide. Butter, which is predominantly saturated fats, does not cause significant morphological changes in the major disulfide and minor thiosulfinate components. Margarine (a.k.a. oleo, for the older generation) is mostly soybean oil, a mixture of solid, saturated oil and liquid, unsaturated oil. From FT-IR data obtained by horizontal ATR, there is significant cleavage of the disulfide linkages to form sulfhydryl functional groups. These are typically not nice things to smell. So, to get the best flavor from your onions and garlic, give 'em each a little pat of butter; maybe your garlic breath won't be as bad the next day.

WHAT'S FOR DESSERT?

Sorry, we only serve main courses at this article. Though the examples shown here have been presented as a leisurely exercise in the applicability of FT-IR to food analysis, the true potentials remain limitless in the areas of nutritional evaluation, reactivity of ingredients, effects of packaging and preparation, shelf life and stability, and even bioavailability if a suitable sampling technique can be implemented. FT-IR has few insurmountable interferences. FT-IR allows automated routines to be executed without any operator intervention, which is ideal for a process environment. FT-IR can be used to detect nanogram contaminants if coupled to the appropriate accessory (i.e., TLC/FT-IR), which is ideal for the high-level researcher. As chemicals—natural and synthetic—become an ever-growing part of the infamous list of "ingredients," the capabilities of food laboratories must also grow to keep in step with the technology of the times. We should not limit the analysis only to the product before the consumer purchases it; we also can evaluate the best ways to prepare the product, documenting the various preparations. The science of culinary spectroscopy can be extremely valuable in the scope and quality of data it can provide. Bon appetit!

ACKNOWLEDGMENTS

My specials thanks to Nani and Mom for the old family traditions, and to my wife Gwen for the new ones.

REFERENCES

1. E. H. S. Bailey, *The Source, Chemistry, and Use of Food Products* (P. Blakiston & Sons, Philadelphia, 1914), Chapters 6, 10, 12, 14, and 18.

2. N. A. M Eskin and R. J. Townsend, *Biochemistry of Foods* (Academic Press, London, 1971), pp. 6–23; 46–62; 70–74; and 130–45.

3. Owen R. Fennema, ed., *Food Chemistry*, 2nd Ed. (Marcel Dekker, New York, 1985), Chapters 4–6, 11, 12, and 15.

4. R. D. King, ed., *Developments in Food Analysis Techniques*, Vol. II (Applied Science Publishers, London, 1980), pp. 185–258.

5. R. Lees, *Analytical and QC Methods for the Food Manufacturer and the Buyer* (Leonard Hill Books, London, 1975), pp. 203–20.

6. D. Pearson, *Developments in Food Analysis Techniques*, 7th Ed. (Chemical Publishing, New York, 1976), Chapters 9, 11, 14, and p. 548.

7. Mrs. Alphonsina DeMenna, private communications (1967).

8. Lily Wallace, *Seafood Cookery* (Crown Publishers, New York, 1968).

CHAPTER II.A.4

HAPPINESS IS A WARM SAMPLE

Greetings all! As I prepare for Pittcon, I have some time to think. (Dangerous, at best.) The best part about writing a column to be distributed at Pittcon is that Spring follows closely upon its heels, and Spring is warmer and brighter than Winter. Because warmth and light are the heart and soul of vibrational spectroscopy, this is a good thing.

I was trying to think of a topic for this edition that I haven't addressed (in my half-decade plus of doing these columns) when the phone rang. I'm always fascinated by phones. Little marvels of modern technology ... except they were invented more than a hundred years ago. Then I started thinking about Alexander Graham Bell and why he worked on sound amplification (his sister was deaf and he was trying to build a hearing aid). We all know how Don Amechi discovered the phone and asked Watson to come to him, but fewer of us are aware that Mr. Bell worked on other inventions as well.

One discovery, published in 1880 (1), was the photoacoustic effect (also referred to as optoacoustic or acousto-optic effect, not to be confused with an AOTF, referred to from time to time in this column). Basically, he found that a sample, sealed in a cell with a microphone, could be made to emit sound waves, which were then detected by the microphone. The overall technique was considered a curiosity until the mid 1970s, when the first instruments began to be built for analyses. These early applications were described in a text by Dewey in 1977 (2). Although much of the first work was performed on gases (3), later work followed on liquids and solids. The applications have also been extended to inorganic, organic, and biological samples (4).

The instrumentation is relatively simple: Light is directed onto a sample that radiates sound waves to a sensitive detector. The essential parts include the following (5–7):

Radiation source. This is most often a xenon arc lamp between 300 and 1000 W. This allows exploration of the UV to near-IR ranges. Lasers, either continuous wave (CW) or pulsed, may also be used. As with the xenon lamp, the required wavelength determines the laser used.

Monochromator. This is most often a standard grating type. The size, blaze angle, and so on are decided by the wavelength range in use. Prisms and acousto-optic tunable filters (AOTF) may also be used for wavelength discrimination.

Modulator. Light impinging on the sample must be "chopped" or modulated for CW sources (CW laser or lamp). A pulsed laser carries its own modulation. Most often this may be a mechanical chopper.

Modulation controller. This is a dual-purpose component; it controls the frequency of the chopper and signals the signal processor with a reference pulse. The frequencies used range between 10 and 100 Hz.

Window. The only transparent portion of the sample cell. It is constructed of the appropriate material for the wavelength used (quartz, glass, low-OH silica).

Sample. The form of the sample varies, but gases are usually in direct contact with the transducers; thus the sample cell is the entire compartment. Liquids may be set up so that they, too, have direct contact with the detector. Solid samples are placed such that they receive the incident light whereas the detector is shaded. Acoustic waves are transmitted through the air or whichever gas is filling the cell.

Detector and preamplifier. The detector is a very sensitive microphone or piezoelectric transducer. In essence, the acoustic wave distorts the surface of the transducer, causing a small electric signal. This minuscule signal is amplified sufficiently to transmit over a wire.

Signal processor. This lockin amplifier compares the signal from the sample with a blank or reference signal, constructing a spectrum. The phase of the photoacoustic (PA) signal lags behind the reference signal that tracks the modulated incident radiation. This allows for "real-time" comparisons.

The theory, when expressed in mathematical terms, is complex, whereas I'm not. The simple explanation is that the sample absorbs wavelengths of light as in any self-respecting spectroscopic technique. However, in photoacoustic spectroscopy (PAS), the sample absorbance is converted to minute amounts of heat (well, in any type, heat is produced, but we assume a constant temperature for Beer's law). This warmth is transferred to the surface of the sample and radiated in pulses of acoustic waves. The transducer picks up these waves, and the rest, as they say, is history.

The sample types analyzed are as varied as in any type of work. Some typical types are:

Gases. One of the simplest types to work with are gases. The sample compartment is used as a sample holder, and the beam is directed through the gas. Multipass mirrors may be included for longer pathlengths for higher sensitivity. Typical applications are usually the detection of trace contaminants in pure gases; for example, methane, NO, NO_2, NH_3, and SO_2. Detection limits, depending on the gas, are in the 0.1–10-ppb (v/v) range. Detection limits are greatly determined by the ratio of blank noise to the response obtained from the sample absorption.

Liquids. Where this technique truly shines (pardon the choice of words) is in highly scattering media such as blood. PAS allows the study of whole blood without prior removal of proteins and lipids (8). HPLC detectors have also been built using PAS with sensitivities over an order of magnitude greater than conventional UV detectors.

Solids. Near and dear to my heart is its ability to work on spots directly on thin-layer plates (9). In the IR region of the spectrum, PAS has a lower S/N than most FT-IR instruments and may be used for cleaner qualitative measurements of pure substances. With highly absorbing solids, the depth of penetration is small enough that sample preparation is unnecessary. When a sample has low absorbencies, a thin wafer (0.01–0.001 cm) may be made.

Another application of PAS is called *photoacoustic Raman spectroscopy* (PARS). With this method, two incident laser beams strike the sample. These cause a given energy state to be populated by a simulated Raman effect. The frequency difference of the two beams is adjusted to equal the frequency of a Raman active transition. Relaxation of the excited molecules generates an acoustic wave detected by the microphone. Because PARS is linear and has concentration detection limits in the ppm range, it is mostly applied to gases. PAS produces spectra generated by the ratio of the PA sample signal to the reference signal at each wavelength.

In dual-beam instruments, the beam is split with one portion deflected to a reference and the other to the sample. These dual-beam instruments separate and detect the signals and continuously provide a ratio to produce a normalized, real-time scan. The time needed to generate a spectrum depends on the wavelength range, type of monochromator, and precision needed.

Detection limits are greatly determined by the ratio of blank noise to the response obtained from the sample absorption. Some noise is source independent, generated as electric noise within the microphone–amplifier combination, Brownian motion of molecules, and mechanical vibrations.

To obviate mechanical noise, the cell is usually separated from the rest of the instrument and acoustic baffles are placed in the sample compartment. The PA signal is proportional to $1/f^n$, where f is the modulation frequency and n is a number between 0.5 and 1.5. The working range for f is between 10 and 1000 Hz. Low frequencies produce a larger magnitude of acoustic waves; however, S/N is better at moderate frequencies because of the $1/f$ character of source-independent noise. In addition, most microphones have a poorer response at low frequencies.

In general, PAS hasn't gotten the press of the more sophisticated techniques, but it, like near-IR, has been around since the nineteenth century and has proven a useful tool. I won't suggest scrapping any other spectrophotometer in lieu of PAS, but, hey, give it a "look-see" at your next conference.

REFERENCES

1. A. G. Bell, *Proc. Am. Assoc. Adv. Sci.* **29**, 115 (1880).
2. C. F. Dewey, Jr., in *Optoacoustic Spectroscopy and Detection*, Y-H Pao, Ed. (Academic Press, New York, 1977).
3. A. Rosencwaig, *Photoacoustics and Photoacoustic Spectroscopy* (Wiley-Interscience, New York, 1980).
4. C. N. Reilley and C. M. Crawford, *Analyt. Chem.* **27**, 716 (1955).
5. J. D. Ingle, Jr., and R. R. Crouch, *Spectrochemical Analysis* (Prentice-Hall, New Jersey, 1988).
6. D. A. Skoog, *Principles of Instrumental Analysis*, 3rd Ed. (Saunders College Pub., New York, 1984).
7. G. D. Christian and J. E. O'Reilly, *Instrumental Analysis*, 2nd ed. (Allyn and Bacon, Boston, 1986).
8. A. Rosencwaig, *Analyt. Chem.* **47**, 596A (1975).
9. *Ibid.*, p. 600A.

CHAPTER II.A.5

NONSTANDARD SPECTROSCOPIC MEASUREMENTS (OR, WHAT TO DO WHEN IT'S NOT A CLEAR SOLUTION OR FILTER ...)

Arthur Springsteen

This selection, dear readers, is a tasty morsel by Art Springsteen of Labsphere, Inc. Art is an interesting fellow, involved in numerous activities centered on reflectance measurements. He works with NIST (National Institute of Standards and Technology, formerly National Bureau of Standards) on numerous projects, all aimed at taking diffuse reflectance from an art (nothing personal, Art) to a science. He is also a fun person to share a wine cooler with at a meeting such as Chambersburg (where I talked him into writing this gem). I trust it will clarify some fuzzy points about reflectance work, integrating spheres, etc. Enjoy!

Typically, spectrophotometers are used for the measurement of nonscattering solutions. Who can forget their days in freshman chemistry slaving away with solutions of potassium dichromate over a Spectronic 20™? Once out in the real world, we realized that spectroscopy was not always the measurement of ideal solutions—there were things to measure that scattered light, or—horrors!—were solid and sometimes couldn't be ground up and placed in Nujol mulls or KBr pellets.

Fortunately, modern spectrophotometers have the versatility to be accessorized to permit the measurement of practically any solid, liquid, or gaseous sample. A wide variety of standard accessories for the measurement of diffuse reflectance and transmittance for practically all current UV-vis, UV-vis-NIR, and FT-IR spectrometers are now produced along with specialized and custom accessories for remote applications, specular reflectance, and fluorescence for a wide range of instruments.

This chapter will describe typical spectrophotometric measurement techniques (with the exception of standard transmittance, of which a knowledge is assumed—or you wouldn't be reading this volume!) and an introduction to accessories and spectrophotometers that will perform that function.

The interaction of light with matter can occur in a number of manners, as outlined in Fig. 1. Each of these interactions requires a certain configuration of instrument to accurately measure the absorbance, transmittance, or reflection of light from the object.

This chapter will deal with all but the regular transmittance (although most sphere-based accessories are configured to measure the regular transmittance of both solid and liquid samples) and retroreflection, the measurement of which is currently not performed by any common commercial spectrometers.

A Treatise on Geometry

No, we're not going to talk about determining the volume of a hypercube or trigonometric functions; so don't let your eyes glaze over. The geometry we'll discuss is that of illumination angle (the angle at which the illuminating beam strikes the object to be measured) and the collection angle (the angle at which the detector is from the measured object). According to ASTM and CIE definitions, the illumination angle is always the first number (or letter, as you will

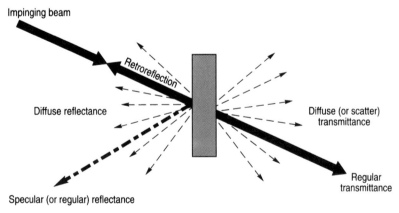

Figure 1. The interaction of light with matter.

see), with the collection angle following separated by a slash (/). Thus, if one illuminates a sample with 4° incident light and has a detector that views the samples from above (normal to the sample or 0°), a geometry that is common in inexpensive color analyzers and compact reflectance measurement instruments, the geometry is properly described as 45/0. If one reverses the detector and light source, the geometry now becomes 0/45. Pretty simple, right? Now suppose you illuminate at 0° but use an integrating sphere as a collector, which catches all forward scatter over 180°. This is called *diffuse collection* (there's also diffuse illumination, but more on that later . . .), abbreviated as *d*. So the geometry is called 0/*d**; the reverse geometry (using the sphere as a light source and viewing the sample at normal incidence is *d*/0).

Fortunately, the previous four geometries describe most of the applications seen (there is the odd 30/*d* geometry specified in one ASTM specification) and multiangle geometries, but remember, it's incidence angle first, collection angle second; you'll have no trouble deciphering geometry nomenclature. [Ah, the tangled web of nomenclature. You will occasionally (especially in European circles) see 8°/*t* and 8°/*d*. The *t* is for total reflectance (specular component included), and the *d* is for diffuse only (specular component excluded).]

Helmholtz reciprocity states (in simple terms) that light doesn't know the direction it is moving: towards or away from a material; thus the 0/45 and 45/0 geometries are equivalent. For directional/directional (that's a number followed by another number) geometry, that is always true. The definitive experiment work was done by Frank Clarke (1) at the National Physical Laboratory in the UK about 20 years ago. This principal also holds in most cases for directional/ hemispherical or hemispherical/directional geometry (that's #/d or d/#). It does NOT hold true if materials are luminescent and the illuminating radiation causes the material being measured to luminance.

DIFFUSE OR SCATTER TRANSMITTANCE

An integrating sphere is an ideal collector for diffuse or scattered transmittance in the 0°/*d* or *d*/0° geometry. The sphere provides the detector with a 180° field of view, which effectively captures all forward scatter. The traditional method of making this measurement, the Scatter-T accessory, simply placed the sample as close to the detector (usually a photomultiplier tube) as possible. This technique is effective for samples up to the size of a cuvette. It is, however, very sensitive for low concentrations of minimally scattering materials. Most integrating sphere accessories for spectrophotometers (with the exception of some downward-looking spheres) will accurately measure this scatter. Consideration must be given when using an integrating sphere accessory on a single-beam instrument or dual-beam instrument that uses only one channel of the two beams to allow for single-beam substitution error (2) inherent in these systems.

DIFFUSE REFLECTANCE

Reflected scattered light can be collected in a number of geometries, as described. Typically, integrating spheres are used to collect reflected radiation, as they collect over all reflected angles and, if small, are extremely efficient collectors. Larger spheres, while not so efficient, offer better integration and higher accuracy of measurement. Depending on the sphere coating, these sphere accessories can be used from the UV to the mid-IR.

So what IS an integrating sphere? Basically, it is a hollow ball with a number of ports for the light beam(s) to enter, sample ports, and places for detectors or sources with a highly reflective coating (or the sphere produced from a highly reflective material) on the inner surface. The coating used depends on which area of the spectrum you wish to measure. In days of old, this coating might be smoked magnesium oxide (and there are hordes of horror stories about making spheres that way!) or a barium sulfate–based paint. Both of these materials limit the range of the spectrophotometer to the very-near UV (>350 nm) to the very-near IR (<1250 nm). Newer materials offer greater flexibility and stability. The author's company, Labsphere, offers a very stable material called Spectralon® (Fig. 2) that gives very high reflectance over almost the entire UV-vis-NIR range (200–2500 nm), is rugged, and is durable. Packed PTFE powder offers similar high-reflectance performance, but spheres made from this material are extremely fragile.

Working in the mid-IR demands a coating that has no IR absorbances, thus ruling out any organics. Two materials are used—packed flowers of sulfur (again, very fragile and usually not suitable for everyday spectroscopic work) and a highly scattering gold coating such as Labsphere's InfraGold® or InfraGold-LF™.

Typically, integrating spheres are used for reflectance measurement of solids, but liquids, pastes, and powders are also easily measured with proper sample handling capability. This can be done either at the accessory level (by using a downward-viewing sphere) or by the use of powder cells that can handle both

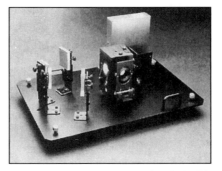

Figure 2. A 60-mm sphere accessory using Labsphere's highly reflective Spectralon.

Figure 3. A biconical accessory.

powders and viscous liquids. Certain high-end instruments are also capable of accepting accessories that measure at 0/45 geometry.

A technique that is often used in IR and becoming more popular in UV-vis-NIR spectroscopy uses what is called *biconical geometry*. (Figs. 3 and 4). This geometry provides not quite perfectly directional illumination and not quite

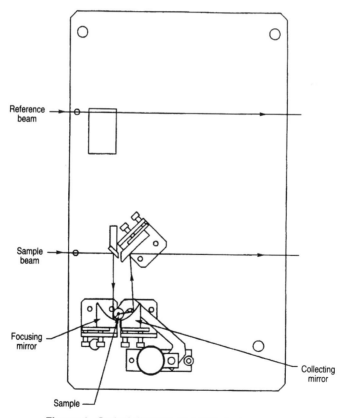

Figure 4. Optical diagram of a biconical accessory.

perfectly hemispherical collection, but the resulting accessories allow users to measure extremely small spots with very high signal-to-noise ratios. The disadvantage is that the results are not as quantitative as one might like, it is but an excellent way of generating qualitative data on materials that might otherwise be impossible to measure. (The author has generated excellent UV-vis-NIR spectra of dinitrobenzene inclusion compounds in cyclodextrins on as little as 5 mg of sample using biconical devices. That you most assuredly *cannot* do with an integrating sphere!) Typical applications would be looking at a small spot of blood on fabric (forensics), a discolored spot on a fabric or plastic blank (quality control), or small tissue or culture samples (medical).

Another technique of diffuse reflectance is the measurement of total hemispherical reflectance (and transflectance, but more on that later) at variable incidence angle and/or at fixed angle in the center of an integrating sphere. This technique is useful for measuring changes in reflectance of a material as the incident angle of the source varies (imagine looking at a painted surface outside as the sun moves across the sky, assuming the sun to be a directional, monochromatic source!), as is commonly done in aerospace and camouflage applications. A similar technique, using a fixed angle center mount, allows for the direct measurement of total absorptance [as $1-(R + T)$] and thus absorbance (as-log of the absorptance). This can be done on solids or, with the proper sample holder, liquids in cuvettes. This technique, known as *transflectance*, is quite useful in measuring scattering liquids such as blood and in the measurement of scattering transmittive samples.

Finally, there's a fairly new method of looking at diffuse reflectance that combines the latter technique with a twist: The sample can be illuminated at different angles by moving the source, AND the detector(s) can be moved. This technique, called *goniospectrophotometry* (Figs. 5 and 6), is really the only way

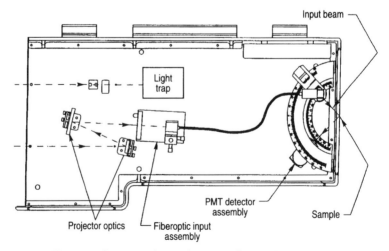

Figure 5. Diagram of a goniospectrophotometry accessory.

Figure 6. Goniospectrophotometry accessory.

to evaluate properly what are called effect pigments (or gonioapparent, to use ASTM terminology) such as mica flake, metal flake, or graphite flake pigments that are used in automotive paints. With these coatings and materials (the pigments easily blend into polymer melts to give the same effect) becoming more and more common in usage, this form of spectrophotometry will become more important to industry. This technique is also useful in measuring holograms in reflectance (no easy feat, as you might imagine) or in gonioapparent inks such as are used in the new $100 bill.

Specular Reflectance

This is the reflectance almost everyone is familiar with, wherein the angle of incidence equals the angle of reflection. There are a number of ways to measure specular (or, as our British colleagues are inclined to call it, "regular" reflectance). These can be broken down into two main categories: absolute and relative. Relative is the measured reflectance relative to an artifact standard, generally another mirror with similar reflectance properties to the sample to be measured. Relative measurements can be done at a number of angles. Typically near-normal is between 6 and 10° and may be measured either on an accessory with this angle built in or with a sphere (so the measurement is for total reflectance at something like $8°/d$, but all the reflectance is assumed specular). Other typical measurement angles are *30°, 45°, and 60° and grazing angle (80°)* (Figs. 7–9). These measurements are quite precise since there is a single bouncing off both the reference and sample mirror in the same orientation; however, accuracy is almost totally dependent on the quality of the reference mirror (3, 4) used.

Absolute measurements are defined as those in which no artifact standard is used and in which the only difference between the baseline (or reference) measurement and the sample measurement is the imposition of the sample in the

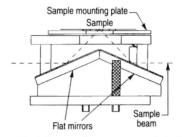

Figure 7. 45° specular accessory (relative).

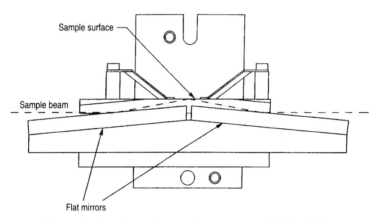

Figure 8. Grazing angle specular accessory (relative 80°).

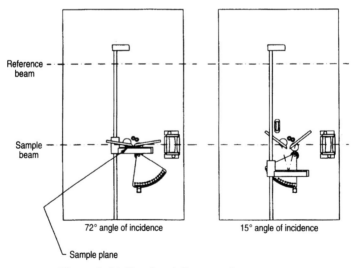

Figure 9. Multiangle relative specular accessory.

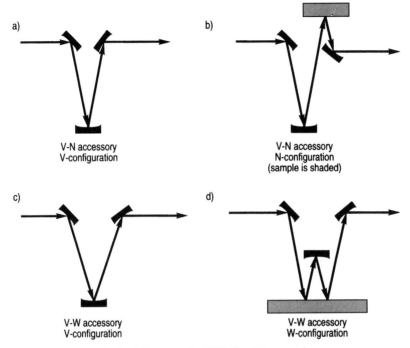

a)

V-N accessory
V-configuration

b)

V-N accessory
N-configuration
(sample is shaded)

c)

V-W accessory
V-configuration

d)

V-W accessory
W-configuration

Figure 10. Optical diagram of a V-W absolute specular accessory.

optical path at the defined angle of incidence. There are typically two geometries used for fixed-angle measurements of absolute specular reflectance—these are the so-called V-W and V-N geometries. In a *V-W accessory* (Fig. 10), the sample is contacted twice at different positions; thus the reflectance measured is actually the square of the actual reflectance. This leads to difficulty in measuring extremely low-reflectance materials such as A-R coatings, where the reflectance is of the order of 0.1% (or about 3 absorbance units). The square of that reflectance would be 0.0001% (or about 6 A), below the noise limits of most spectrophotometers. In a V-N design, the sample is only contacted a single time. The V-N is certainly the favored device but is much more difficult to design and produce from an optical standpoint.

A more recent design allows for measurement of *absolute reflectance at variable incidence angle*. The variable angle absolute specular accessory (Fig. 11) allows measurement of absolute reflectance on nonscattering samples from about 10° to about 70° by use of a moving integrating sphere collector. This collector travels to follow the reflected beam as the angle is changed. This is a single-bounce accessory, much like a V-N, but the sphere collector does attenuate the signal to the detector somewhat. Very low-reflectance coatings may require reference beam attenuation.

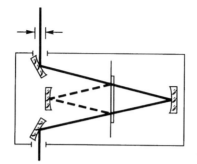

Figure 11. Optical diagrams of V-N and V-W accessories.

CONCLUSIONS

It becomes apparent from the preceding that a large number of sample measuring problems may be overcome by reflectance spheres. It is also apparent that there are more than one approach for any particular sample. The common knowledge that sample preparation is 90% of any spectroscopic measurement certainly applies to the samples described. It is hoped that this explanation of the physics of integrating spheres will allow more spectroscopists to be successful in their work.

SELECTED REFERENCES

1. F. J. J. Clarke and D. J. Parry, "*Helmholtz Reciprocity: Its Validity and Application to Reflectometry*," *Lighting Research and Technology* **17**, 1–11 (1985).
2. A. Springsteen, S. Blanchard, T. Ricker, and K. Dwyer, "*Quantitation of Single Beam Substitution Error*," Labsphere Technical Note, 1996.

GENERAL REFERENCES

3. K. Mielenz and C. Burgess (eds.), *Advances in Standards and Methodology in Spectroscopy* (Elsevier Press, 1987).
4. C. Burgess and D. Jones (eds.) *Spectrophotometry, Luminescence, and Colour, Science and Compliance* (Elsevier Press, 1994).
5. F. Billmeyer, E. Carter, and D. Rich, ISCC Technical Report 89–1, "*Guide to Material Standards and Their Use in Color Measurement*" (Intersociety Color Council, 1989) (currently in revision).
6. A. Springsteen, "*A Guide to Reflectance Spectroscopy*," Labsphere Technical Guide (1993, 1995 revision).

CHAPTER II.B.1

PART 1: ARTERIAL ANALYSIS WITH A NOVEL NEAR-IR FIBER-OPTIC PROBE

Robert A. Lodder, Lisa Cassis, and Emil W. Ciurczak

This month, we provide a double-barreled installment of the Molecular Spectroscopy Workbench. I have turned the first few pages over to Robert A. Lodder and Lisa Cassis, both of the University of Kentucky's Chandler Medical Center, Rose Street, Lexington, Kentucky 40536. Lodder and Cassis have been developing a diffuse-reflectance fiber-optic probe to collect near-infrared spectra from the intimal surfaces of arteries. The colorful results of their experimentation are shown here. In Part 2 of this month's column, we present the findings of the instrument repair and maintenance survey I initiated (see p. 84). This informal poll provides a glimpse at how spectroscopists view their instruments and the technicians they must call into action when things go wrong.

Our recent work at the University of Kentucky Medical Center has been directed towards the development of a diffuse-reflectance fiber-optic probe to collect spectra from the intimal surfaces of arteries. A prototype fiber-optic probe was used in experiments that mapped the lipoprotein composition of the thoracic aorta in rats. The near-IR fiber-optic probe is intended for use in studies of the growth of atherosclerotic lesions and in the chemical examination of arterial endothelium (the portion of the artery now thought to be a major factor in the development of arterial blockages as well as in arterial blood-pressure control). Until now, it has not been possible to determine location and quantities of substances such as high-density lipoprotein (HDL), low-density lipoprotein (LDL), and apolipoproteins in living arterial tissue.

The ability to perform such analyses opens up the possibility of performing kinetic experiments in which the quantities of these materials are studied over time in a lesion, both as the lesion grows and as cholesterol-lowering drugs are used in attempts to shrink the lesion. Theories intended to describe what may cause lesion shrinkage, such as reverse cholesterol transport (a theory in which HDL acts to remove cholesterol from tissues to the liver, where it is excreted in the form of bile acids), will be tested with the fiber-optic device.

A compound parabolic concentrator (CPC), similar to those used for solar power concentration (1), was used to compress the beam from a transmitting optical fiber onto a small spot on the artery surface. The CPC was molded from a polymer and had a polished aluminum lining. Nonimaging CPCs have been shown to improve light gathering by a factor of four or more over lens-based focusing designs. The aperture at the distal end of the fiber-optic probe was ~0.74 mm^2 and could be moved across the arterial surface in increments as small as 10 μm with a micropositioning stage. Near-IR light across a wavelength range from 1100 to 2500 nm was transmitted through the concentrator. Collimated source light was directed on the artery by the concentrator. Slightly skewed transmitted rays were still focused into the aperture by the shape of the CPC. However, scattered light from the artery tends to travel in all directions from the point of scattering and therefore produced a substantial amount of scatter at the proximal end of the CPC. This scattered radiation was detected at the proximal end of the CPC by lead sulfide detectors located off the axis of the incident beam.

The false-color "maps" that appear in Figs. 1 and 2 represent the lipoprotein composition of the thoracic aortas obtained from two laboratory rats. The aortas were each excised and partially denuded, removing the endothelium from a portion of the vessel wall of each rat. One of these aortas (Fig. 1) was incubated

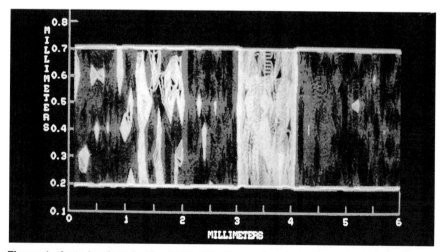

Figure 1. Aorta incubated without LDL. The "normal" aorta is shown in shades of gray.

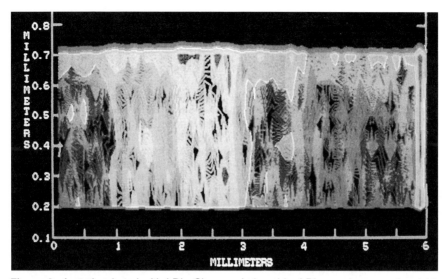

Figure 2. Aorta incubated with LDL. Changes induced by LDL are shown on the right-hand side in shades of gray.

for 2 h in Krebs physiological salt solution. The other aorta (Fig. 2) was incubated for 2 h in a Kreb's physiological salt solution that also contained LDL (500 μg/ml).

The arteries were washed following incubation and passed beneath the concentrator while the concentrator was held fixed. The micropositioning stage allowed spectra to be collected along the axis of the artery when the artery was opened to expose its intimal surface. Both segments of the thoracic aorta were scanned along a track 1 × 6 mm in dimension. Figure 3 depicts the spectral vector in the principal axis transformation matrix corresponding to the spectral change observed in the artery when the artery was incubated in LDL. The major spectral changes were observed near 1560 nm and between 1700 and 1870 nm.

Figure 4 summarizes the process by which colors were assigned to pixels in the arterial images. The "O"s represent sample spectra obtained from the aorta incubated without LDL, and the "+"s represent sample spectra obtained from the aorta incubated in the LDL. Distances are measured in standard deviations (SDs) in spectral hyperspace between the center of the aorta spectral points incubated without LDL and each test-sample spectral point (which may or may not come from an aorta incubated in LDL). A parallel assimilation algorithm (2) was implemented on IBM 3090–600J supercomputers at the Cornell National Supercomputer Facility and the University of Kentucky Center for Computational Sciences to calculate the distances and to assemble the results into the final color images.

The center of the aorta spectral points incubated without LDL are coded dark blue. Vertical motions along the solid lines in Fig. 4 (representing baseline shifts

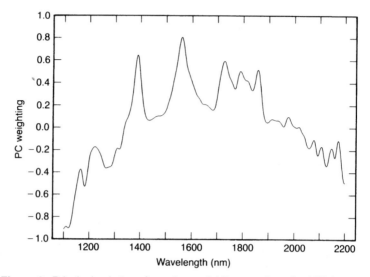

Figure 3. Principal-axis transformation weighting spectrum for LDL in aorta.

that probably correlate to vessel wall thickness) are coded green. Horizontal motions along the dotted lines in Fig. 4 (apparently representing uptake of LDL) are coded pink and red. Red represents a larger movement in SDs than pink. The contour lines in the arterial images are drawn every 0.1 SD, and a color change occurs approximately every 3 SDs.

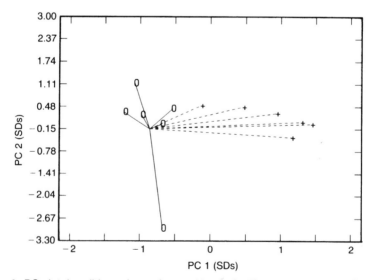

Figure 4. PC plot describing color assignment to pixels. 0's represent spectra from aorta without LDL, and +'s represent spectra from aorta with LDL.

The image in Fig. 1 is predominantly blue and green, indicating that the aorta is similar to spectra obtained from such tissues incubated without LDL. The green portion is caused by a baseline shift that may be the result of a thickening of the vessel wall that brings the sample closer to the optical aperture. The image in Fig. 2 shows nearly orthogonal changes that correspond to the uptake of LDL by the aorta wall. The red color represents regions of maximum uptake of LDL. The exact form in which this LDL resides in the arterial wall is still unknown.

In the future, we will be constructing an improved fiber-optic catheter that will be used to conduct *in vivo* imaging of arterial lesions. The system is expected to increase existing knowledge of arterial endothelium, permit kinetic studies to be performed on lesions both as they grow and as they shrink in response to drug therapies, and clarify the role of lipoproteins in the artherosclerotic disease process.

REFERENCES

1. P. Gleckman, J. O'Gallagher, and R. Winston, *Nature* **339**, 198–200 (1989).
2. R. A. Lodder and G. M. Hieftie, *Appl. Spectrosc.* **42**, 1351–65 (1988).

PART 2: INSTRUMENT REPAIR SURVEY RESULTS

The following includes the results of an instrument repair poll.

I wouldn't want to color readers' interpretations of the numbers I am about to present; so I will withhold my comments on the results until the latter part of the column.

THE PURCHASE

During your last purchase, was the sales representative helpful in choosing the proper model and/or options?

Yes 87%
No 13%

Were you ever sold an instrument that did not perform the task for which it was purchased?

Yes 32%
No 68%

If yes, do you think it was because:

The product was misrepresented . . 43%
I assumed I knew the
  capabilities of the instrument . . . 29%
The rep didn't know how to
  answer my questions 14%
The rep withheld information 14%

THE INSTRUMENTS

What types of spectrometers do you use frequently?

FT-IR 27%
NMR 20%
Mass 18%
UV/Vis 16%
Grating near-IR 7%
Dispersive IR 4%
Filter fluorometer 2%
Scanning fluorometer 2%
Other 2%

Which suppliers do you patronize for these instruments?

Baird, Beckman, Bio-Rad/Digilab, Bomem, Bruker, Coleman/Hitachi, Delsi-Nermag, Finnigan MAT, General Electric, Guided Wave, Hewlett-Packard, Hitachi, JEOL, Jobin-Yvon/Coherent, Kratos, Mattson, Nicolet, NIRSystems, Perkin-Elmer, Shimadzu, Varian

In general, is durability or versatility the main characteristic you seek in a new instrument?

Durability 59%
Versatility 27%
Both 14%

From your experience, is your most used instrument:

Versatile 14%
Durable 18%
Both 68%

Overall, how would you rate your instruments in terms of dependability and serviceability?

Excellent 44%
Good 44%
Fair 4%
Poor 8%

QUALITY OF SERVICE

What kind of training have you received from the supplier?

Good 23%
Fair 36%
Poor 23%
None 18%

The last time you ordered a part, how long did it take to arrive?

Overnight 14%
<1 week 27%
1 week 14%
2–3 weeks 27%
Don't know 9%

When an instrument breaks down, what is the average response time before:

(a) Your call is returned?
No wait 4%
1–2 hours 17%
Same day 35%
1–2 days 30%
3–4 days 4%
Never calls 4%

(b) You receive the needed part?
Overnight 38%
~2–4 days 13%
1 week 19%
1–2 weeks 19%
Never (used third party) 4%

(c) A service rep calls?
Overnight 6%
1–2 days 33%
<1 week 39%
1–2 weeks 17%
3 weeks 6%

When a service technician calls on your facility, does he/she:

Have all needed parts 19%
Know the model in question 30%
Have the knowledge to repair . . . 30%
Explain how to avoid problems . . . 21%

When the repair is made, does the instrument work as well as before the problem occurred?

Yes 100%

GENERAL PERCEPTIONS

How much importance do you place on service when making a purchase?

A great deal 65%
Some 35%
Not much 0%
None 0%

Do you believe your instrument reflects the general reliability of the supplier's other instruments?

Yes 82%
No 18%

Overall, how would you rate the repair service from your principal instrument manufacturer or supplier?

Excellent 25%
Good 50%
Fair 20%
Poor 5%

On what do you base your judgment for this rating?

Personal experience 80%
Experience of colleagues 16%
Surveys 0%
Manufacturer claims/reputation 4%

SERVICE NEEDS

On average, how often must you replace expendable parts?

Every 0–6 months 32%
Every 12 months 26%
Every 1–2 years 15%
Every 2–3 years 21%
Less often 5%

Do replacement parts last as long as the originals?

Yes 82%
No 12%

Are service calls usually necessary?

Yes 24%
No 76%

Was the operator's manual any help the last time you attempted to fix an instrument or did you ask someone else for help?

Manual 42%
Other person 35%
Luck/skill 23%

Most responses were from spectroscopists working in private industry, although academic institutions and crime labs were also well represented. Respondents reported that their facilities employed anywhere from 8 to 2500 souls, with budgets ranging from "the Lord will provide" to $3 million per year. There seemed to be no correlation between site size and the perceived quality of service; big customers were sometimes unhappy while the "little guy" was well taken care of.

Some findings I found particularly interesting included the 100% completion rate of repairs; when the instrument was fixed, it stayed fixed. The wide range of time for calls to be returned was also interesting. The one comment that decried a company that "never calls back" was directed toward a company that got overall good grades from other customers.

One thing I might remind readers is that my 20+ years in industry taught me that instrument manufacturers are staffed by human beings: They make mistakes, and they can be excellent. Like any other service operation (including police and fire departments), they only hear problems. Often they are personally blamed for problems that are not their fault, nor indeed anyone's fault. To bruisingly paraphrase a common quote, "Hit me and I hurt, prick me and I bleed, kvetch too much and I hang up!"

This is not to imply that there is no room for improvement in the way repair service is provided. Often the instrument company's sales-driven management looks only to the bottom line for the quarter (month, week) and considers service to be an overhead expense rather than a good business practice. This philosophy is not much different than a car dealer who forgets your name the day after you drive off with your new car. For another example, how many large chemical or pharmaceutical firms think of R&D and QC as expenses, but not the $200 million spent in advertising every year?

In its brevity, this survey has taken some things for granted. As many readers will be quick to point out, there really is no comparing the service requirements of an NMR spectrometer with those of, say, a UV/vis instrument. The expendable products for various molecular spectroscopy instruments cover a vast range of technical complexity, operational lifespan, cost, and ease of acquisition. The data presented here don't attempt to compare apples to oranges, and so most questions were formulated for generic application to all methods. Accordingly, readers are encouraged to view these findings as a rather broad indicator of their colleagues' opinions.

As a final comment, I wish to thank all the service technicians who taught me more about my "toys" over the years than any university or manufacturer could ever have done. I think these individuals do a super job in the face of ever more complex and less user-fixable instruments that now cover the scientific landscape.

CHAPTER II.B.2

IDENTIFICATION OF ACTIVES IN MULTICOMPONENT PHARMACEUTICAL DOSAGE FORMS USING NEAR-INFRARED REFLECTANCE ANALYSIS

Emil W. Ciurczak and Thomas A. Maldacker

Three methods of data treatment—spectral subtraction, spectral reconstruction, and discriminant analysis—were applied to the near-infrared spectra of a dosage form containing three actives. In two cases, the individual spectra of the actives were reconstructed (via a computer algorithm). In the third case, the sample was categorized as containing all actives (in acceptable proportions) or as missing one or more actives on the basis of near-infrared spectral data. I am thrilled that, after 10–12 years, the FDA and the rest of the industry are "discovering" some of this earlier work!

In traditional applications of spectroscopy by the pharmaceutical industry, the near-infrared region of the spectrum has rarely been used. Structural information typically has been acquired with "real" infrared spectroscopy, while quantification has been accomplished with UV/vis analyses of drug substance solutions. The near-infrared region—the range of wavelengths between 800 and 2500 nm—has been considered less functional because only overtones and combination bands exist in the region and extinction coefficients are up to 1000 times weaker than those in the mid-range IR region (1). New instrumentation, however, has been developed specifically for precise near-IR work (2–4). These instruments have high-powered quartz halogen lamps with precision gratings and optics and use larger samples and diffuse reflectance sample holders to overcome most former handicaps. Instrument noise that is only in the microabsorbance unit range (5) has allowed for precise quantitative work on agricultural samples (6–8).

While some near-infrared reflectance (NIR) spectroscopy has been conducted in pharmaceutical applications (9), only simple quantitative methods have been reported. Advances in chromatography have made quantification of complex mixtures almost trivial. Prior to this development precise identification of the active ingredients within a dosage form has been an arduous task. Usually, some type of extraction was required, followed by a colorimetric or spectrophotometric technique for simple dosage forms. In addition, two or more actives often require a separation step. Thus the qualitative identification of the actives can take much longer than their quantification.

This chapter will describe three approaches used to attempt to identify each active ingredient within a three-component dosage form using computer-controlled near-infrared reflectance analysis (NIRA).

EXPERIMENTAL

Instrumentation

All spectra were generated on a model 500C InfraAlyzer (Technicon Instrument Corporation. Tarrytown, New York) equipped with a Hewlett–Packard model 1000/A600 minicomputer and software from Technicon.

Reagents and mixtures

All active drug substances and excipients were USP grade or better. True placebos were made by mixing all relevant substances (with the exception of the one active to be identified) in the exact proportions specified in the master manufacturing procedure. The samples for the third experiment were prepared differently: The amounts of active were independently varied between 90% and 110% of the recommended proportions so that they would correspond to allowed USP limits. To ensure homogeneity, the mixtures were ground for 10–15 s in a powder mill and scanned in the sample cup.

RESULTS AND DISCUSSION

The simplest and most direct approach to identify active ingredients involves the use of standard software to perform a spectral subtraction. The true placebo

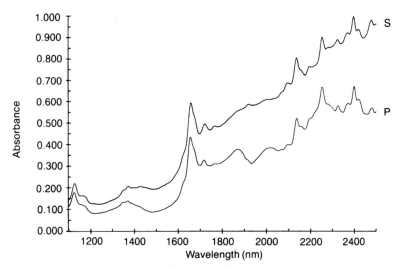

Figure 1. Spectral subtraction to identify aspirin. S = subtracted spectrum; P = pure component spectrum.

spectrum for aspirin (inactive ingredients plus butalbital and caffeine) was subtracted from the spectrum of the total tablet. Figure 1 compares the resultant spectrum with a spectrum of pure aspirin. Similar subtractions are shown for caffeine in Fig. 2 and for butalbital in Fig. 3. In each case, the resultant spectra were similar. Although subtraction for comparative purposes is a rapid substi-

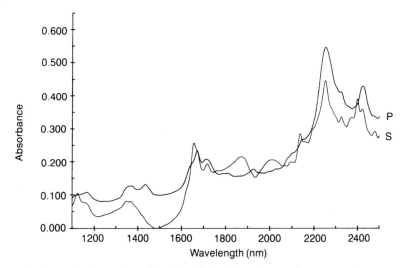

Figure 2. Spectral subtraction to identify caffeine. S = subtracted spectrum; P = pure component spectrum.

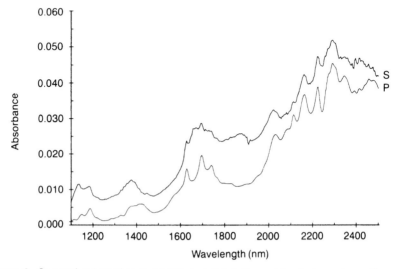

Figure 3. Spectral subtraction to identify butalbital. S = subtracted spectrum; P = pure component spectrum.

tute for existing methods, other techniques to identify actives using NIRA are available that may be quicker.

The second approach used a commercially available program named RECON (available from Technicon), which is based on the Ph.D. dissertation by Honigs (10) and expanded by Hirschfeld, Honigs, and Hieftje (11). The program examines a series of mixtures with known values. Using modified multiterm linear regression equations, the program correlates the frequently changing portions of the spectra with the values given to each spectrum. The program then reconstructs the spectrum of the changing component. Although this program ideally needs more than three or four spectra, the reconstructed spectrum of aspirin in Fig. 4 was the result of only two spectra. Spectra of caffeine (Fig. 5) and butalbital (Fig. 6) were also reconstructed in the same manner. Greatly assisted by software, this technique also allows the archived placebos and the dosage form spectra to be placed in the same file. Despite the advantages of this approach, it still depends on the active's position within a narrow range for a perfect spectral reconstruction.

The third approach proved to be the most rugged and the least labor intensive for the analyst. A discriminant analysis program was introduced (12) at the Seventh NIRA Symposium. The program uses the absorbance values (log $1/R$) of 2–19 wavelengths from the total spectrum. Using the group means of these data in multidimensional space (wavelengths versus absorbances) to define the location of each grouping, the inclusion in, or proximity to, each location is defined by a unit distance vector called the Mahalonobis distance. Mark and Tunnel (13) have written an extensive review of this subject.

Briefly, the Mahalonobis distance can be described by an ellipse around the

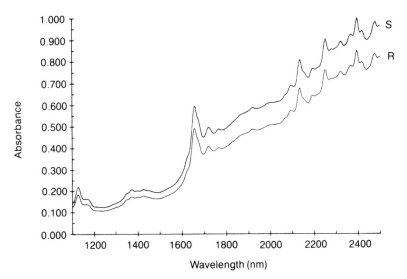

Figure 4. Reconstruction of spectrum of aspirin. R = RECON spectrum; S = standard spectrum.

group means, and discrimination between unknowns is made on the basis of whether or not the mean absorbances fall within the group. This method initially was used to distinguish between types of wheat, soy meal, barley, and other natural products that only differed because of small constituent changes such as hardness, moisture, or particle size. The initial pharmaceutical application

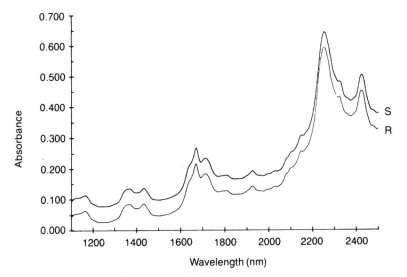

Figure 5. Reconstruction of spectrum of caffeine. R = RECON spectrum; S = standard spectrum.

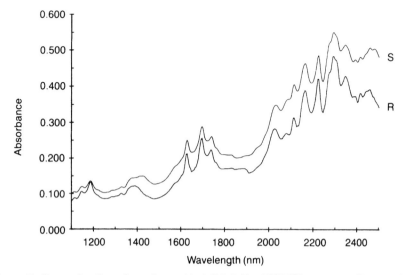

Figure 6. Reconstruction of spectrum of butalbital. R = RECON spectrum; S = standard spectrum.

of discriminant analysis was for the qualitative identification of pure drug substances and excipients (12). This use was trivial, because of the frequently large differences between substances tested. The discriminant analysis program was used in the studies described in this chapter—the identification of dosage forms.

The groupings for the discriminant analysis program (DSCRIM) included the following: dosage form, containing all actives ranging independently from 90% to 110% of label content; no caffeine, containing no caffeine, with aspirin and butalbital ranging from 90% to 110%; no butalbital; and no aspirin. Each of the 34 samples was scanned between 1100 and 2500 nm at 2-nm intervals, and the program was allowed to determine the best three wavelengths for analysis after scanning all the spectra. After approximately seven hours of computer time (*the same program now takes about 10–15 sec; the initial work was on a pre-"anything"-86 chip, actually a dual-floppy system "jury-rigged" with a Winchester 9 MByte hard drive; some applications took over 72 hours to crunch!*) for the initial data analysis, the best wavelengths chosen were 2258, 2314, and 2290 nm. These wavelengths were chosen to define the areas in space that would produce the greatest separation of groupings on the plot. The two-dimensional graphic representation of these data can be seen in Fig. 7.

Because the program saw the range between 2200 and 2350 nm as the most significant, the samples were rescanned in this range at 1-nm intervals and new wavelengths were calculated. At the completion of the process, the wavelengths chosen by DSCRIM differed slightly from the original calculation (2266, 2306, and 2294 nm). Plotting mean absorbance values at these new wavelengths yielded Fig. 8, in which the domains in space are more clearly defined. Finally, several samples were put through the program, includ-

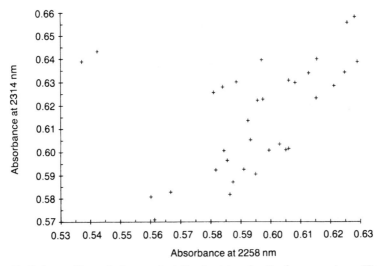

Figure 7. Data used in preliminary calculation to determine the four groupings. (The randomness of data suggests that different wavelengths would be more suitable for discriminant analysis.)

ing perfect tablets (100%, 100%, 100%), borderline tablets, and mixtures that lacked one component. All samples were identified correctly (data not shown). Although this approach is rapid and apparently precise, the spectra are almost indistinguishable, and one must depend on the software's performance—rather than spectral comparison—to make identifications.

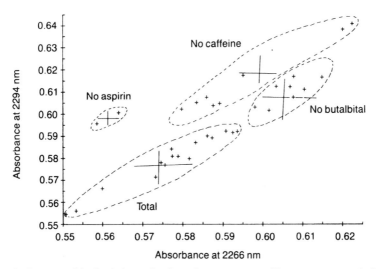

Figure 8. Data used in final determination of groupmeans. (Groupmeans are indicated by +; ellipses are sketched in to indicate areas of each grouping.)

Near-infrared spectroscopy can be used to identify individual dosage-form components in a complex mixture. With computer assistance, the nondestructive procedure can be more rapid than either chromatographic or chemical techniques for identification of dosage-form components.

ACKNOWLEDGMENTS

The authors thank Nicholas Del Medico for his assistance with sample preparation and Drs. Honigs and Mark for their helpful discussions concerning the software used in our instrument.

REFERENCES

1. O. H. Wheeler, *Chem. Rev.* **59**, 629–666 (1959).
2. R. F. Goddu, in *Advances in Analytical Chemistry and Instrumentation*, Vol. 1, C. N. Reilley (ed.) (Interscience Publishers, Inc., New York, 1960), p. 347.
3. J. S. Shenck, I. Landa, M. R. Hoover, and M. O. Westerhaus, *Crop Sci.* **21**, 355–358 (1981).
4. T. M. Long, *Anal. Proc.* **20**, 69–83 (1983).
5. T. Hirschfeld, D. Honigs, and G. Hieftie, *Appl. Spectrosc.* **39**, 430–34 (1985).
6. D. Baker and K. H. Norris, *Appl. Spectrosc.* **39**, 618–21 (1985).
7. B. L. Bruisma and G. L. Rubenthaler, *Crop Sci.* **18**, 1039–42 (1978).
8. M. J. Allison, I. A. Cowe, and R. McHale, *J. Inst. Brew.* **84**, 153–55 (1978).
9. J. J. Rose, T. Prusik, and J. Mardekian, *J. Parent. Sci. and Tech.* **36**, 71–78 (1982).
10. D. Honigs, Ph.D. Dissertation, Indiana University (1984).
11. D. Honigs, G. M. Heiftje, and T. Hirschfeld, *Appl. Spectrosc.* **38**, 317–22 (1984).
12. E. W. Ciurczak, paper presented at the Seventh NIRA Symposium, 17–19 July 1984, Tarrytown, New York.
13. H. L. Mark and D. Tunnel, *Analyt. Chem.* **57**, 1449–56 (1985).

COMMENT

Donald A. Burns

In the preceding chapter, the authors very effectively demonstrated the value of the discriminant analysis software. The reader, however, may infer that computation time is excessive. The statement "After approximately seven hours of computer time for the initial data analysis ..." probably should have been followed by one that distinguished between calibration and actual analysis.

Admittedly, the calibration process can be time consuming. One must "pay one's dues" by performing good work up front—allowing the system to look at all possible combinations of absorbances at each wavelength and known composition of the "teaching set." Thereafter, however, the analysis is frequently as fast as 20–30 s. This calibration procedure is only performed once for any given collection of up to 100 groups of materials with several samples per group.

Second, Fig. 8 is a two-dimensional plot of wavelength #1 versus wavelength #2, obviously choosing the best pair from among the three possible ones (1&2, 1&3, 2&3). Unless the reader understands that there is additional separation in the third dimension, it may erroneously be concluded that the separation of some groups from one another is marginal. A pseudo-three-dimensional plot would have made this point clearer, as would a table of the Mahalanobis distances.

Figure 9 is derived from Fig. 8. The absorbance plane λ_1–λ_2 depicts the shadows of the actual clusters of data points in space. While distance J (the separation in two dimensions) appears small, distance K (the actual separation in three dimensions) is more than adequate for discrimination with high confidence.

While three-dimensional plots are difficult to draw in two dimensions and are not always easily understood, a table of Mahalanobis distances is a relatively easy way to quantitate the discrimination. Moreover, these distances are computed and printed in the output of the DSCRIM program. A typical form

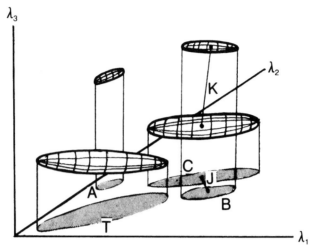

Figure 9. A pseudo-three-dimensional plot derived from Fig. 8.

of presentation is as follows:

	A	B	C	T
A	0			
B	40	0		
C	30	42	0	
T	25	60	35	0

The table is not unlike the distance table found on many road maps. Note, however, that distances are non-Euclidian and cannot be measured in the usual manner. For a good explanation of this, consult Ref. 13 by Mark and Tunnel.

CHAPTER II.B.3

EVALUATION OF A NEAR-INFRARED DETECTOR FOR HIGH-PERFORMANCE LIQUID CHROMATOGRAPHY

Emil W. Ciurczak and Frances M. B. Weis

Fran Weis was my first research student as well as the first person to work with me on the idea of using NIR for HPLC detection. The topic became the research project for my doctorate at Seton Hall University.

This chapter examines the effectiveness of using a near-IR detector for high performance liquid chromatography applications. The near-IR detector's performance is judged for drift and noise levels, sensitivity, response time, linear dynamic range, dead volume, selectivity, and nondestruction of the sample.

The vast majority of (*recently*) reported work in the near-IR region has been performed on solid materials (1–5). Articles concerning a number of uses for static liquid cells with dairy products (6–8) or with liquid pharmaceuticals (9, 10) also have been published. Several manufacturers now are offering remote sensing capabilities for the near-IR region using fiber optics. These devices usually are designed to monitor effluent streams of chemical reactors or other large applications.

The ability to quantify small amounts of organic substances in aqueous solutions (10) and the low noise and the long-term stability of near-IR instrumentation (5, 11, 12) seem to offer another potential tool for high-performance liquid chromatography (HPLC) detection.

Another motivation for a modeling study was the minicomputer capability for both data processing and spectral manipulation. Many types of LC detectors are already on the market: UV/vis, refractive index, electrochemical, infrared, conductance, and fluorescence, to name a few. It would have been interesting, but not very useful, to duplicate present detector capabilities with a near-infrared detector. Instead, we chose to examine two areas where currently available detectors have not addressed all the problems involved: the detection of solutes without chromophores and preparative-level concentrations.

Whichever detector type is used in HPLC applications, there are several general criteria that describe an ideal LC detector (13); It should have low drift and noise levels, high sensitivity, fast response, wide dynamic range, and low dead volume. It should also be insensitive to solvent type, flow rate, and temperature; and it should exhibit a selectivity of wavelength (or potential). It should be nondestructive, and finally, it should be able to detect all materials (that is, it should be a "universal detector"). All these criteria were checked with our near-IR detector, and the detector's performance in each category was examined.

EXPERIMENTAL

Instrumentation   All spectra were generated with a model 500C Infra-Alyzer (Technicon Instrument Corporation, Tarrytown, New York), which was equipped with a model 1000/A600 minicomputer (Hewlett–Packard, Palo Alto, California) and operated with software from Technicon. The liquid samples were scanned in the thermostatted liquid cell from Technicon. The cell was replumbed with 1/16-in. tubing to reduce dead volume to a minimum. The cell has a nominal volume between 150 and 400 μl depending on the thickness of the separating O ring (Fig. 1). A low-pressure reciprocating pump (Fluid Metering, Oyster Bay, New York) was used to introduce the samples and to check for flow rate effects.

Sample preparation.   For the linearity study, food-grade sucrose was dissolved in distilled water until a saturated solution was achieved (about 2.5 M).

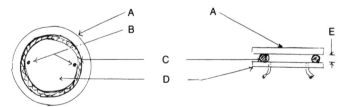

Figure 1. Schematic of a near-IR flow cell. A = low OH quartz cover; B = entrance and exit ports; C = spacer O-ring; D = reflective ceramic plate; and E = 0.1–3.0-mm path length.

This was serially diluted to a lower concentration of 0.1 mg/ml. This series represented a solute without significant chromophores at concentrations ranging from an analytical level to beyond the preparative level.

Glycine and valine were dissolved in trichloroacetic acid (adjusted to pH 4.0) to mimic a typical separation of amino acids. The concentrations ranged from 0.1 to 10.0 mg/ml. This series represented very similar compounds—neither of which contained a chromophore—that were to be examined both quantitatively and qualitatively.

Aqueous solutions of methanol, tetrahydrofuran, and acetonitrile were prepared from 10% to 90% (v/v) to determine the effects of mobile-phase composition changes both quantitatively and qualitatively.

Finally, 25%, 50%, and 75% (v/v) solutions of methanol, ethanol, propanol, and tetrahydrofuran (all of which are normal LC solvents that were used because of their transparency) were injected by themselves as solutes to determine the precision of the technique for normally undetectable solutes. The mobile phase also was subtracted to show how near-IR analysis could be used as an identification tool for LC solutes.

RESULTS AND DISCUSSION

Low drift and noise levels. With a high-intensity (1000-W) quartz halogen lamp and a low-noise lead sulfide (PbS) detector, near-IR instruments routinely have noise levels below 20 μAU. With absorbances that routinely are above 1.0 AU (Fig. 2), the signal-to-noise ratio for major peaks was 50,000:1. For some of the minor peaks, such as 0.1 mg/ml valine with an absorbance of 0.002 from the mobile phase, the S/N ratio is 100:1, which is still quite respectable.

The drift was less than 0.5% over the course of one day. This was largely

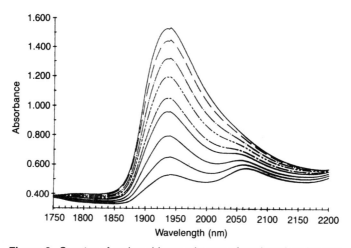

Figure 2. Spectra of various binary mixtures of methanol and water.

due to the cell being thermostatted to compensate for the high-energy output of the light source.

High sensitivity. The extinction coefficients are inherently lower in the near-IR than in, for instance, the IR range. In some cases, the values are as low as 0.001. As mentioned, however, the exceedingly low noise enhances this sensitivity, resulting in a working range that usually is beyond 100:1 S/N.

Fast response. It is well documented that the PbS detector used in near-IR analysis is comparable with any other spectrophotometric detector on the market (5).

Wide linear dynamic range. Sucrose was used to evaluate the dynamic range. Fig. 3a shows the linear response to aqueous solutions over a 0.00–2.48 M range of concentration, while Fig. 3b shows the linearity over the 0–50 mM range. Tables I and II show the linearity at two wavelengths, while Table III

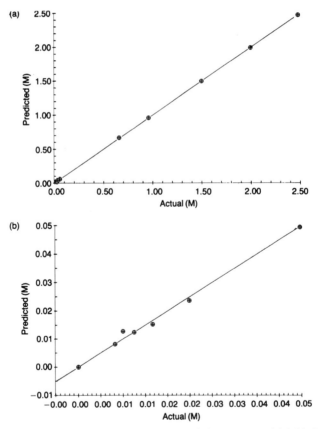

Figure 3. Linear response of sucrose to aqueous solutions over an (a) 0.00–2.48 M range and a (b) 0–50 mM range of concentrations.

Table I. Calibration data for sucrose at (a) 0.008–2.48 M with an intercept of −12.205 and a multiple correlation coefficient of 0.99999 and (b) 0.008–0.050 M with an intercept of −6.003 and a multiple correlation coefficient of 0.9964.

	Wavelength (nm)	
(a)	1740	1504
Regression coefficient	−38.808	−3.897
t value of coefficient	724.548	−2.942
Partial F for deletion	524969.000	8.656
(b)	1740	1500
Regression coefficient	30.698	−9.768
t value of coefficient	9.788	−4.368
Partial F for deletion	95.803	19.076

Table II. Calibration data for sucrose at 0.008–2.48 M: best five combinations.

1st	2nd	3rd	4th	5th
1982	2038	1996	1436	1968
2136	2108	2136	2262	2150

Correlation coefficients:

1.000	1.000	1.000	1.000	1.000

Wavelength (nm)	Regression coefficient	t value of coefficient
1982	−1.565	−158.812
2136	20.092	958.086

Multiple correlation coefficient = 1.00000; intercept = −10.355.

Table III. Calibration data for sucrose at 0.008–2.48 M with an intercept of −14.302.

	1740 nm
Regression coefficient	38.917
t value of coefficient	665.668
Partial F for deletion	443126.690

shows the linearity of a single wavelength (as would be the case for a simple filter or a detector grating instrument kept at a single wavelength). This range of 2.7–848 mg/ml represents a 315-fold linear range. Rather than indicating the limit of detection, the low end was simply the lowest dilution attempted. These data show the ability of a single cell to be effective at virtually any range of solute concentration from dilute to saturated. This is important because virtually all detector designs are aimed at detecting smaller and smaller sample quantities, and thus linearity is often lost above the 1.0 mg/ml range.

Figures 4a and 4b show responses to the 25%, 50%, and 75% v/v methanol and ethanol solutions when these were used as samples. The linearity data of these materials are not as straightforward to obtain—one cannot just tune in on the maximum and measure the absorbance. Because the mobile phase contains mostly water, the baseline shift as the water percentage falls from 75% to 25% is considerable. Figure 5 shows a breakpoint in the linearity of the

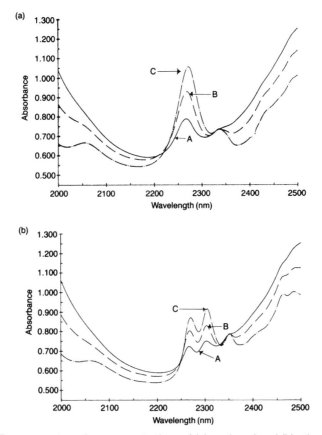

Figure 4. Responses to various concentrations of (a) methanol and (b) ethanol in water. In both figures, A = 25%, B = 50%, and C = 75% concentration.

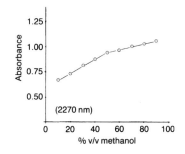

Figure 5. Spectrum shows a breakpoint in the linearity of the methanol–water solution at approximately the 50% mark.

methanol–water solution at about the 50% mark, which is at the greatest density, but the molal properties are the subject of another study (14). When using methanol, for example, merely subtracting the absorbance value of the preceding minimum (2180 nm) from the absorbance maximum (2270 nm) gives a linear relationship for a methanol range from 10% to 90% (Figure 6a). Similar

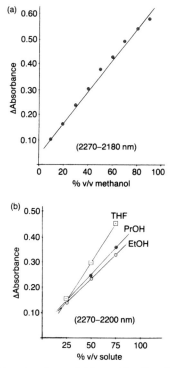

Figure 6. Linear relationships are shown (a) by subtracting the absorbance value of the preceding minimum from the absorbance maximum for a methanol range from 10% to 90% v/v. Similar results are obtained (b) with mixtures of water and ethanol, propanol, and tetrahydrofuran.

results were obtained with ethanol, propanol, and tetrahydrofuran water mixtures (Figure 6b).

Low dead volume. Our detector cell failed miserably in this category. The cell has a volume of 150–400 μl (Fig. 1). In most analytical runs, this would be disastrous. In a semipreparative or a preparative separation, however, this volume would be of no great consequence. Thus it is apparent that this design would only be useful for low-pressure column work with larger sample sizes or for preparative work.

Insensitivity. Concerning insensitivity to flow rate, the extremely low pressure drop within the cell and its relatively thick quartz window renders the cell indifferent to flow variations from 0.2 to 5.0 ml/min.

Temperature control. This is extremely important in near-IR analysis. Figures 7a and 7b show the two major peaks of pure water at 20 and 35°C. These

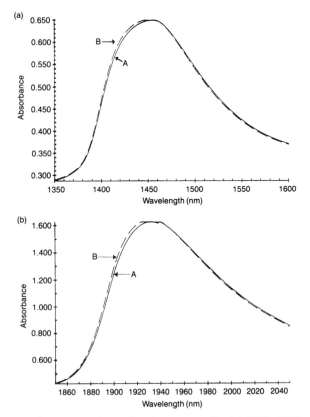

Figure 7. Spectra show two major peaks of water over the (a) 1350–1600-nm range and the (b) 1860–2040-nm range. In both figures, curves at A are at 20°C and curves at B are at 35°C.

shifts may be even more significant for organic compounds. Because hydrogen bonding is quite significant in the near-IR, temperature control should be maintained to +0.5°C for most runs. In our experiment, this was fairly simple to accomplish because of the large thermal mass of our cell and water recirculation.

Selectivity of wavelength. Because a high-resolution grating system was used in this experiment, wavelength selection was not a problem. In the case of near-IR analysis, however, reproducibility is more important than exact selectivity. The very wide bands can be precisely scanned with 10–20 nm bandpasses. In this case, a filter-type dedicated instrument could easily be as sensitive and accurate as a grating-type instrument.

Nondestructive quality. This characteristic is self-evident in a pure spectrophotometric detector. Unlike a post-column reaction or an electrochemical detector, the sample remains as unchanged as it does in UV/vis detectors.

Detects all materials. Like the universal solvent for which no container can be constructed, near-IR analysis does indeed "see" all materials. The immediate problem, then, for a 1 mg/ml solution is the presence of 999 mg/ml of solvent. This huge ratio is exacerbated in the case of water, which has a phenomenal extinction coefficient in the near-IR.

The immediate options were, first, to use as-is absorbances. This worked well in the case of sucrose over a wide range of concentrations. The second option was to use a two-wavelength ratio. As in the case of the alcohols and tetrahydrofuran, this is still a simple yet effective method. The third option was to use spectral subtraction. If the samples fall within a very narrow range, spectral subtraction followed by a second derivative seemed to be quite effective. This is shown in Figs. 8a and 8b for valine and glycine in the 0.1–10 mg/ml range (from 99.9% to 99% mobile phase). The increasing concentration (in this case as the second derivative) is noticeable enough to be quantified. The spectra also are different enough to be used for identification purposes.

Because so many LC separations are performed using gradient elution, the results of our tests were quite interesting. The tremendous change in the spectra for a 10–90% change in the organic phase can be seen in Figs. 9a and 9b. In addition to multiwavelength data manipulation and subtraction of mobile-phase spectra, another possibility becomes apparent: the use of isosbestic points as elution windows. These are naturally occurring points of constant absorbance that can be controlled with temperature and buffer changes. Figures 10a and 10b show the effects of small salt concentrations on both the wavelength and the absorptivity of water maxima. This same control can be useful in manipulating both solute and solvent peaks to create useful data. The effect is also valid in isocratic runs and must be taken into account in LC methods development using near-IR as a detection source.

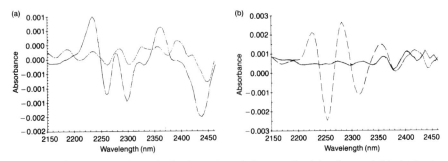

Figure 8. Spectra are shown in the 0.1–10 mg/mL range for (a) valine and (b) glycine in trichloroacetic acid.

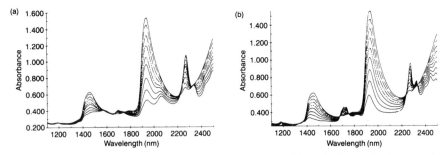

Figure 9. Figure shows the change in spectra for a 10%–90% v/v change in the organic phase for (a) methanol and water and (b) tetrahydrofuran and water.

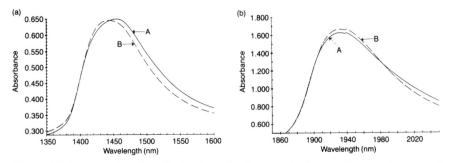

Figure 10. Spectra show the effects of small salt concentrations on the wavelength and the absorptivity of water maxima (20°C water versus a 3% NaCl solution).

CONCLUSIONS

In conclusion, we believe that the advantages of using any near-IR liquid chromatography detector far outweigh the disadvantages. We predict the most immediate success in preparative work and in the identification of natural products that have little or no conventional absorbance properties.

ACKNOWLEDGMENTS

The authors would like to thank Jeanne Sze for allowing them to use some of her near-IR spectra of binary liquids for illustrative purposes. They also would like to thank Technicon (*now Bran + Leubbe*) for the use of a liquid cell and technical advice concerning this project.

REFERENCES

1. C. H. Loftier and R. H. Busch, *Crop. Sci.* **23**, 167 (1983).
2. P. C. Williams, K. H. Norris, and W. S. Zarowski, *Cereal Chem.* **59**, 473 (1982).
3. I. Ben-Gera and K. H. Norris, *J. Food Sci.* **33**, 64 (1968).
4. S. Ghosh, Proceedings from the 8th International Symposium on NIRA (Technicon, Tarrytown, New York, 1985).
5. P. Rotolo, *Cereal Foods World* **24**, 94 (1979).
6. K. Luchter, Proceedings from the 4th International Symposium on NIRA (Technicon, Tarrytown, New York, 1983).
7. I. Ben-Gera and K. H. Norris, *Israel J. Agri. Res.* **18**, 117 (1968).
8. L. Rademacher, Proceedings from the 8th International Symposium on NIRA (Technicon, Tarrytown, New York, 1985).
9. J. J. Rose, T. Prusik, and J. Mardekian, *J. Parenter. Sci. Technol.* **36**(2), 1240 (1966).
10. E. W. Ciurczak and R. P. Torlini, *Spectroscopy* **2**(3), 41 (1987).
11. I. Landa, *Rev. Sci. Instrum.* **50**(1), 34 (1979).
12. M. S. Day and F. R. B. Fern, *Lab. Pract.* 439 (1982).
13. A. M. Krstuloyic and P. R. Brown, *Reversed Phase High Performance Liquid Chromatography: Theory, Practice, and Biomedical Applications* (John Wiley & Sons, New York, 1982).
14. E. W. Ciurczak and J. Sze, College of St. Elizabeth, unpublished data, 1987.

CHAPTER II.B.4

PARAMETERS OF A NEAR-IR HPLC DETECTOR, PART I: LIMITS OF ANALYTICAL SENSITIVITY FOR REVERSED-PHASE CHROMATOGRAPHIC SYSTEMS

Emil W. Ciurczak and Tracey A. Dickinson

In this chapter I have the chance to have my cake and eat it too: I am featuring myself as a "guest" author. My coauthor is Tracey Dickinson, who is now pursuing a graduate degree at the University of Virginia at Charlottesville. Tracey worked with me for two years at the College of St. Elizabeth and presented much of this work at a Pittsburgh Conference. I trust you will like this offering, because I continue to crank out research on this topic.

The editing of the conclusions and/or observations for space constraints kept us from pointing out how well the technique actually worked for so many things. Hopefully, some other soul will do some of the experiments we dabbled in.

The feasibility of a near-infrared (near-IR) detector for high-performance liquid chromatography (HPLC) has been reported for several years (1–4). To date, we have shown the possibility of using a near-IR detector for aqueous preparative-level work for nonchromophore-bearing solutes (5) and described what might be a workable design for the detector itself (6). Here, we would like to expand on the technique by introducing the concepts of real-time solvent extraction and apparent extinction coefficient.

Our earliest work was performed on a transmission–reflectance ("transflectance") flow cell. With this cell, it was seldom possible to achieve an accurate calibration using only one wavelength due to nonlinearities with changing refractive index, density, and concentration. Two or more wavelengths were required to describe the constituent and physical properties. Often the ratio or difference between two wavelengths was needed to perform quantitative analyses.

In utilizing a transmission instrument, we noted that a binary mixture (for example, methanol and water) displayed numerous wavelengths suitable for a single-wavelength calibration equation.

EXPERIMENTAL

All spectra were generated on either an NIRSystems (Silver Spring, Maryland) model 6250 or 6500 near-IR spectrophotometer equipped with an IBM (Groton, Connecticut) PC/AT or IBM PS/2 model 50Z minicomputer, utilizing commercially available NSAS (NIRSystems) software. A 0.5-mm pathlength flow cell (200-μl volume), thermostated to 25°C, was used for HPLC simulations.

Solvent systems were chosen from references in Kirkland and Snyder's classic text as well as applications notes from column and instrument manufacturers (7, 8).

RESULTS AND DISCUSSION

In a normal transmission spectrum of water, the major peaks (~1430 and 1940 nm) are the only noticeable features. If an organic constituent appears, the water spectrum still dominates the overall spectrum, making analytical wavelength selections difficult. The second derivative of the spectrum allows the analyst visually to select wavelength ranges where major absorbances might occur. For example, Figs. 1a and 1b show second-derivative spectra of water and a 50% water–methanol mixture versus an air reference.

Because our instruments are single beam, the reference scan (subtracted from the sample scan) can be of the mobile phase as well as of air. Figure 2 shows the resultant spectrum of water mobile phase minus itself. Because so little light was transmitted in the 1900- and 2500-nm regions, the noise at these points was exaggerated. Through the remaining spectral range, it was still only 10^{-4} AU at this "pseudozero absorbance." Overall, the working signal-to-noise ratio (S/N) was usually above 1000.

Scans of solutions representing pure solvent through 100% methanol

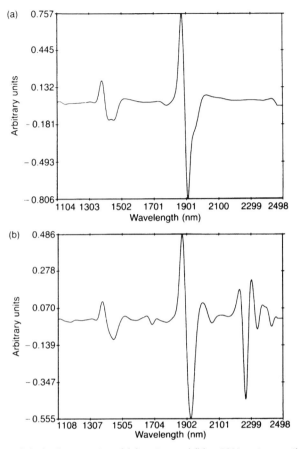

Figure 1. Second-derivative spectra of (a) water and (b) a 50% water–methanol mixture, both versus an air reference.

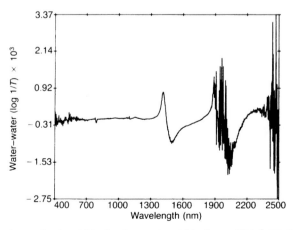

Figure 2. Spectrum of solvent blank minus solvent blank resulting from reference scan.

(MeOH) are shown in Fig. 3. Several significant wavelengths are highlighted, especially in the 2200–2300 nm range. We allowed the software to choose the best single wavelength with which to quantify MeOH and, excluding the 1940-nm water peak, the wavelength chosen was 1440 nm, the other major water peak.

The correlation coefficient of -1.00 is excellent correlation, but in the negative direction, indicating disappearance of solvent. This could still be a usable wavelength because in well-behaved HPLC there is only a single analyte mixed with mobile phase in the detector at any one time. Thus the disappearance of the mobile-phase signal would serve to mark the elution of a solute. However, in method development, where the purity of the peak has not already been demonstrated, more specific wavelengths may be required.

When a plot of correlation coefficients versus wavelength is examined, several regions of high correlation may be identified. Peaks at 1600, 1702, and 2274 nm, attributable to the alcohol, coincide with positive correlation coefficients. The negative peaks at 1446 and 1926 nm are from the disappearance of water and are associated with negative correlation coefficients.

Specifying the 2265–2285 nm region allowed the algorithm to generate the best equation based on the solute. The 2276 nm wavelength chosen is seen in Fig. 4. The equation generated obtained a multiple correlation coefficient of 0.9983. This equation was tested for reliability against a series of MeOH-water solutions. The results of the analysis are shown graphically in Fig. 5.

Because the system has a noise level of $\sim 1 \times 10^{-4}$ AU, an analytical or quantifiable amount should give a signal difference of $\sim 1 \times 10^{-3}$ AU. This would represent a 10× noise level; this lowest quantifiable amount is also commonly referred to as the *limit of detection* or the *sensitivity*.

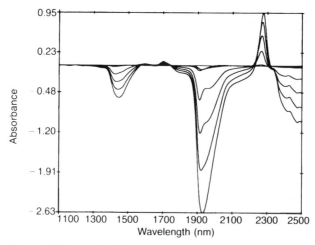

Figure 3. Scans of solutions from 0–100% methanol (MeOH).

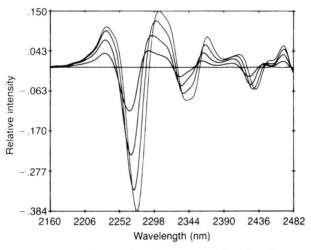

Figure 4. Second-derivative spectra of methanol.

The measured or "apparent" extinction coefficient for methanol in this system at 2276 nm is ~7.68×10^1 cm^{-1} mol^{-1}. Thus a signal of 1×10^{-3} AU would represent an MeOH concentration of 0.026 M or about 0.08% (80 ppm). This value would be valid for this mobile phase at this temperature. Unlike electronic absorbances in ultraviolet (UV) and visible detectors, near-IR is a vibrational mode and more susceptible to solvent and temperature effects.

Hydrogen bonding may cause apparent peak shifting in most polar solvents. In the example used, the pure methanol exhibits a major peak at 2280 nm. This

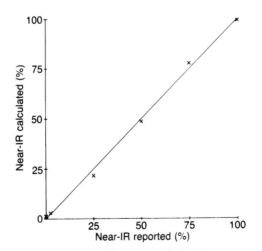

Figure 5. Plot of the calculated near-IR spectra of MeOH–water solutions versus the actual near-IR spectra at 2276 nm.

apparent peak shift to 2274 nm occurs in a 25% MeOH–water mix, with no further shifting after that point. After 25%, the MeOH is most likely completely solvated.

This technique was also used on several common amino acids. For example, solutions of tyrosine ranging from 0.05 to 3.0% were adjusted to pH 8.0 and scanned, and a spectrum of water with a pH of 8.0 was used as a blank (Fig. 6 shows tyrosine versus tryptophan spectra). Again, the (disappearing) water peak at 1440 nm yielded the best correlation coefficient (-0.9946). The regression coefficient was almost exactly the same as for methanol, based on the molarity. This is important when the analyst considers that all well-resolved chromatographic peaks are, in reality, binary mixtures of the solute and mobile phase (of which, usually, only the water and polar modifier are NIR absorbing).

Focusing on the water peak, one would see a mole-for-mole diminution of water by the solute. While the solutes might have different extinction coefficients, water (within the same solvent system) would maintain the same approximate value for most solutes. While corrections might be necessary for certain strong hydrogen-bonding materials, this approach could be construed as a universal detector.

Table I shows the results of a second set of tyrosine solutions predicted using the single-wavelength equation. Adding a second wavelength (2204 nm) gave a somewhat stronger equation (results in Table II), bringing the correlation coefficient to 0.9995. An abbreviated series of tryptophan solutions displayed similar spectra and gave comparable results.

The linearity of aqueous solutions of sugars was discussed previously (5). We also examined sugars in alcoholic–aqueous solutions. This was done because the OH groups on the sugars are spectrally similar to the alcoholic OHs. Figure 7 shows the spectra of 1.5 and 3.0% sucrose in 50:50 MeOH–water minus the

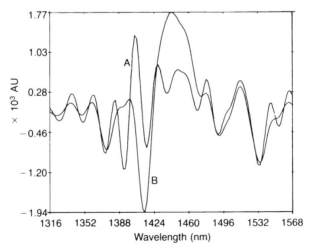

Figure 6. Spectra of (a) tyrosine versus (b) tryptophan.

Table I. Comparison of actual and predicted tyrosine solutions (using single-wavelength equation).

Actual (%)	Predicted (%)	Residual (%)
0.000	0.013	0.013
0.047	0.010	-0.036
0.188	0.182	-0.006
0.375	0.439	0.064
0.750	0.723	-0.027
1.500	1.489	-0.011
3.000	3.003	0.003

Table II. Comparison of actual and predicted tyrosine solutions (using two-wavelength equation).

Actual (%)	Predicted (%)	Residual (%)
0.000	0.010	0.010
0.047	0.018	-0.028
0.188	0.174	-0.013
0.375	0.392	0.017
0.750	0.767	0.017
1.500	1.507	0.007
3.000	2.991	-0.009

solvent (second-derivative spectra). Several wavelengths are indicated for equations. Figure 8 shows differences between lactose and sucrose in solution (using the second derivative).

CONCLUSIONS

For materials that are not amenable to fluorescence, electrochemical, or UV detection, near-IR could be a rapid, nondestructive alternative to pre- and post-column reaction detectors. For preparative-level work, the qualitative nature of near-IR spectroscopy might allow the chromatographer to ascertain the purity of the compound of interest on a real-time basis in addition to generating a traditional chromatogram. The majority of preparative detectors now are refractive-

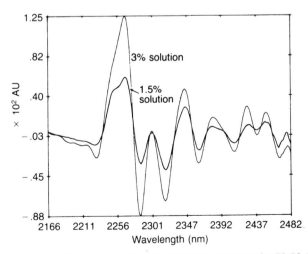

Figure 7. Second-derivative spectra of 1.5 and 3.0% sucrose in 50:50 MeOH–water minus solvent.

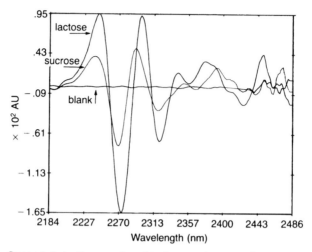

Figure 8. Second-derivative spectra of lactose and sucrose in water minus water.

index types that offer little or no qualitative discrimination. We believe that near-IR has great potential in the field of HPLC.

REFERENCES

1. E. W. Ciurczak and F. M. B. Weis, *"Use of a NIR Detector in HPLC for Detection of Solutes without Chromophores,"* paper presented at PittCon, Atlantic City, New Jersey, March 1987.

2. E. W. Ciurczak and I. M. Vance, *"Use of a NIR Detector for Analytical and Preparative Scale LC of Amino Acids and Polymers,"* paper presented at PittCon, New Orleans, March 1988.

3. E. W. Ciurczak and T. A. Dickinson, *"Application of a NIR Detector in the HPLC of Amino Acids, Proteins, and Drug Substances,"* paper presented at PittCon, Atlanta, March 1989.

4. E. W. Ciurczak and T. A. Dickinson, *"The Use of NIRS in Normal Phase and Non-Aqueous Size Exclusion Chromatography,"* paper presented at PittCon, New York, March 1990.

5. E. W. Ciurczak and F. M. B. Weis, *Spectroscopy* **2**(10), 33 (1987).

6. E. W. Ciurczak and I. M. Vance, *Spectroscopy* **3**(9), 56 (1988).

7. J. J. Kirkland and L. R. Snyder, *"Introduction to Modern Liquid Chromatography"* (2nd Ed., John Wiley & Sons, New York, 1979).

8. R. Pfiefer et al., *Am. Lab.* **15**(3), 86 (1983).

CHAPTER II.B.5

PARAMETERS OF A NEAR-IR HPLC DETECTOR, PART II: APPLICATION TO NONAQUEOUS SYSTEMS

Emil W. Ciurczak, Tracey A. Dickinson, and Wyatt R. Murphy

In this chapter, I am presenting the second of two papers. The original work was performed at the College of St. Elizabeth with former student Tracey Dickinson, now at the University of Virginia at Charlottesville. I received a great deal of input from my graduate mentor, W. R. (Rory) Murphy at Seton Hall University, on content and layout. This was critical, as I had to forsake several figures from the original to fit space constraints. I trust that the information provided remains clear and useful.

In Part 1 (1, preceding chapter), we discussed the use of near-infrared (near-IR) spectroscopy as a tool in reversed-phase high-performance liquid chromatography (HPLC) (2). This was the culmination of several years' work, which was reported elsewhere (3–6). Since that time, a new cell has been constructed, allowing flows of up to 25 ml/min to better simulate semipreparative and preparative systems.

In this work, we examined nonaqueous systems common to size exclusion and normal-phase separations. "Chromatograms" were created from carefully controlled flowing systems and did not involve the use of a column. Also, in aqueous systems, inherent hydrogen bonding causes deviations from Beer's law (for a single-wavelength calibration). When using organic solvents, the vast majority of absorption bands increase without significant shifting (caused by solvent–solute interactions).

The "richness" of near-IR spectra was examined to pinpoint regions where one solute absorbed and a second had little or no absorptivity to quantify over-lapping peaks.

EXPERIMENTAL

All spectra were generated on a NIRSystems (Silver Spring, MD) model 6500 near-IR spectrophotometer equipped with an IBM (Groton, CT) PS/2 model 50Z computer. Commercially available NSAS (NIRSystems) software was used for all calculations. A 1.0-mm thermostated flow cell was used for all sample measurements and simulated chromatograms. Path-length variations were performed in standalone cuvettes with indicated path lengths.

Solvents were HPLC grade, supplied by several manufacturers, and polymers were donated by several producers. Solvent systems were determined by consulting Kirkland and Snyder's classic text (8).

RESULTS AND DISCUSSION

Because the largest portion of the light was absorbed by the solvent (see Fig. 1), real-time solvent spectral subtraction was routinely performed. Thus all spectra displayed are the solute–solvent mix minus the solvent. In addition, at preparative-level concentrations (~4–7% w/v), there are serious refractive index (RI) changes. I previously reported on adjustment methods implemented for

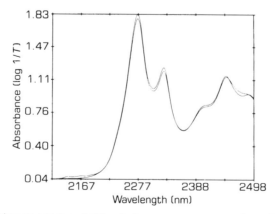

Figure 1. Absorbance spectra of 5% polystyrene versus pure solvent (tetrahydrofuran). Note the little difference and almost no spectral features.

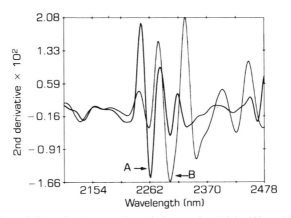

Figure 2. Portion of the polyester spectrum (minus solvents) in (A) acetonitrile and (B) tetrahydrofuran.

RI (and temperature) changes (9). For this work, merely taking the second derivative compensates for virtually all the baseline shifting caused by RI effects.

One of the more common (*commercial*) polymers, polyester, was examined. Figure 2 shows the spectra of acetonitrile and tetrahydrofuran, which demonstrate the effect of using different solvents. Although the analyte does not shift with increasing concentration, the solvent may be chosen to aid in the detection of various solutes in, for example, size-exclusion work. Theoretically, absorption and adsorption do not influence size exclusion; therefore, the analyst is free to choose a solvent that will aid detection.

In a case where the peaks are well resolved, a wavelength at which both absorb may be used to develop a chromatogram. In the case of overlap, a common peak may not be successful. Figure 3 shows polyester and 1,4-butanediol

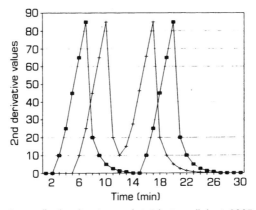

Figure 3. "Chromatogram" of polyester and 1,4-butanediol at 2305 nm. First peak = polyester; second peak = 1,4 butanediol. ■ = separated, + = overlap.

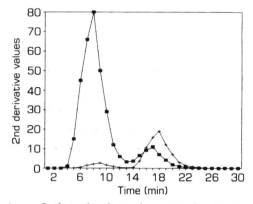

Figure 4. "Chromatogram" of overlapping polyester and 1,4-butanediol at two wavelengths: each component is maximized by one while the other peak is minimized. ■ = 2130 nm, + = 1572 nm.

in both situations at 2305 nm. One of two routes may be taken at this point: Assume equal molar extinction coefficients for both materials, or find wavelengths where one material is maximized while the other is minimized.

Classic size-exclusion software, often designed for RI determination of peaks, frequently chooses the former path. With near-IR spectra, the latter may be used. Figure 4 shows the same "chromatograph" of polystyrene and 1,4-butanediol when 2130 nm is chosen to maximize the polyester, and 1572 nm is chosen to emphasize the butanediol. Utilizing both wavelengths in a common multiterm equation should give a precise determination of one component in the presence of the second. This is truly the strong point of near-IR spectroscopy: chemometrics! A comparison chromatogram of the three wavelengths for a "well-resolved" system is shown in Fig. 5.

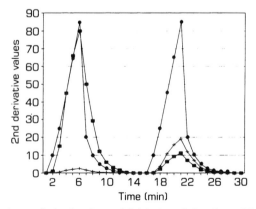

Figure 5. "Chromatogram" of polyester and 1,4-butanediol at three different wavelengths: ■ = 2130, ● = 1572, + = 2305.

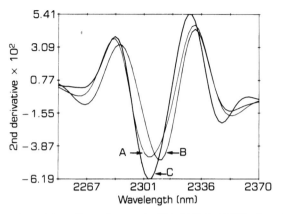

Figure 6. Intrasystem interactions in nearly identical systems: A = 1,3-butanediol, B = 1,4-butanediol, C = 1,6-hexanediol.

Even without the strong influence of water–hydrogen bonding, intrasystem interactions in nearly identical systems show differences. Figure 6 reveals the differences between three diols: 1,3-butanediol, 1,4-butanediol, and 1,6-hexanediol. The absorbance around 2305 nm may be used for all three, but examination of the precise maximum aids in identification of the material. When there is baseline separation of materials, choosing a wavelength isn't a problem. As the spectrum of polystyrene shows (see Fig. 7), there are numerous wavelengths in which a calibration and quantification may be made.

Because organic solvents have much smaller absorptivities than water, longer path lengths may be used to observe lower concentrations of solute. A constant amount of polyester was measured at path lengths varying between 2 and 20 mm. As Fig. 8 shows, there is a clear peak around 2048 nm. When put into

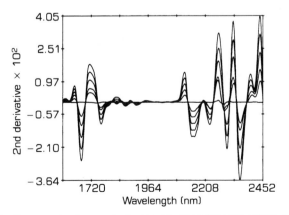

Figure 7. Calibration spectra for a polystyrene equation (spectra minus solvent, second derivative).

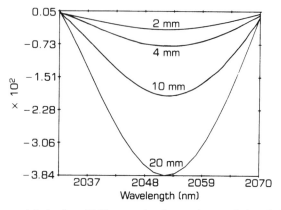

Figure 8. Second-derivative, 2048-nm polyester peak as path length is increased.

a linear-regression equation, the peak was shown to be linear with path length (Fig. 9). This is common to ultraviolet/visible applications, but vibrational spectroscopy doesn't always follow Beer's law as a result of frequent solvent–solute interactions.

CONCLUSIONS

Liquid chromatographic nonaqueous systems may be more manageable than the more common water-based systems with less hydrogen bonding and lower absorptivities associated with the solvents. With the longer time frames associated with size exclusion and preparative chromatography, even existing near-IR spectrophotometers may be used. We have shown data points at 1-min intervals. These may be shortened, and, with minor instrument adjustments, true analog signals may be produced.

With the higher selectivity of the near-IR spectra, more accurate allocations of materials may be made in the case of overlapping peaks. The ability to use spectral identification software (10) makes the method even stronger. We believe that a commercial system is not too far off.

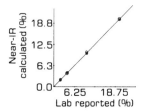

Figure 9. Plot of the linearity of polyester absorbance with varying path length.

REFERENCES

1. E. W. Ciurczak and T. A. Dickinson, *Spectroscopy* **6**(2), 12 (1991).

2. E. W. Ciurczak and F. M. B. Weis, *"Use of a NIR Detector in HPLC for Detection of Solutes without Chromophores,"* paper presented at PittCon, Atlantic City, New Jersey, March, 1987.

3. E. W. Ciurczak and I. M. Vance, *"Use of a NIR Detector for Analytical and Preparative Scale LC of Amino Acids and Polymers,"* paper presented at PittCon, New Orleans, March, 1988.

4. E. W. Ciurczak and T. A. Dickinson, *"Application of a NIR Detector in the HPLC of Amino Acids, Proteins, and Drug Substances,"* paper presented at PittCon, Atlanta, March, 1989.

5. E. W. Ciurczak and T. A. Dickinson, *"The Use of NIRS in Normal Phase and Non-Aqueous Size Exclusion Chromatography,"* paper presented at PittCon, New York, March, 1990.

6. E. W. Ciurczak and F. M. B. Weis, *Spectroscopy* **2**(10), 33 (1987).

7. E. W. Ciurczak and I. M. Vance, *Spectroscopy* **3**(9), 56 (1988).

8. J. J. Kirkland and L. R. Snyder, *Introduction to Modern Liquid Chromatography*, 2nd Ed. (John Wiley & Sons, New York, 1979).

9. E. W. Ciurczak and D. A. Honigs, *"The Measurement of the Refractive Index of Liquids via NIRS Utilizing a Single Fiber Optic Probe,"* paper presented at PittCon, New York, 1990.

10. E. W. Ciurczak and D. M. Mustillo, *Spectroscopy* **6**(3), 26 (1991).

CHAPTER II.B.6

THE DESIGN PARAMETERS OF A NEAR-INFRARED DETECTOR FOR HIGH-PERFORMANCE LIQUID CHROMATOGRAPHY

Emil W. Ciurczak and Ina M. Vance

Once again, a research student of mine was interested in my ongoing HPLC project. Ina not only was a bright young lady, but sang like a bird. She was last heard singing in her husband's band. Brains and beauty!

This article discusses efforts to increase the efficiency of using a near-IR detector in routine high-performance liquid chromatography applications. An enhanced cell design for preparative and analytical LC detection is discussed, as well as the detector output and physical components of the cell.

The feasibility of using a near-infrared detector for high-performance liquid chromatography (HPLC) was investigated by Ciurczak and Weis (1). At that time, a scanning instrument was used with a macro-liquid cell. With this setup, parameters such as limits of detection, linearity, and specificity were determined for non-chromophore-bearing solutes such as alcohols, sugars, and simple amino acids. The previous work was performed on an instrument equipped with an integrating sphere that employed a reflectance/transmittance feature wherein the light passes through the sample twice. For many reasons, this is not practical for routine HPLC analyses. In this chapter, we attempt to determine a more advantageous design for a cell to be used for preparative and analytical LC detection. We have proposed a model cell and described the detector output and physical components.

EXPERIMENTAL

All spectra used in this work were generated on a Bran + Leubbe (Elmsford, New York) model 500c InfraAlyzer equipped with a Hewlett–Packard (Palo Alto, California) model 1000 computer. The software was the standard program Bran + Leubbe supplies with the firm's commercial instruments. The cell was a Bran + Leubbe Chemical Cell modified to be fed by an FMI pump (Fluid Metering Corporation, Long Island, New York). All reagents were ACS or USP grade or better. The HPLC-grade acetonitrile, methanol, tetrahydrofuran, and propanol were all obtained from Fisher Scientific (Chicago, Illinois).

The spectra were generated on IBM PCs equipped with a 20-Mbyte hard-disk drive, utilizing ASYSTANT+ software (MacMillan Software Company, New York, New York).

The solutions were made at the indicated maximum concentrations and then serially diluted to produce absorbances for the model LC peaks. The cell was flushed thoroughly between readings, filled with the solutions, and scanned from 1100 to 2500 nm. As was done previously (1), the maxima for each solute were determined from the spectra, and apparent minima were similarly determined. Absorbance (log I/R) values were obtained from the appropriate print-outs.

The minimum (attributed to "pure" mobile-phase absorbance) was then numerically subtracted from the absorbance maximum attributed to the solute(s).

Reasonable retention times were estimated from values published in the literature and in manufacturers' applications notes (2–6); this approach was also used for determining peak widths and elution orders of all materials. The absorbance differences were then plotted against the time points chosen to render pseudo-chromatograms. These simulated chromatograms were plotted and then used for further data analysis. First derivatives were simulated by first estimating the points between the experimentally determined values (the "dots were connected"), then simply subtracting each data point from the one immediately following (for example, point "1" was subtracted from point "2" point "2" was subtracted from "3," etc.). These resultant points were then plotted versus the predetermined time points.

RESULTS AND DISCUSSION

The spectra of 10 to 90% (v/v) water–methanol mixtures are shown in Fig. 1a (a 100% water spectrum is shown in Fig. 1b). From this, the 2170- and 2270-nm lines were chosen as "baseline" and "peak" wavelengths, respectively. The plot of the change in absorbance from 2270 to 2170 nm as a function of concentration for methanol is shown in Fig. 2. Figure 3 shows the spectra of all the solvents used as solutes with the water spectrum subtracted. From this, it was determined that the change in absorbance from 2270 to 2170 nm would give a suitable difference chromatogram for all the solvents used as solutes.

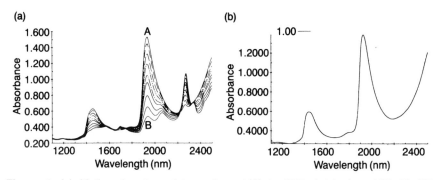

Figure 1. (a) Methanol water mixtures from 10% to 90% (v/v). A = 10% MeOH, B = 90% MeOH; (b) a 100% water spectrum.

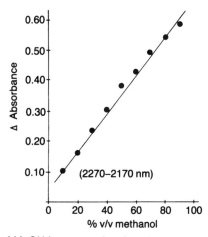

Figure 2. Linearity of MeOH in water using the 2270-nm – 2170-nm absorbance.

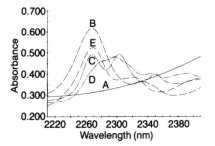

Figure 3. Spectra of "solutes" minus the mobile phase. A = mobile phase – 90% mobile phase; B = methanol; C = ethanol; D = propanol; E = tetrahydrofuran.

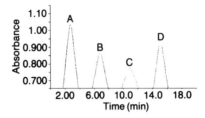

Figure 4. Pseudochromatogram of (A) methanol, (B) ethanol, (C) propanol, and (D) tetrahydrofuran.

A generated chromatogram of the four "solutes" is shown in Fig. 4. Since the v/v percents of all the "solutes" were identical, the simulated chromatogram also shows the relative extinction coefficients for each material at the wavelengths chosen. To generate a difference chromatogram, one must use either a cell in which two wavelengths can be measured simultaneously or two cells, each of which monitors a single wavelength with a minimum of dead volume between them.

A two-cell design is shown in Fig. 5. The light source is a 1000-W quartz halogen lamp with a carefully controlled power supply. The filters are commercial interference types, and the detectors are PbS of the type used in virtually all commercial near-IR instruments. The flow cells are "off-the-shelf" 1 × 1 cm cuvettes with a 1-cm pathlength (the pathlength may need to be shorter

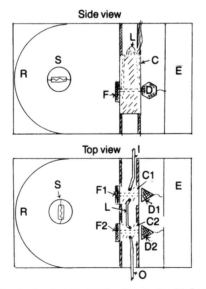

Figure 5. Schematic of a dual-cell detector. F = filters, D = PbS detectors, C = flow cells, S = light source, L = connecting tubing, R = back reflector, E = electronics, I = incoming liquid, and O = flow to waste.

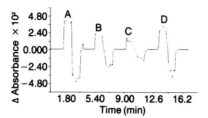

Figure 6. Pseudo first derivative of chromatogram in Figure 4.

for strongly absorbing solvents). Although this is a rough schematic, it is quite close to the actual prototype that we are now evaluating. The electronics are also essentially the same as in standard instruments. Because commercial instruments (double-beam spectrophotometers, in any case) already subtract the reference from the sample, no modification was needed. Thus, if the two filters were chosen from a scanned spectrum (as is done routinely in HPLC to determine the wavelength of maximum absorbance) to represent the maximum and minimum for the solute in question, plotting the resultant output as a function of time would produce a chromatogram.

If the connecting tube between the two flow cells is lengthened, and the two filters are the same wavelength, then a "first derivative" is generated and may be plotted as a function of time. This is shown in Fig. 6 for the four solvents. Electronically, this is done by weighting the outputs of the detectors "−1" and "+1" and adding the signals; for a pseudo-second derivative, a third cell is required with weightings of "+1 ," "−2," and "+1" for the three. This, however, is outside the range of the current work. The point at which the derivative crosses the baseline will then represent the highest point of the peak, that is, the retention time. Figures 7 and 8 show the absorbance and derivative chromatograms, respectively, for five simple sugars "separated" in aqueous solution.

One immediate consequence of this approach is that if the wavelength monitored for a derivative chromatogram is the peak absorbance for water, the system can be used to detect almost any chromatographic analyte. The rate of change of the water spectrum will be greater than for almost any other material

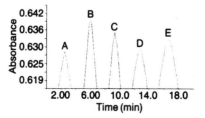

Figure 7. Pseudochromatogram of (A) sucrose, (B) fructose, (C) arabinose, (D) maltose, and (E) galactose.

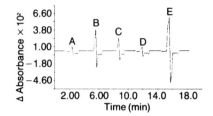

Figure 8. Pseudo first derivative of Figure 7.

in the near-IR region (water has virtually the largest extinction coefficient in the near-IR range). The linearity of such a plot is shown in Fig. 9. The slope of this Beer's law plot is steeper, and the actual absorbance values higher, than for the solute in question (methanol). Thus the difference and derivative chromatograms will be of the disappearance of water (solvent) and will be proportional only to the amount of mobile phase displaced as each peak elutes. This very nearly amounts to a universal detector. A little physical chemistry must be called into play here, taking into account the apparent molal volumes. However, for similar compounds, similar chemistry may be assumed in an isocratic run.

CONCLUSION

In conclusion, we believe that near-IR detection may be used successfully in HPLC analysis, even for those solutes that have little or no conventional absorbances in the near-IR range. As was seen in previous work (1), the linearity of the near-IR detector encompassed several orders of magnitude (0.1–850 mg/ml for sucrose). This makes it particularly attractive as a preparative LC detector. Most commercial preparative LC detectors are nonspecific (usually

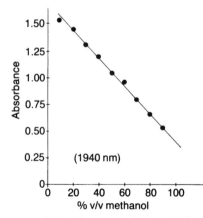

Figure 9. Linearity of water peak absorbance at 1940 nm in binary water/MeOH mixtures.

they use the refractive index as the method for detection); however, the electronics of our detector could be modified to provide wavelength ratios, in which case the outputs of the two detectors would simply be divided rather than subtracted. This procedure might be used for preparative peak-purity determinations, much as is now done in applications involving UV analytical detectors. Our future work will include the building of a prototype cell that will be based upon the findings of this chapter.

REFERENCES

1. E. W. Ciurczak and F. M. B. Weis, *Spectroscopy* **2**(10), 33 (1987).
2. Alltech brochure #84 (Alltech Associates, Deerfield, Illinois, 1986), p. 9.
3. Alltech brochure #84 (Alltech Associates, Deerfield, Illinois, 1986), p. 39.
4. Perkin–Elmer Product Journal, Chromatography International (Perkin–Elmer Ltd., Beaconsfield, United Kingdom, 1986).
5. "Primary Amino Acids," Gilson Application Notes (Gilson Medical Electronics, Middleton, WI, 1984).
6. "Sugars," Gilson Application Notes (Gilson Medical Electronics, Middleton, WI, 1984).

CHAPTER II.B.7

DETERMINATION OF ENANTIOMERIC PURITY OF VALINE VIA NEAR-INFRARED REFLECTANCE SPECTROSCOPY

Bruce R. Buchanan, Emil W. Ciurczak,
Arthur Q. Grunke, and David E. Honigs

This was a classic collaboration: I had an idea, David Honigs (at that time a Professor at the University of Washington) agreed to work on it, his grad student, Bruce, supervised the project, and a senior, Art, actually did the work. The most people occupied on a small project not *employed by the government!*

Often, racemates of amino acids exhibit very different infrared spectra compared with optically active forms. A reflectance technique that exploits the spectral differences between D-, L-, and DL-valine was used to predict the isomeric ratio of a series of recrystallized mixtures. Predicted residual error sum of squares (PRESS) values were calculated to determine the number of wavelengths that should be used to develop the calibration. A spectral reconstruction based on the ratios of one valine form to the other was performed and used to partially explain the wavelengths chosen.

Pharmaceutical companies desire a fast and reliable method of analysis to determine optical isomeric purity of a drug for process control in drug manufacture. These determinations are of prime importance because compounds with different optical isomers often exhibit different biological activity. While one configuration might have the desired biological activity, the other configuration could have vastly different activity or even be inactive. One such pair is quinine and quinidine, used, respectively, to treat malaria and arrhythmia. (*A second, more tragic example is thalidomide, a tranquilizer discovered in the 1960s. It was responsible for many birth defects in Europe. The problem was the wrong optical isomer, not the "proper" isomer. Routine screenings were not common in those days!*) For this reason, pharmaceutical companies control their products within strict limits, leading to the recent interest in the near-IR investigation of pharmaceuticals (1–5).

This chapter explores the use of near-infrared reflectance spectroscopy to determine the ratio of D- and L- isomers of valine within a sample. Analytical wavelengths were chosen using multiple linear regression techniques.

The PRESS method was used to determine the optimal number of wavelengths to be used in the calibration. An all-possible-combinations wavelength search was performed to find the best set of wavelengths to use. The wavelengths were then justified using spectral reconstruction.

THEORY

The crystal structures of L-valine and DL-valine have been studied extensively (6–9). While the structure and spectra of both D- and L-valine are identical, the crystal consisting of both isomers shows a definite infrared shift. This shift is related to the change in hydrogen bonding that occurs because of the difference in the orientation of the groups in the mixture crystal (Table I). There probably also exists a rotation of the isopropyl group, allowing the two optical isomers in the mixed crystal to form the most favorable packing. The formation of these "rotational isomers" also affects the midinfrared spectra (10). These spectral changes occur only when the mixture of valines has formed racemic crystals. A conglomerate (or a mechanical mixture of crystals of the two enantiomers) will have a spectrum equivalent to either the pure D or L form. Because the near-IR spectrum

Table I   Crystal data of valine

	L-valine[a]	DL-valine[b]
Space group	$P21$	$P21$
Crystal system	Monoclinic	Monoclinic
Cell dimension		
(a)	9.71 ± 0.01	5.21 ± 0.02
(b)	5.27 ± 0.02	22.10 ± 0.04
(c)	12.06 ± 0.02	5.41 ± 0.02
	90.8 ± 0.20	$90°$

[a] See Ref. 5.
[b] See Refs. 6–8. There is agreement between Refs. 8 and 10, but Ref. 9 shows triclinic, (a) = 5.25, (b) = 11.05, and (c) = 5.43.

of a compound is dominated by overtones and combinations of the fundamental bands, any change in the mid-IR region should be detectable in the near-IR. Furthermore, the anharmonic nature of hydrogen bonding causes the near-IR to be especially sensitive to small changes in crystal structure.

Diffuse reflectance is typically used in the near-IR region to obtain the spectra of solids. Because the near-IR has weak absorptivities, radiation can penetrate a considerable length into a solid and still yield an analytical signal. However, diffuse reflectance destroys any optical polarization present in the incident beam (11). Thus it might appear impossible to measure the difference between optically active compounds using diffuse reflectance. In the present case, optical activity is not the property measured; rather, what is measured is the change in the crystal that arises when the one enantiomer is incorporated into the crystal structure of the other.

For the calibration, the ratio of weight fractions, which is the weight fraction of the lesser constituent, x, divided by the weight fraction of the larger constituent, $(1 - x)$, was used in place of the more conventional weight-percent composition. The calibration is found in this manner because the spectra that have equivalent ratios but different enantiomers in excess should be the same. As a result, the predicted value is only an indication of the concentration of the minor constituent. No information as to the identity of the minor constituent is contained in the ratio; however, this is not a fatal disadvantage because the identity of the bulk is usually known in a manufacturing process. The analytical wavelengths are chosen so that the sums of the squares of the errors in equation 1 are minimized:

$$R = [A]B + E \tag{1}$$

where B is the coefficient vector; $[A]$, the pseudoabsorbance matrix ($-\log$ reflectance); R, the valine ratio vector; and E, the error vector.

A limited sample set was available to develop the calibration; so some method was necessary to determine the number of wavelengths needed to give the best correlation without overdetermining the solution. The predicted residual error sum of squares was used (12). Simply stated, the PRESS calculation consists of using stepwise multiple linear regressions (SMLRs) without one sample present. The ith sample is then predicted using the equation developed with the rest of the samples. This procedure is continued until all samples have been held out. A PRESS value is calculated for the predicted value using p number of wavelengths, according to:

$$\text{PRESS} = \left(\sum_{l=1}^{n} (Y_{pi} - Y_{pi})^2 \right)^{1/2} \tag{2}$$

where p is the number of wavelengths used; Y_{pi}, the actual value of the ith sample; and Y_{pi}, the predicted ith value.

There is no clearcut method of choosing the perfect PRESS value, but the usual practice is to choose the correlation that gives the lowest PRESS. The wavelength subsets for the PRESS calculations were chosen using an SMLR algorithm. After the optimal number of wavelengths needed was determined, an all-possible-combination algorithm was used.

The "spectrum" of the valine ratio was found using a reconstruction program (RECON, Bran + Leubbe Analyzing Technologies, Elmsford, New York). The program is based on a cross-correlation technique (13). By cross-correlating the spectra with the isomeric ratio, the pure component spectrum is extracted from the mixture spectra. In this study, the reconstruction process should yield a spectrum of the effect of the change in the crystal.

EXPERIMENTAL

Mechanical mixtures of the USP-grade D-, L-, and DL-valine were made, and the weight percentages were computed. The samples were in a solid form; so diffuse reflectance was used to collect the data. The measurements were made on a Technicon InfraAlyzer 500c instrument (also available from Bran + Luebbe) with a scanning monochromator, and data were collected using an HP1000/A600 minicomputer (Hewlett–Packard, Palo Alto, California). Wavelengths from 1100 to 2500 nm were scanned in 2-nm increments. A fixed slit of 1 mm was used. The spectra of the conglomerates (mechanical mixtures) were obtained only for the first eight samples because the spectra showed no apparent differences. Next, the mixtures were recrystallized from doubly distilled water. Finally, diffuse reflectance spectra of the mixtures recrystallized from water were collected under these same instrumental conditions.

RESULTS AND DISCUSSION

Table II lists concentrations and ratios used in these analyses. The ratio is $x/(1 - x)$, where x is the weight fraction of the minor constituent. The spectra of the recrystallized samples are shown in Fig. 1a. The spectra are limited to the region of wavelengths used in the correlation. For comparison purposes, the conglomerate spectra are shown in Fig. 1b. There is little change in the conglomerate spectra, as was expected in light of earlier work done by Draper and Smith (11). The major difference in the spectra is baseline offset caused by changes in the particle size and packing of the samples. Although the recrystallized spectra show numerous changes from the physical mixtures, choosing analytical wavelengths visually would be extremely difficult.

Table III shows the results of the PRESS calculation. The three-wavelength case was chosen as having the best prediction ability. The individual residual errors (E) values show that the samples at the extremes (samples 1 and 2, 100%

Table II   Ratios and mole fractions of valine

Mole fraction scan number	L-valine	D-valine	Ratio
1	1.0000	0.0000	0.000
2	0.0000	1.0000	0.000
3	0.4986	0.5014	0.994
4	0.2327	0.7673	0.303
5	0.4760	0.5240	0.908
6	0.2696	0.7304	0.369
7	0.7464	0.2536	0.340
8	0.6133	0.3867	0.631
9	0.1507	0.8492	0.177
10	0.3590	0.6410	0.560
11	0.8208	0.1793	0.218
12	0.8726	0.1274	0.146
13	0.4212	0.5788	0.720
14	0.9387	0.1613	0.065

L-valine and 100% D-valine, respectively) are apparently not well predicted. These ends of the scale should be further investigated.

Regression techniques are used to choose analytical wavelengths in the near-IR because of confusing band overlap. It is not readily apparent from Fig. 1a why the marked wavelengths are of analytical interest. These points are not of particular importance in the spectrum of valine. However, the reconstructed spectrum (Fig. 2) indicates that the two wavelengths at 1562 and 1576 nm are a baseline correction, whereas the 1660-nm peak is the major peak of interest. At 1660 nm, there is a large negative correlation to the isometric ratio. It is not surprising that the wavelength chosen is not directly at the absorbance minimum; by choosing a wavelength off the minimum, small shifts in the spectra have large effects. The fit of the data is shown in Fig. 3. The data are fit well

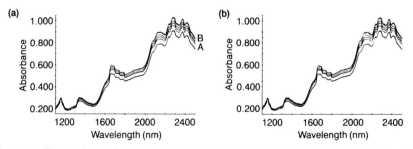

Figure 1. Figure shows (a) near-infrared spectra of recrystallized valine samples. A = 100% L; B = 100% D; (b) near-infrared spectra of valine conglomerates.

Table III   PRESS calculations on valine samples.
All numbers are multiplied by 10^{-2}

Sample removed	RE[a] 2 wavelengths	RE 3 wavelengths	RE 4 wavelengths
1	4.558	2.500	4.818
2	3.201	2.544	3.073
3	2.132	0.629	1.178
4	0.164	<0.001	0.010
5	0.712	1.626	1.381
6	0.630	−0.457	0.213
7	0.238	0.147	<0.001
8	0.554	0.027	0.033
9	0.806	1.000	0.276
10	0.031	1.430	1.395
11	0.153	0.120	0.002
12	0.251	0.023	0.095
13	1.000	0.340	1.641
14	0.013	0.686	0.372
PRESS	14.443	11.538	14.470

[a]RE = residual error.

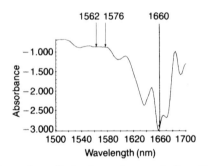

Figure 2. Reconstruction using ratio in place of concentration. Marked wavelengths are the regression wavelengths.

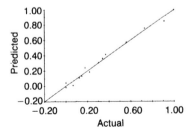

Figure 3. Fit of the regression values using the calibration given in Table IV.

Table IV Regression results for valine (all possible combinations)

Wavelength	Coefficient
Intercept	1.562
1562	−294.228
1576	330.922
1600	−44.606

F value = 432. R = 0.996. Degrees of freedom = 10.

by the regression mathematics with an R value of 0.996 (Table IV). The fact that the individual PRESS values are small would indicate that the correlation should have good prediction ability.

CONCLUSIONS

It might appear that near-IR reflectance could not distinguish the amount of one enantiomer of an amino acid present in another. On the surface, this would be a correct assumption, because the spectrum of D-valine is exactly the same as that of L-valine. However, this is not the case. It was shown that the crystal structure of DL-valine differed from those of D- and L-valine in a way detectable by near-IR. This change was linear with respect to the ratio of the valine enantiomeric mole fractions. While one would not be able to tell from the ratio which enantiomer was the impurity, the major species in a process system is usually known. If one is producing L-valine for marketing and desires to know the amount of D-valine contaminating the sample, the ratio measurement technique appears to work well. The near-IR method would appear to be satisfactory for process measurements of amino acids. Other optically active pharmaceuticals such as vitamins should logically be amenable to this approach. Near-IR may well be used as a rapid, nondestructive technique for measuring the enantiomeric purity of pharmaceutical actives.

REFERENCES

1. J. J. Rose, T. Prusik, and J. Mardekian, *J. Parent. Sci. Tech.* **36**, 71 (1982).
2. E. W. Ciurczak and T. A. Maldacker, *Spectroscopy* **1**(1), 36 (1986).
3. E. W. Ciurczak, R. Torlini, and M. T. Demkowicz, *Spectroscopy* **1**(7), 36 (1986).
4. E. W. Ciurczak and R. Torlini, *Spectroscopy* **2**(3), 41 (1987).
5. R. G. Whirfield, M. E. Gergor, and R. L. Sharp, *Appl. Spectrosc.* **41**, 1204 (1987).
6. K. Torii and Y. Litaka, *Acta Crystallogr.* **B26**, 1317 (1970).
7. B. Dawson and A. M. Mathieson, *Acta Crystallogr.* **4**, 475 (1951).

8. M. Mallikarjunar and S. Thyagaraja Rao, *Acta Crystallogr.*, **B25**, 296 (1969).

9. M. Tsuboi and T. Takenishi, *Bull. Chem. Soc. Jpn.* **32**, 726 (1959).

10. P. Kubelka and F. Munk, *Zh. Tech. Fiz.* **12**, 593 (1931).

11. N. R. Draper and H. Smith, *Applied Regression Analysis* (John Wiley & Sons, NY, 1966).

12. D. E. Honigs, G. M. Hieftje, and T. B. Hirschfeld, *Appl. Spectrosc.* **38**, 317 (1984).

13. J. R. Lacher, V. D. Croy, A. Klanpour, and J. D. Park, *J. Phys. Chem.* **58**, 206 (1954).

CHAPTER II.B.8

ANALYSIS OF SOLID AND LIQUID DOSAGE FORMS USING NEAR-INFRARED REFLECTANCE SPECTROSCOPY

Emil W. Ciurczak and Ronald P. Torlini

Near-infrared spectroscopy (NIRS) was used to analyze pharmaceutical solid and liquid dosage forms. The purpose of this study was to compare NIRS analysis in terms of speed and accuracy with other analytical techniques.

Although uses of near-IR spectroscopy (NIRS) are plentiful in agriculture and in the textile and polymer industries, few applications have been published for the pharmaceutical industry. And of these pharmaceutical applications, very few have been used for the analysis of actives in dosage forms. One such assay was for a parenteral solution with minimum matrix interference (1). Other analyses of dosage forms were performed either after extraction (2, 3) or on a model system containing only one drug and a single excipient (4). Quantitation of raw materials has been performed (5, 6) and several papers (7–9) have appeared on qualitative applications for raw materials and dosage forms, but none has been published about the analysis of more than one active substance in a complex matrix. One reason for the scant use of near-IR spectroscopy has been the lack of suitable data-handling capabilities for available instrumentation. Even pure organic materials have spectra that are rather difficult to interpret in the near-IR region, because they consist of combinations or overtones of the fundamental bands originating in the mid-IR region. Mixtures are even harder to work with. The complete use of the near-IR region awaited the development of instruments that could be coupled with micro- and minicomputers.

The software developed by all major manufacturers of near-IR instruments since the mid-1970s is rather sophisticated. When combined with computer power, modern near-IR spectrometers offer excellent wavelength reproducibility and exceedingly small baseline noise levels that are routinely in the microabsorbance range. These developments now allow for precise quantitative work to be performed with complex mixtures.

The absorbance for a single substance in a nonabsorbing solvent, according to Beer's law, is simply:

$$A = \epsilon bc \tag{1}$$

where A is the total absorbance (at a given wavelength); ϵ, the molar extinction coefficient; b, the path length; and c, the molar concentration of the absorbing material.

Analyzing for a substance in a matrix in which every component absorbs is a little more difficult. In solid mixtures and liquid solutions, several wavelengths and a series of standards must be scanned and their entire spectra utilized. A multiple-wavelength regression method was first introduced by BenGara and Norris (10) in the 1960s for use in meat products:

$$C = b_0 + b_1 \times A_1 + b_2 \times A_2 + \cdots + b_n \times A_n \tag{2}$$

This equation relates a given concentration (C) of a particular component with the absorbances measured (A_n) through factors (b_n) determined from previously analyzed calibration sets. For natural products, calibration sets can only be obtained from analyses performed by a reference method (such as Kjeldahl nitrogen for protein or Karl Fischer for moisture determinations). For pharmaceutical dosage forms, however, the matrix is well defined and can be synthetically reproduced to create a calibration set. The accuracy and reproducibility of a near-IR analysis are only as good as the calibration data. These data are usually obtained from standard methods such as Kjeldahl, Karl Fischer, or HPLC; thus synthetic samples can produce extremely accurate calibrations.

In this work, we used calibration techniques that are common in agriculture and in textile manufacturing for analysis of pharmaceutical liquid and solid dosage forms. The purpose was to compare near-infrared reflectance analysis with classical techniques—such as high-performance liquid chromatography (HPLC)—for speed and accuracy of results.

EXPERIMENTAL

Instrumentation. All spectra were generated with a model 500C InfraAlyzer (Technicon Instrument Corporation, Tarrytown, New York) that was equipped with a model 1000/A600 minicomputer (Hewlett–Packard, Palo Alto, California) and operated with software from Technicon. Solid samples were ground in a Tecator (Herndon, Virginia) model A-10 powder mill. Liquid samples were scanned in a thermostatted liquid "transflectance" cell from Technicon.

Sample preparation. All materials used were USP-grade reagents. The caffeine, chlorpheniramine maleate, acetaminophen, and codeine phosphate were assayed in the authors' laboratory. The starch, lactose, magnesium stearate, 95% ethanol, sorbitol, and alginic acid were from production stock.

Solid samples were made from a stock supply of weighed excipient mixture with the appropriate amount of actives added by weight. These materials were placed in the powder mill and milled for 10 s. The resultant powder was placed into a closed sample cup and scanned with the near-IR spectrometer. The calibration was performed on approximately 80% of the mixtures made. The remaining samples were considered unknowns and analyzed as such. The caffeine–acetaminophen tablet mixture was examined with and without milling, while the acetaminophen tablet mix was only analyzed after milling.

Liquid samples were prepared in a 25% ethanolic (v/v), 3% sorbitol (w/v) aqueous vehicle. This was sufficiently similar to commercial syrups and provided a highly absorbing background. One percent (w/v) solutions of chlorpheniramine maleate, codeine phosphate, and acetaminophen were prepared in the syrup vehicle and analytically metered into volumetric flasks with burettes. These were brought to volume with the solvent to give a series of cough syrups with varying strengths (the ingredients were varied randomly). Again, most of the solutions were used for the calibration, while some were set aside to be unknowns.

The solutions were measured in a thermostatted liquid cell provided by Technicon. This cell has a fixed path length of about 2–3 mm and works in a mixed transmittance–reflectance mode.

RESULTS AND DISCUSSION

Table I shows the calibration data generated by the regression algorithm for the acetaminophen–caffeine tablet mixture. The wavelengths, regression coefficients, partial Fs for deletion (a measure of the importance of any number to a given equation: the larger the value of partial F, the better), and the multiple correlation coefficients are tabulated therein. Table II shows the prediction data for the unknown samples. Tables III and IV present the calibration and prediction data, respectively, for the 325-mg acetaminophen tablet mixtures. Tables V and VI show the calibration and prediction data for acetaminophen, codeine phosphate, and chlorpheniramine maleate in the cough syrup.

Experience with sample runs leading to the reported data shows that, depending on the particular component, milling might not appreciably improve the calibration. This can only be determined experimentally by an individual scientist.

Table I. Calibration data from acetaminophen–caffeine tablets.

| | Acetaminophen | | |
| | Regression | t value of | Partial F for |
Wavelength (nm)	coefficient	coefficient	deletion
Intercept	0.281		
1172	− 695.676	−6.380	40.710
1136	655.791	6.600	43.565
2016	495.116	5.330	28.410
1960	− 455.666	−4.490	20.158

Multiple correlation coefficient = 0.937.

	Caffeine		
Intercept	5.766		
1172	2198.188	8.680	75.335
1400	144.020	4.088	16.709
1188	−2947.448	−7.772	60.401
1216	613.225	3.549	12.596

Multiple correlation coefficient = 0.926.

Table II. Prediction data from acetaminophen–caffeine tablets.

| | Acetaminophen | | |
Sample	Theory	Found	Percent difference
1	4.47	4.47	0
2	4.56	4.59	0.66
3	4.69	4.60	−2.00
4	4.69	4.74	1.00
5	4.44	4.52	1.89

	Caffeine		
1	4.50	4.45	− 1.11
2	4.40	4.43	0.68
3	4.25	4.35	2.35
4	4.25	4.19	− 1.46
5	4.63	4.58	− 1.08

Table III. Calibration data from 325-mg acetaminophen tablets.

Wavelength (nm)	Regression coefficient	t value of coefficient	Partial F for deletion
Intercept	75.281		
1964	−5921.586	−5.311	28.217
1872	1757.113	6.146	37.775
1924	1302.130	2.329	5.423
1972	2897.631	1.770	3.133

Multiple correlation coefficient = 0.999.

Table IV. Prediction data (% wt) from 325-mg acetaminophen tablets.

Sample	Theory	Found	Percent difference
1	71.864	71.787	−0.11
2	76.604	76.759	0.20
3	70.155	69.985	−0.24
4	73.196	73.244	0.07
5	74.718	74.785	0.09

In the case of the caffeine–acetaminophen tablets, the acetaminophen measurements improved significantly while the caffeine data were virtually unchanged (data not shown). The relatively large (up to 2%) deviations from theory are most likely caused by the sample handling involved in measuring both unmilled and milled samples.

Because the acetaminophen (APAP) tablet formulation contained 70–82 wt % active, it is to be expected that the prediction data should be better than those

Table V. Cough syrup calibration data.

Wavelength (nm)	Acetaminophen Regression coefficient	t value of coefficient	Partial F for deletion
Intercept	55.420		
1328	512.095	9.657	93.263
2392	−273.382	−7.333	53.776
1148	−646.779	−10.034	100.679
2408	228.559	6.738	45.405

Multiple correlation coefficient = 0.984.

	Codeine phosphate		
Intercept	47.884		
1284	−215.103	−17.178	295.073
1808	88.780	9.494	90.145
2060	17.904	11.014	121.317
2228	−61.819	−12.098	146.367

Multiple correlation coefficient = 0.993.

	Chlorpheniramine maleate		
Intercept	4.567		
1824	−65.535	−25.241	637.144
1144	125.591	42.525	1808.406
1296	−37.640	−14.852	220.571
2472	−1.041	−4.371	19.105

Multiple correlation coefficient = 0.999.

Table VI. Cough syrup prediction data.

	Acetaminophen (mg/mL)		
Sample	Theory	Found	Percent difference
1	0.629	0.618	−1.75
2	0.579	0.569	−1.73
3	0.634	0.638	0.63
4	0.604	0.608	0.66
	Codeine phosphate (mg/mL)		
1	0.2920	0.2908	−0.41
2	0.2900	0.2902	0.07
3	0.3110	0.3117	0.23
4	0.2980	0.2981	0.03
	Chlorpheniramine maleate (mg/mL)		
1	0.1210	0.1212	0.17
2	0.1300	0.1298	−0.15
3	0.1230	0.1230	0
4	0.1200	0.1203	0.25

obtained for actives in the 5% range. This was the case. In fact, the average difference was approximately 0.25%, which is quite competitive with HPLC determinations (11). While not within the scope of the experiment, it might be inferred from the results that a solid dosage form containing one active as a major percentage of the total weight might be assayed in its tablet form without requiring grinding. (*This is exactly what most manufacturers are claiming today. I like to think that I predicted this.*)

This factor, although it remains to be proven, might be useful in a quality control setting. With the speed of analysis inherent in NIRS, large numbers of tablets from a production or pilot run could be assayed rapidly and nondestructively. This information could be essential in process design and control. Secondary tests, such as dissolution, hardness, disintegration, and friability, could be run on the actual tablets from which the assay results were derived.

In a process validation study, hundreds of granulation or tablet samples might be submitted for assay. These usually are assayed by chromatographic or standard spectrophotometric methods. The sample preparation time for several hundred samples is considerable. Using near-IR spectroscopy, even if milling is required, the time savings (not to mention savings in solvent purchases and the cost of solvent disposal) would be significant. Typical HPLC suitability requirements contain the condition that replicate samples from the same vial must reproduce within 2% RSD. Because the results presented herein easily meet this criterion, a well-designed near-IR assay should be considered at the onset of methods development.

The data from the liquid samples are interesting for more reasons than the accuracy of the prediction data that were achieved. Most manufacturers of near-IR equipment recommend the technique for determination of components present in amounts of 1% or greater for the best results. The components we used were in the range of 0.01 to 0.06% (w/v). Several recent articles suggested that charged molecules might well have a greater effect upon the near-IR spectra of solutions than uncharged molecules. An article by Hollenberg and Ifft (12) that describes the use of near-IR spectroscopy to determine the hydration numbers of amino acids suggests the same proposition. In light of this, the apparently more accurate data for increasingly small concentrations in the same solutions can be explained. Acetaminophen (1.1% variation) is a neutral molecule, codeine (0.3% variation) can have a single charge, while chlorpheniramine (0.1% variation) can possess a double charge in solution. These data might point the way to trace analysis of charged molecules where only major components were analyzed before. *[I explained these data in 20/20 hindsight using Debye radii and a little physical chemical thoughts at a later date (13, 14).]*

REFERENCES

1. J. J. Rose, T. Prusik, and J. Mardekian, *J. Parenter. Sci. Technol.* **36**(2), 71 (1982).
2. A. F. Zappala and A. Post, *J. Pharm. Sci.* **66**, 292 (1977).
3. S. Sherken, *J. AOAC* **51**, 616 (1968).
4. J. K. Becconsall, J. Percy, and R. F. Field, *Analyt. Chem.* **53**, 2037 (1981).
5. J. E. Sinsheimer and A. M. Keuhnelian, *J. Pharm. Sci.* **55**, 1240 (1966).
6. J. R. Rose, paper presented to the Second NIRA Symposium, August 1982, Tarrytown, NY.
7. E. W. Ciurczak and T. A. Maldacker, *Spectroscopy* **1**(1), 36 (1986).
8. E. W. Ciurczak, R. P. Torlini, and M. P. Demkowitz, *Spectroscopy* **1**(7), 36 (1986).
9. E. W. Ciurczak, paper presented at the Seventh NIRA Symposium, July 1984, Tarrytown, NY.
10. I. Ben-Gara and K. H. Norris, *J. Food Sci.* **33**, 64 (1968).
11. S. J. van der Wal, LC, *Liq. Chromatogr. HPLC Mag.* **3**(6), 488 (1985).
12. D. Hollenberg and M. Ifft, *J. Phys. Chem.* **36**, 1939 (1982)
13. E. W. Ciurczak, *"Analysis of Multi-Component Liquid Pharmaceutical Preparations Using NIRS,"* paper presented at 27th EAS, New York, November, 1987.
14. E. W. Ciurczak, *"Effect of Dielectric Constant, Salt Concentrations, and Ionization of Weakly Dissociating Molecules on the NIR Spectra of Solutes in Aqueous Solutions,"* presented at the 8th International Diffuse Reflect. Conf., Chambersburg, PA, August, 1996.

CHAPTER II.B.9

DETERMINATION OF PARTICLE SIZE OF PHARMACEUTICAL RAW MATERIALS USING NEAR-INFRARED REFLECTANCE SPECTROSCOPY

Emil W. Ciurczak, Ronald P. Torlini, and
Michael P. Demkowicz

Near-infrared (NIR) reflectance spectroscopy was used to determine the mean particle size of pure, granular substances. Using low-angle laser light scattering (LALS) to determine particle size distributions, calibration graphs were constructed versus NIR reflectance values (log 1/R). These calibration graphs could be used to assess the particle size of pure samples in quality control applications.

When a material cannot be analyzed by spectrophotometric methods using transmitted light, it may be convenient to use reflectance methods. The theories pertaining to diffuse reflectance have been derived and refined since Stokes (1) first presented them in the 1860s. Since then, a number of other workers (2–9) also have studied the optics of diffusely reflected light, but the most widely quoted formulas are from Kubelka and Munk (5). Depending on whether the material is light, dark, poorly scattering, or intensely scattering, one form of the Kubelka–Munk theory covers the expected reflectance. In all cases, the scattering coefficient S and absorption coefficient K are taken into account.

Equation (1) is a simplified and therefore usable statement of the

Kubelka–Munk equation:

$$(1 - R)^2/2R = K/S \tag{1}$$

From this relation, it is apparent that reflectance R increases as the mean particle size decreases (indicating more scatter), while R decreases as the absorptivity increases. If the absorptivity were constant, then upon reaching a small enough particle size, one would obtain a perfect reflector. Because this implies that very small particles do not absorb radiation, one might naively presume that spectroscopic phenomena cannot be real. Because spectroscopy is practiced as a science, the theory cannot be so simply adhered to, and the value of K must change rather than remain constant. Felder (10) developed a system of spherical monodisperse particles to explain the required change of K as a function of transmission T:

$$K_T = \frac{3\Phi(P_m)}{2d} \ln\left(1 - \frac{p(1 - T_d)}{\Phi(P_m)}\right) \tag{2}$$

where d is the particle diameter; P, the packing density of particles $[(N/V)$ $(\pi d^3/6)]$; and $\Phi(P_m)$, the function of a maximum possible packing density P_m.

Equation (2) implies that the absorbance for a constant concentration (either a mixture or a pure substance) increases with diminishing particle size. The limit will correspond to the absorbance of the molecularly dispersed dissolved material. At the other end of particle sizes, where scattering diminishes with increased d, radiation penetrates farther into the powder, which effectively increases the path length that the light travels, therefore increasing the absorbance.

Because of the recent increased use of NIR reflectance instrumentation, the problem of light scattering has required addressing. Most published accounts have discussed negative correlations or described ways to avoid interferences. Watson and colleagues (11) stated only that they found no correlation between classes of wheat and particle size. Looking for minimal interference, Hunt and co-workers (12) used a series of mills to determine the optimum particle size and grinder to be used for grain analyses. Osborne and colleagues (13) performed some useful measurements on milled flour. They observed diminishing absorbance with diminishing particle size and compared the sieve size (a reciprocal of particle size) to the logarithm of $1/R$. The observed correlation of 0.95 may be a result of using sieve sizes instead of actual distributions on any sieve. Norris and Williams (14) used normalized second derivatives to attempt to eliminate most particle size effects rather than to measure them.

An attempt to correlate particle size to near-infrared reflectance measurements was presented by Ciurczak (15). This chapter is an attempt to expand on that work and to glean useful data from what was formerly considered a nuisance.

EXPERIMENTAL

Instrumentation. All the spectra were generated on a model 500C InfraAlyzer (Technicon Instrument Corporation, Tarrytown, New York) that was equipped with a model 1000/A600 minicomputer (Hewlett–Packard, Palo Alto, California) using software from Technicon. Mean particle sizes of the samples were determined with a model 3600E low-angle laser light-scattering (LALS) particle sizer (Malverne Instruments, Malverne, United Kingdom).

Sample preparation/reagents used. USP-grade reagents were used. Aspirin and ascorbic acid were assayed in the authors' laboratories; aluminum oxide and ammonium phosphate, dibasic, were used as purchased (Fisher Scientific, Pittsburgh, Pennsylvania). Sieving was performed on standard brass and/or stainless steel sieves (Newark Wire Cloth Company, Newark, New Jersey). The actual sieves used in each case depended upon the sample.

Sieve cuts were measured using near-IR reflectance spectroscopy and analyzed with the Malverne particle sizer. The particle size distribution, mean particle size, and absorbance values ($\log 1/R$) at selected wavelengths were recorded.

RESULTS AND DISCUSSION

Tables I–IV contain the data gathered from the particle sizer and the near-infrared spectrophotometer. Figure 1 shows representative near-infrared reflectance spectra for aspirin, ascorbic acid, and aluminum oxide. Figure 2 illustrates graphs of the reciprocal particle size versus absorbance ($\log 1/R$) at various wavelengths for aspirin, ascorbic acid, aluminum oxide, and ammonium phosphate, dibasic. The most striking feature of the graphic depiction of the data is the linearity in all experiments for particle sizes from greater than

Table I. Data from particle size and near-infrared analyses of aspirin.

Particle size (average)		Absorbance					
$\bar{x}\,(\mu m)$	$1/\bar{x}\,(\times 10^3)$	1800 nm	2000 nm	2200 nm	2300 nm	2400 nm	2500 nm
247	4.04	0.670	0.830	1.036	1.161	1.341	1.285
209	4.80	0.629	0.784	0.986	1.107	1.287	1.234
146	6.84	0.560	0.698	0.882	0.992	1.161	1.113
143	6.99	0.570	0.713	0.906	1.023	1.206	1.151
112	8.97	0.493	0.619	0.788	0.894	1.060	1.009
94	10.65	0.435	0.543	0.691	0.784	0.928	0.886
78	12.87	0.403	0.501	0.635	0.719	0.850	0.813
62	16.20	0.384	0.480	0.614	0.698	0.832	0.793

Table II. Data from particle size and near-infrared analyses of ascorbic acid.

Particle size (average)		Absorbance				
$\bar{x}\,(\mu m)$	$1/\bar{x}\,(\times 10^3)$	1900 nm	2100 nm	2250 nm	2400 nm	2450 nm
222.3	4.50	0.868	0.994	1.211	1.080	1.230
153.4	6.52	0.812	0.938	1.169	1.024	1.182
139.4	7.17	0.798	0.923	1.160	1.010	1.172

Table III. Data from particle size and near-infrared analyses of aluminum oxide.

Particle size (average)		Absorbance					
$\bar{x}\,(\mu m)$	$1/\bar{x}\,(\times 10^3)$	1800 nm	1950 nm	2100 nm	2300 nm	2400 nm	2500 nm
110.6	9.04	0.165	0.258	0.238	0.298	0.324	0.377
102.5	9.76	0.159	0.241	0.227	0.284	0.306	0.353
100.6	9.94	0.157	0.235	0.223	0.281	0.305	0.351
91.2	10.76	0.148	0.219	0.207	0.261	0.281	0.324
84.5	11.83	0.146	0.211	0.202	0.256	0.274	0.314
77.7	12.87	0.144	0.208	0.197	0.250	0.267	0.307
67.8	14.75	0.141	0.197	0.189	0.241	0.257	0.293
61.1	16.37	0.138	0.192	0.183	0.233	0.247	0.282

Table IV. Data from particle size and near-infrared analyses of ammonium phosphate, dibasic.

	Particle size (average)		Absorbance			
Mesh	$\bar{x}\,(\mu m)$	$1/\bar{x}\,(\times 10^3)$	1700 nm	1900 nm	2200 nm	2500 nm
40	380	2.63	1.254	1.546	2.028	1.867
70	200	5.00	1.084	1.360	1.979	1.695
90	160	6.25	0.995	1.246	1.889	1.566
110	130	7.69	0.914	1.148	1.818	1.456
140	107	9.35	0.783	0.986	1.635	1.268

200 μm down to about 85 μm. In the cases in which smaller particle size data are available, a second, less steep, sloping line is observed.

The authors believe that the rapidly diminishing absorbance between 250 and 85 μm is controlled almost entirely by the S factor of the Kubelka–Munk equation, while the K factor exerts a larger influence on data from particles smaller than 85 μm.

The most obvious consequences of these data are twofold. Particle sizes of pure materials can be measured most accurately over the larger particle size range (above 85 μm); conversely, although they still are feasible, measurements of the size of particles smaller than 85 μm are somewhat less accurate. A less

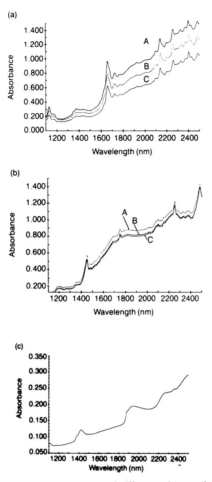

Figure 1. Near-infrared reflectance spectra of different size particles of aspirin, ascorbic acid, and aluminum oxide. (a) Aspirin: A = 40, B = 100, and C = 200 mesh material; (b) ascorbic acid: A = 40, B = 90, and C = 140 mesh material; (c) aluminum oxide, 100 mesh.

obvious, but perhaps more important, result concerns the slope change below 85 μm. Instrument manufacturers have recommended grinding samples to prevent the occurrence of particle size effects, but no size range has been definitely specified. From the data presented, it may be inferred that reducing solid samples to sizes less than 80 μm would greatly reduce particle size effects encountered in diffuse reflectance near-infrared spectroscopic measurements. This result was empirically observed in the sample preparation process of a previous experiment (16).

In an attempt to determine if particle size effects could be avoided, the second-derivative spectra of different size aspirin samples were obtained. Figure 3 shows these spectra; it is apparent that the second-derivative spectra of

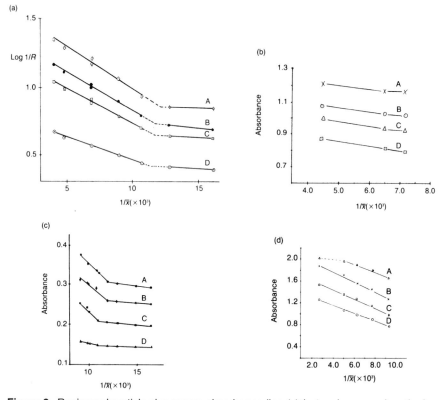

Figure 2. Reciprocal particle size versus absorbance (log $1/R$) at various wavelengths for aspirin, ascorbic acid, aluminum oxide, and ammonium phosphate, dibasic. (a) Aspirin: A = 2400, B = 2500, C = 2300, and D = 1800 nm; (b) ascorbic acid: A = 2250 (peak), B = 2400 (shoulder), C = 2100 (minor peak), and D = 1900 nm (baseline); (c) aluminum oxide: A = 2500, B = 2400, C = 1950, and D = 1800 nm; (d) ammonium phosphate, dibasic: A = 2200, B = 2500, C = 1900, and D = 1700 nm.

different size particles are very similar, indicating that the effect of particle size is greatly reduced but not eliminated.

Figure 4 illustrates a different manner of presenting the data. The absorbance values for each spectrum at 1658 nm (the major peak) are plotted against the absorbance values at 1784 nm (the baseline). The resultant graph is linear, with a correlation coefficient of 0.99999. Samples that represent the largest and smallest acceptable particle size ranges could be used to generate these data. It is not necessary, however, to know absolute particle sizes to calculate the ratio of absorbances (1658 to 1784 nm). The resultant plot could be then used as a quality control tool when new materials are received. Subsequent work found (data not shown) that this property was valid at any two wavelengths, which means filter, rather than dispersive instruments, could be used on quality control applications.

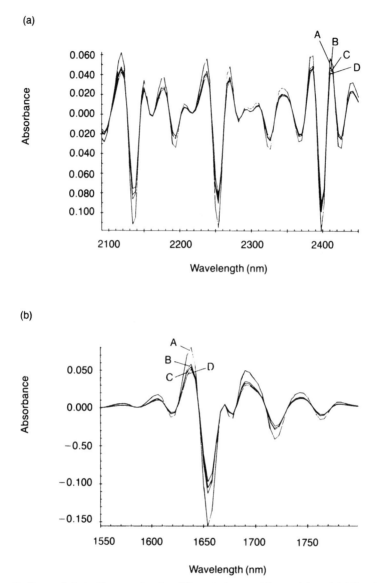

Figure 3. Second-derivative spectra for different size aspirin particles. (a) Wavelength region: 2100–2450 nm; (b) wavelength region: 1550–1800 nm. Sizes: A = 115, B = 200, C = 230, and D = 270 mesh.

This chapter has discussed the application of near-infrared reflectance spectroscopy to measure the mean particle size of a pure pharmaceutical substance. In addition, the data suggest a range to which solid mixtures should be reduced in order to minimize particle size effects when analyzing with near-infrared reflectance spectroscopy.

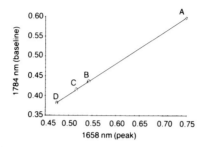

Figure 4. Plot of aspirin absorbances (log $1/R$)—1658 nm (peak) versus 1784 nm (baseline)—for different size particles. Sizes: A = 115, B = 200, C = 230, and D = 270 mesh material.

REFERENCES

1. G. G. Stokes, *Proc. Royal Soc. London* **11**, 545 (1860–1862).
2. H. J. Channon, F. F. Renwick, and B. V. Storr, *Proc. Royal. Soc. London* **94**, 222 (1918).
3. H. D. Bruce, Technical Paper 306 (NBS, Gaithersburg, MD, 1926).
4. M. Gurevic, *Physik Zeits.* **31**, 753 (1930).
5. P. Kubelka and F. Munk, *Zeits. E. Tech. Physiks* **12**, 593 (1931).
6. B. D. Judd, *J. Research Nat. Bur. Stand.* **12**, 345 (1934).
7. B. D. Judd, *J. Research Nat. Bur. Stand.* **13**, 281 (1934).
8. L. Amy, *Rev. Optique* **16**, 81 (1937).
9. T. Smith, *Trans. Opt. Soc. (London)* **33**, 150 (1931).
10. B. Folder, *Heir. Chim. Acta* **47**, 488 (1964).
11. C. A. Watson, W. C. Shuey, O. J. Banasik, and J. W. Dick, *Cereal Chem.* **54**, 1264 (1977).
12. W. H. Hunt, D. W. Fulk, and T. Thomas, *Cereal Foods World* **22**, 143 (1978).
13. B. G. Osborne, S. Douglas, and T. Fearn, *J. Sci. Food Agric.* **32**, 200 (1981).
14. K. H. Norris and P. C. Williams, *Cereal Chem.* **61**, 158 (1984).
15. E. W. Ciurczak, presented at the Eighth NIRA Symposium, August 1985, Tarrytown, NY.
16. E. W. Ciurczak and T. A. Maldacker, *Spectroscopy* **1**(I), 36 (1986).

CHAPTER II.B.10

APPLICATIONS OF NEAR-IR SPECTROSCOPY IN THE PHARMACEUTICAL INDUSTRY

James K. Drennen and Emil W. Ciurczak

This month I asked Jim Drennen (Associate Professor of Pharmaceutics, Graduate School of Pharmaceutical Sciences, Duquesne University, Pittsburgh, PA 15282) to coauthor a chapter on two subjects near and dear to us both: near-IR spectroscopy and pharmaceutical analysis. Jim earned his degree under Rob Lodder at the University of Kentucky a few years back, and I've known him since he was a graduate student (I even knew Rob as a grad student ... perhaps the gray hair isn't premature after all). I feel that Jim's academic credentials and his abilities with computers are a good addition to my more "empirical" approach. In fact, if the truth be known, we are collaborating on a book.

The subject of near-infrared (near-IR) spectroscopy is well known. For this reason, our overview of near-IR spectroscopy in the pharmaceutical industry will focus on the applications rather than the theoretical aspects. The order in which we discuss the applications here will roughly approximate the path of a product being either developed or produced.

RAW MATERIALS

Current methods of raw material analyses are often based on the "tried and true" methodologies listed in the United States Pharmacopoeia (USP). These are mostly wet chemical or time-consuming instrumental methods. Methods were introduced as early as 1984 for the qualitative use of near-IR spectroscopy for determining the identity of raw materials (1, 2). Torlini (3) showed the technique to be a potential method for purity determinations in drug substances, while Whitfield (4) used a discriminant technique as a precursor to the quantitative analysis of a finished product. These and other uses were described by Ciurczak (5) in 1987. Polymeric packaging materials were tested using near-IR and the results presented in a paper by Shintani-Young (6) at the 1985 Pittsburgh Conference.

One of the most often reported analyses for raw materials is water: Sinsheimer and Poswalk (7) in 1968, Warren and others (8) in 1970, and Torlini and Ciurczak (9) in 1987, just to name a few.

PREFORMULATION/ FORMULATION

The homogeneity of a dry blend was determined in a mixing study discussed by Ciurczak (10) in 1991, and similar work was reported (11) using fiber optics in 1992. White (12) reported measuring the drying of a granulation in a fluid-bed drier in real time using a reflectance probe in the vessel wall connected by a fiber-optic bundle. The ability to measure the particle size of pure materials as well as granulated formulations was demonstrated in 1986 (13).

Methods were introduced as early as 1984 for the qualitative use of near-IR spectroscopy for determining the identity of raw materials.

In 1990, Drennen and others (14) described a method for determining the homogeneity of ointments. Concentrations of active ingredients were also determined. The method was touted as a noninvasive means of finding the end point of a mixing process. This method is now being used for making qualitative decisions about the homogeneity of powder mixtures.

ASSAY/CONTENT UNIFORMITY

Numerous reports of finished product assays have been reported over the years: Zappala and Post (15) in 1977 for Meprobamate, Rose and co-workers (16) in 1982 for multicomponent parenterals, and Ciurczak and Torlini (17) in 1987 for multicomponent parenterals and solids. More recently, intact dosage forms have been analyzed. In 1987, Lodder and Hieftje (18) used uniquely designed holders for the analysis of tablets and capsules. Individual intact tablets have been analyzed for the determination of degradation product and moisture content (19), hardness (20), and active ingredient–excipient concentrations. Studies are now under way at the Duquesne University Graduate School of Pharmaceutical Sciences to prove the possibility of predicting numerous tablet properties with one rapid and nondestructive near-IR scan.

A 1990 study (19) describing determination of degradation product and moisture content in individual intact tablets demonstrated low prediction errors for both the mass of water absorbed (±0.04% of tablet mass) and the mass of degradant formed (±0.04% of tablet mass). Drennen (20, 21) has shown that tablet hardness variations appear as a nonlinear baseline shift. Apparently, as tablet hardness increases, the tablet's surface becomes smoother, more light penetrates into the tablet because less light is scattered from the surface (softer tablets with rougher surfaces scatter more light), and absorbance is greater. Ciurczak (22, 23) has also performed content uniformity studies of intact tablets.

Researchers at the University of Kentucky (24) have achieved good results predicting the dissolution rate of carbamazepine tablets. They performed quantitative tests to predict the dissolution rate of individual tablets from near-IR spectra ($R^2 = 0.985$) and qualitative tests to identify tablets with unacceptable dissolution rates (20). The bootstrap error adjusted single-sample technique (BEAST) was shown to have a significant advantage for identifying "bad" tablets when compared to other algorithms used for discriminant analysis (20). The BEAST also permits information from all spectral wavelengths to be incorporated into the qualitative prediction model, thereby eliminating the preliminary data reduction step required by other qualitative calculations and increasing the accuracy of qualitative decisions.

MISCELLANEOUS

Numerous other pharmaceutical applications have been reported. Noninvasive measurement of residual moisture in freeze-dried materials has been performed by measuring directly through the vial (25, 26). Kamat and co-workers (25) described the noninvasive determination of residual moisture in lyophilized sucrose, a cryoprotectant material. More traditional methods of moisture analysis such as gas chromatography, thermogravimetry, or the commonly used Karl Fischer titration are all destructive tests. Kamat's results suggest that the near-IR method is at least as sensitive as the Karl Fischer titration. Good correlation between the Karl Fischer reference test and the near-IR method ($R^2 = 0.97$) was obtained. Near-IR spectroscopy has been used for determining polymorphism of drugs in formulations (27), as a detector for quantitative and qualitative analysis of drug substances on thin-layer chromatographic plates (28, 29), and for determining the enantiomeric purity of optically active materials (30). Galante and co-workers (31) developed a near-IR assay for determining the sterility of parenteral formulations. Detection of bacteria, mold, and yeast contamination was proven possible. Near-IR spectroscopy should be valuable for the quality control of parenteral products that will not withstand terminal sterilization.

CONCLUSION

Being insiders, we are aware of the rapid acceptance of near-IR as a tool in pharmaceutical analyses. In less than 10 years, near-IR spectroscopy has matured from merely a curiosity to a full partner in analytical work. We would strongly recommend that those who are not familiar with this technique watch for upcoming articles on the subject. The Pittsburgh Conference and the Eastern Analytical Symposium have been averaging four or five sessions on near-IR spectroscopy at recent meetings, and the trend seems to be the presentation of more, rather than fewer, papers on this topic.

REFERENCES

1. E. W. Ciurczak, "Discriminant Analysis of Pharmaceutical Components Using NIRA," paper presented at the 7th International NIRA Symposium, Tarrytown, NY, July 1984.
2. D. E. Honigs, G. M. Hieftje, and T. Hirschfeld, *Appl. Spectrosc.* **38**, 317 (1984).
3. R. P. Torlini, "*Use of Near-IR Spectroscopy in Determining Purity Levels of USP Grade Aspirin*," paper #401 presented at The Pittsburgh Conference, Atlanta, GA, March 1989.
4. R. G. Whitfield, *Pharm. Manuf.* **4**, 31–40 (1986).
5. E. W. Ciurczak, *Appl. Spec. Rev.* **23**(1&2), 147 (1987).
6. T. Shintani-Young and E. W. Ciurczak "*Identification of Polymeric Packaging Materials via NIRS*," paper presented at The Pittsburgh Conference, New Orleans, LA, March 1985.
7. J. E. Sinsheimer and N. M. Poswalk, *J. Pharm. Sci.* **57**(11), 2007 (1968).
8. R. J. Warren, J. E. Zarembo, C. W. Chong, and M. J. Robinson, *J. Pharm. Sci.* **59**(1), 29 (1970).
9. R. P. Torlini and E. W. Ciurczak, "*Determination of Moisture in Pharmaceutical Raw Materials via NIRA*," paper presented at The Pittsburgh Conference, Atlantic City, NJ, March 1987.
10. E. W. Ciurczak, *Pharm. Tech.* **15**(9), 87 (1991).
11. E. W. Ciurczak, "*Use of Fiber Optics to Follow the Homogeneity of a Mixing Study*," paper presented at Process Analysis '92 (SPIE Conference), New Brunswick, NJ, March 1992.
12. J. White, "*Following the Drying Profile of a Pharmaceutical Mixture in a Fluid-Bed Drier Using NIRS*," paper presented at the Eastern Analytical Symposium, Somerset, NJ, November 1990.
13. E. W. Ciurczak, R. P. Torlini, and M. P. Demkowicz, *Spectroscopy* **1**(7), 36–39 (1986).
14. J. K. Drennen, P. P. DeLuca, and R. A. Lodder, "*Near-Infrared Determination of Ointment Homogeneity*," paper presented at the AAPS National Meeting, Las Vegas, NV, November 1990.
15. A. F. Zappala and A. Post, *J. Pharm. Sci.* **66**(2), 292 (1977).
16. J. J. Rose, T. Prusik, and J. Mardekian, *J. Parent. Sci. Technol.* **36**(2), 71 (1982).
17. E. W. Ciurczak and R. P. Torlini, *Spectroscopy* **2**(3), 41–43 (1987).
18. R. A. Lodder and G. M. Hieftje, *Appl. Spectrosc.* **42**(4), 56 (1987).
19. J. K. Drennen and R. A. Lodder, *J. Pharm. Sci.* **79**(7), 622 (1990).

20. J. K. Drennen, "*A NOISE in Pharmaceutical Analysis: Near-Infrared Outside/Inside Space Evaluation*," Ph.D. dissertation, University of Kentucky (1990).

21. J. K. Drennen, "*Near-Infrared Analysis of Individual Intact Tablets and Capsules*," paper presented at the 30th Eastern Analytical Symposium, Somerset, NJ, November 1991.

21. E. W. Ciurczak, "*Analysis of Intact Tablets and Capsules by NIRS*," paper presented at the 30th Eastern Analytical Symposium, Somerset, NJ, November 1991.

23. E. W. Ciurczak, "*Non-Destructive, Single-Tablet Analysis via Near-Infrared Spectroscopy*," paper presented at the AAPS National Meeting, Washington, DC, November 1991.

24. P. N. Zannikos, W. Li, J. K. Drennen, and R. A. Lodder, *Pharm. Res.* **8**(8), 974 (1991).

25. M. S. Kamat, P. P. DeLuca, and R. A. Lodder, *Pharm. Res.* **6**(11), 961–65 (1989).

26. P. Brimmer, "*Determination of Moisture Levels in Freeze Dried Antibiotics Using NIRS*," paper presented at the 30th Eastern Analytical Symposium, Somerset, NJ, November 1991.

27. R. Gimet and A. T. Luong, *J. Pharm. Biochem. Anal.* **5**(3), 205 (1987).

28. E. W. Ciurczak, L. J. Cline-Love, and D. M. Mustillo, *Spectroscopy* **5**(8), 38–42 (1990).

29. E. W. Ciurczak, W. R. Murphy, and D. M. Mustillo, *Spectroscopy* **6**(3), 34–40 (1991).

30. B. R. Buchanan, E. W. Ciurczak, A. Q. Grunke, and D. E. Honigs, *Spectroscopy* **3**(5), 54–56 (1988).

31. L. J. Galante, M. A. Brinkley, J. K. Drennen, and R. A. Lodder, *Anal. Chem.* **62**, 2514 (1990).

CHAPTER II.C.1

CHEMOMETRICS: A POWERFUL TOOLBOX FOR UV/VIS SPECTROSCOPY

Jerome Workman, Jr.

When I spoke with Jerry Workman recently, he tried to explain a little more about chemometrics. It sounded so good that I asked him to put it down on paper. In fact, Jerry does do more than statistics. In spite of his busy schedule, he graciously agreed to write this opus for us. Either I'm getting used to the terms he uses or Jerry did a good job writing this chapter. I think the latter.

With the continued proliferation of papers describing mathematical or applied statistical methods for IR and near-IR spectroscopy, one might ask, "Do all these great (or potentially great) techniques also apply to UV/vis spectroscopy?" One might also ask:

- Would it be desirable to routinely calibrate and predict analyte concentrations for complex mixtures with many components using UV/vis spectroscopy?
- Are both quantitative and qualitative measurements important with multicomponent mixtures?
- Is real-time monitoring for at-line or on-line spectroscopy ever important to UV/vis users?
- Are ideal calibration sets for UV/vis quantitative analysis ever incomplete or unavailable?

- Do we sometimes lack pure component spectra when working with multicomponent mixtures?
- Are matrix interactions—such as band shifts, impurities, and background scattering—ever a problem in UV/vis measurements?
- Does low digitization of data in overlapping bands ever cause difficulty in determining band shapes?
- Would a complete spectral search–spectral matching and archiving database system be useful in UV/vis spectroscopy?

The answers to these questions and solutions to such problems can be found in the proper understanding and application of chemometrics.

The rapid expansion of the use of chemometrics, or the formal application of statistics and mathematics to chemistry, has caught the non-IR/NMR spectroscopic community slightly unawares. The power of these techniques, when properly applied (that's a big when), can provide a rich toolbox for gleaning real information from a maze of digital chaos. One might make the analogy that chemometrics is to digital data what a spectrometer is to electromagnetic radiation.

THE NOSTALGIA OF STATISTICS

Many of us can recall the fascination we experienced when fitting our first linear least-squares line to a scatter plot of x–y data using a ruler. This simple technique allowed us actually to visualize a complex solution of $Y = b_1 x + b_0$ in two-dimensional space. We felt comfortable with the notion of measuring an absorbance for our sample, moving our ruler to the line fit, and then estimating our concentration from this point—how elegant. We could then move conceptually into two-dimensional space using, say, a quadratic or quintic fit, still feeling comfortable.

Our next step might have been multiple linear least-squares regression. If we did the computations for this technique manually, we used a calculator and discovered that those alien-looking summation signs (Σ) involved a reasonable amount of work when $n \rightarrow 25$. With multiple regression we still may have felt comfortable, being able to visualize that a change of absorbance at one wavelength multiplied by a factor then added to a factor multiplied by the absorbance at another wavelength could represent a concentration. Of course, the notion of a simple line was gone, and we were left with the description of a calibration plane in three-dimensional space (Fig. 1).

But what about principal components regression, the K-matrix (1), P-matrix, partial least squares, and acronyms like SQUAD, SIMCA, WIZARD, SELEX, and the like; what a leap from our ruler line fit to functions such as these! What can be done for people requiring real information for real bosses and real regulatory agencies?

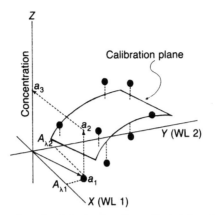

Figure 1. A sample spectrum is plotted using the optical data at two wavelengths ($A_{\lambda 1}$ and $A_{\lambda 2}$). The vector sum of the two wavelength line segments results in a point (a_1). A point directly above a_1 is found on the calibration plane (a_2). From a_2 the concentration is found (a_3) for this sample. The calibration plane (like a calibration line) is determined by analyzing known samples (dots) and minimizing the residuals. The residuals in this case are distances to a plane rather than a line.

WHAT IS REAL?

A quality assurance director probably would be uneasy knowing that all of the quality measurement results for several product runs were reported directly as obtained from the company's "Mister Wizard" computer program. So how do we as serious users of chemistry make the leap from plastic rulers to artificial intelligence (AI) with stops at advanced chemometric methods along the way?

Our first step is to realize that Monte Carlo simulation has nothing whatsoever to do with a travel video. In other words, we must learn the benefits and limitations of these mathematical tools and especially the correct ways to apply them in routine as well as research applications. In 1986, the Association of Official Analytical Chemists (AOAC) released this statement following a symposium on Chemometrics in Scottsdale, Arizona: "The general consensus of the symposium was that while chemometrics can greatly improve analytical data collection and interpretation, some degree of *caveat emptor* must be used in its application. In methods development, and particularly in interlaboratory validation of methods, consideration must be given to potential problems introduced by improperly applied chemometrics."

This statement could be interpreted to mean "stay away from," or "beware of," chemometrics until complex calibration techniques are made push-button or foolproof, which will not happen until a computer can compose a calibration set by itself. By then, the counterpart to AI, that is, AS (artificial stupidity), may itself be incorporated into many algorithms. Caution is a powerful motivating force, but knowledge of the truth is even more powerful. There is no magic here, only mathematics and physics. Thus each method and result can be understood in fundamental terms.

MULTICOMPONENT TRIVIA

One aspect of acute interest to practitioners of pure and applied analytical chemistry is multicomponent analysis. The first fundamental review bearing the title "Chemometrics" appeared in *Analytical Chemistry* in 1980 (2). This review by Bruce Kowalski came at a time when researchers using multivariate techniques were searching for standard terminology and for answers to the confusing problems encountered in day-to-day measurements using advanced mathematical techniques. Except for a few papers, the multivariate methods used in the literature were related to infrared techniques.

General-purpose UV/vis spectroscopy was limited to single-, two-, three-, or at most four-component mixtures. The rapid availability of multivariate calibration software and faster and less expensive microcomputer hardware has brought powerful mathematics to even the most modest laboratory. Capabilities that required software that took months to write and hardware worth many tens of thousands of dollars in 1980 now require only an AT clone at $1000–2000 and, in some cases, free public-domain software. Multiple linear regression (MLR), principal component regression (PCR), and partial least squares (PLS) are common buzz words to "calibrationists." But few can answer such questions as, "What is a principal component?" or "What is the difference between principal components and partial least-squares dimensions?" or "What similarities are there between Fourier analysis and principal components analysis?" These concepts can be confusing to those who just require quantitative results and have a need to be assured their results are correct and proper.

Well, let's quit wandering and get on with it. Simple multicomponent measurements are sometimes used when the spectra of two or more chromophores in a mixture are found to be marginally or severely overlapped across an entire spectral region. When performing multivariate analysis on these mixtures, the chemist can either select evenly spaced wavelengths across the entire spectral region for calibration or determine specific wavelengths where the molar absorptivities for the components are very different. Both of these wavelength selection techniques might seem simple to understand, but undoubtedly they often result in a nonoptimum calibration. Another possible approach involves all-possible-combinations (APC) search techniques. For these techniques, all the wavelengths are tested in prespecified wavelength group sizes to determine which wavelengths correlate most closely to changes in concentration.

Although the use of APC-MLR produces the optimum wavelength sets specific to any set of calibration samples, the computational time increases as k^3, where k is the number of wavelengths used for the final calibration equation. Thus an APC calibration of four wavelengths requires greater computational time than a three-wavelength calibration (64/27). Of course, even if wavelength selection were instantaneous, in complex cases it is impossible to know if the ideal set of wavelengths has indeed been chosen for all cases. In all MLR techniques, the assumption is made that absorptivities are additive and that the addition of wavelengths in a calibration equation can compensate for some noise, interferences, nonlinearity, background scattering, and other deviations from Beer's law (3–8).

PCR has been applied in a variety of spectroscopic techniques with reported success (9–13). It has the obvious benefit of not requiring wavelength searches. The best solution to the calibration problem is always available; the user only has to decide how many components to use in the regression. Of course if a poor set of calibration standards is used, a poor calibration will result, irrespective of the mathematics used. PLS is a technique that has more recently been brought to the attention of chemists. PLS brings added dimensionality to PCR and is thus a closer modeling technique. PLS has been compared directly to PCR in an excellent article from the *Journal of Chemometrics* (14). Tutorials and special articles describing PLS are, for most users, easier to find than to understand (15–20).

HOW MANY DIMENSIONS?

Visualizing a calibration plane, such as Fig. 1, allows us to feel much more comfortable about moving into multidimensional space. Likewise, if we look at principal components in two dimensions we can see that a spectrum reduced to two wavelengths can be plotted in this space. Four two-dimensional spectra can then be easily plotted in space (Fig. 2). Now if we take the average wavelength values at each axis for the four samples, we can plot a centroid (Fig. 3). If we next center our axes on the centroid and draw a vector to describe the maximum variance of our four spectra, the result is Fig. 4.

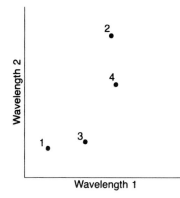

Figure 2. Four two-dimensional spectra plotted in wavelength space. Each data point represents a two-wavelength spectrum; thus four sample spectra are represented.

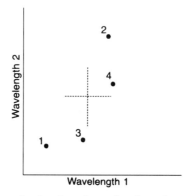

Figure 3. This figure shows the four two-dimensional spectra with a centroid plotted. The centroid represents the average optical data in both axes.

At this point, we can define B as the angle of the principal component vector and a distance S for each individual spectrum point from the centroid. We have just given a visual description of a principal component score S and its corresponding loading (which is equal to the cosine of B). Descriptions like this can be very helpful for users and assist in breaking down barriers of misunderstanding or mathematical malaise.

Computationally, we can provide the equations for the PCA:

$$S_1 = L_{11}X_1 + L_{12}X_2$$
$$S_2 = L_{21}X_1 + L_{22}X_2$$
$$S_3 = L_{31}X_1 + L_{32}X_2$$
$$S_4 = L_{41}X_1 + L_{42}X_2$$

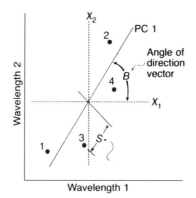

Figure 4. The four two-dimensional spectra showing the X_1 and X_2 axes centered on the data. The figure also illustrates the relationship of the principal component scores (S) and loadings ($\cos B$) to the plot of the four spectra. The first principal component (PC 1), which describes the maximum amount of variance in the four spectra, is also illustrated.

where S_i is the ith principal component score (Fig. 4); X_k, the optical data at the kth wavelength; and L_{ik}, the PC loading of the ith component for data at the kth wavelength. PCR then relates the scores to the dependent variable using the well-known calibration equation obtained using MLR:

$$y = B_0 + B_1 S_1 + B_2 S_2 + \cdots + B_n S_n$$

So we see that whereas standard MLR relates the optical data as the predictor variables (independent variables) and the concentration as the response variables (or dependent variable), PCR uses the principal component scores as the predictor variables as related to concentration as the response variable. All calibration techniques decompose the spectral data and use this decomposition factor as a predictor variable.

When learning basic chemometrics, hopeful students must familiarize themselves with the fundamentals of linear algebra, statistics, and calculus. Serious practitioners of chemometrics must be able to attack problems from three main approaches: spacially (visualization), computationally, and theoretically. To my knowledge, no tutorials walk the user through the more complex techniques starting from basic algebra to spacial, computational, and theoretical levels. These types of tutorials are essential to provide "a warm and fuzzy" feeling to practicing analysts and regulatory groups, in order to break down barriers to using more advanced methods. As Steve Brown et al. stated recently, "Many texts titled 'An Introduction to . . .' are far from introductory!" (21).

This is precisely the complaint of the majority of would-be users. This problem combined with the abundant misuse of statistics in sales or media presentations has left many understandably miffed.

A SOLUTION

It has been said, "That which you have freely received, also freely give." It behooves those who have seen the real value of chemometrics to continue to publish and present as much as possible in tutorial materials to bolster the foundations of computational chemistry for all practitioners of chemistry.

An "Additional Sources" section at the conclusion of this chapter is a good place for would-be chemometricians to start.

ACKNOWLEDGMENT

The author thanks Dr. Don Burns for the discussions regarding the need to improve chemometrics tutorials.

REFERENCES

1. H. J. Kisner, C. W. Brown, and G. J. Kavarnos, *Analyt. Chem.* **54**, 1479–85 (1982).
2. B. R. Kowalski, *Analyt. Chem.* **52**, 112R–22R (1980).
3. J. Sustek, *Analyt. Chem.* **46**, 1676–79 (1974).
4. D. E. Honigs, J. M. Freelin, G. M. Hieftje, and T. B. Hirschfeld, *Appl. Spectrosc.* **37**, 491–97 (1983).
5. M. A. Maris, C. W. Brown, and D. S. Lavery, *Analyt. Chem.* **55**, 1694–703 (1983).
6. H. J. Kisner, C. W. Brown, and G. J. Kavarnos, *Analyt. Chem.* **55**, 1703–07 (1983).
7. M. Otto and W. Wegscheider, *Analyt. Chem.* **57**, 63–69 (1985).
8. S. D. Frans and J. M. Harris, *Analyt. Chem.* **57**, 2680–84 (1985).
9. D. Metzler, C. M. Harris, R. L. Reeves, W. H. Lawton, and M. S. Maggio, *Analyt. Chem.* **49**, 864A–74A (1977).
10. R. N. Cochran and F. H. Home, *Analyt. Chem.* **49**, 846–53 (1977).
11. I. A. Cowe and J. W. McNicol, *Appl. Spectrosc.* **39**, 257–66 (1985).
12. S. Kawata, H. Komeda, K. Sasaki, and S. Minami, *Appl. Spectrosc.* **39**, 610–14 (1985).
13. H. Mark, *Analyt. Chem.* **58**, 2814–19 (1986).
14. S. Wold, P. Geladi, K. Esbensen, and J. Ohman, *J. Chemometrics* **1**, 41–56 (1987).
15. P. Geladi and B. Kowalski, *Analyt. Chem. Acta* **185**, 19–32 (1986).
16. D. Haaland and E. Thomas, *Analyt. Chem.* **60**, 1193–202 (1988).
17. D. Haaland and E. Thomas, *Analyt. Chem.* **60**, 1202–08 (1988).
18. W. Lindberg, J-A. Persson, and S. Wold, *Analyt. Chem.* **55**, 643–48 (1983).
19. I. E. Frank, J. H. Kalivas, and B. R. Kowalski, *Analyt. Chem.* **55**, 1800–04 (1983).
20. A. Lorber, L. E. Wangen, and B. R. Kowalski, *J. Chemometrics* **1**, 19–31 (1987).
21. S. Brown, T. Barker, R. Larivee, S. Monfre, and H. Wilk, *Analyt. Chem.* **60**, 256R (1988).

ADDITIONAL SOURCES

1. S. R. Seafie, *Matrix Algebra Useful for Statistics* (John Wiley & Sons, New York, 1982).
2. N. Draper and H. Smith, *Applied Regression Analysis* (John Wiley & Sons, New York, 1981).
3. S. A. Muhammad, D. L. Illman, and B. R. Kowalski, *Chemometrics* (John Wiley & Sons, New York, 1986).
4. D. L. Massart, B. G. M. Vandeginste, S. N. Deruing, Y. Michotte. and L. Kaufman, *Chemometrics: A Textbook* (Elsevier, New York, 1987).

CHAPTER II.C.2

A REAL "COOL" TECHNIQUE AND OTHER MUSINGS
Comments and Ideas for Better Sample Handling

As is my habit, being eclectic, at best, this column is part of two distinct sections. The second half, being concerned with fiber optics, is Chapter II.G.3 in addition to appearing under the UV/vis portion of the text. I hope my readers are sophisticated enough to follow the bouncing ball.

This chapter discusses the effects of low temperatures on luminescent spectroscopy.

FLUORESCENCE

To expand on what I believe is a useful discussion of the fundamentals of fluorescence and phosphorescence, I have a few more comments about the two techniques. Because sample handling is so important in all methodologies, it deserves some comment here.

The luminescent signal can be reduced by various types of quenching. One type, *dynamic quenching*, is also called collisional quenching. It happens when the excited lumophore collides with a quenching species. This is a kinetically (or diffusion)-controlled process; so it might be mitigated somewhat by lower solvent temperatures and higher solvent viscosities (1).

Although higher temperatures generally decrease luminescence through dynamic quenching—phosphorescence is even more greatly affected—some instances of delayed fluorescence are increased due to the thermal assist of the $S_1 \rightarrow T_1$ intersystem crossing. (Consult a Jablonski diagram if you are interested in the fine details of intersystem crossings.) In general, however, the lower the temperature, the more sensitive the analysis.

Choice of the solvent should be based not only on solubilities, but its quenching proclivities as well. Characteristics such as polarity, viscosity, and hydrogen-bonding ability can disrupt luminescing of analytes. The higher viscosities slow the collisions of excited state and quencher, while high polarities can cause red-shifting and weaker fluorescence.

The effects of solvents are even subjects of undergraduate instrumentation courses (2). Simple differences, such as substitution of deuterium for hydrogen in water, allow the Eu^{3+} ion to fluoresce more intensely.

The pH of a solution is also important because protonated *versus* unprotonated forms are either active or not! Oxygen is also an effective quencher and, because gases are more soluble at lower temperatures, must be excluded in low-temperature work. One also must be aware of the effect of solvents or solutes containing heavy atoms, such as bromine. Some salts of halides and halogenated solvents can increase the rate of intersystem crossing, increasing phosphorescence yields.

The thing that got me started on this topic in the first place was seeing a fused silica cell for low-temperature work. When I asked about the choice of fused silica, I was told that cells fashioned from individual faces pull apart at liquid nitrogen temperatures. To quote one of my students, "Duh, I knew that!"

DIPPING FIBER-OPTIC PROBES

As a master of *non sequiturs (I served as the "sequitary" of the National Non-Sequitur Society for a brief time. Our motto: "We may not make much sense, but we do like pizza!")* I am reminded of fiber optics at this point. The use of fiber optics has increased almost exponentially in recent years. Having used them to no small degree myself, there is one thing I have not seen addressed in the design of dipping probes: adjustment for differing refractive indexes.

As a quick reminder of how the probe works, the light from the fiber(s) emerges roughly at the acceptance angle at which it entered the fiber. This diverging beam is collimated with a convex lens, allowing a roughly parallel beam to emerge from the tip of the probe (3).

The light travels through the sample, doing what a good beam does (absorbing, scattering, etc.) and is picked up by a second fiber or is reflected back to the bifurcated fiber from which it emerged. In either case, the convex lens focuses the light to within the acceptance angle of the fiber(s), allowing transmission back to a detector.

What I have found is that although the probes are used in varying solutions, they are calibrated in a uniform manner. That is, the beam is calibrated to be collimated in air or water. Most samples are neither pure air nor pure water;

so some loss of signal might be expected through divergence. The problem is exacerbated when a mirror is used to reflect the light back to the source. There is one more solid–liquid interface to diverge the light.

Some work I performed with David Honigs in 1990 (4) showed that this loss of light was sufficient to calibrate for either refractive index of different solutions or temperature of a single solvent. (*Wow! Light follows the rules of physics! Sometimes I amaze myself with the simplicity of my discoveries. This is yet one more case where an interference is used as a measurement. The old Ann Landers bromide was never truer: "When life hands you a lemon, make lemonade!"*)

In an industrial setting, a probe is often permanent and used for a limited number of products; so it should be possible to make slight adjustments of either the lens or the distance of the fiber-optic tip from the lens.

Fiber optics, especially single fibers, have a limited throughput of light; so it seems to me that simple adjustments based on the solvent system, operating temperature, and solids content of a system could be made *to maximize the amount of light retrieved from the solution.* Perhaps spacers, such as those used to "gap" spark plugs, could be used to adjust the distance from the fiber to the lens *for each specified solvent system.* In any case, you heard it here first.

REFERENCES

1. J. D. Ingle, Jr., and S. R. Crouch, *Spectrochemical Analysis* (Prentice Hall, Englewood Cliffs, NJ, 1988).
2. U. A. Hofacker, *Chemical Experimentation* (W. H. Freeman, San Francisco, 1972).
3. F. A. Jenkins and H. E. White, *Fundamentals of Optics* (McGraw–Hill, New York, 1957).
4. E. W. Ciurczak and D. E. Honigs, "The Measurement of the Refractive Index of Liquids via NIRS Utilizing a Single Fiber Optic Probe," paper presented at PittCon, New York, 1990.

CHAPTER II.C.3

TO DESCRIBE IN
GLOWING TERMS

This chapter was written in the spring. In thinking about the crocuses, tulips, and other spring flowers we tend to take for granted, the array of colors is most striking. And thinking of colors always makes me think of light absorption. (I know, I need to get out more!)

I tend to write about the vibrational/rotational absorbences of molecules frequently. As the number of daylight hours increases, electronic absorbences don't seem to be out of the question for a topic.

THEORY

Although the lion's share of UV/vis spectroscopy is done in the absorbence mode, an important technique is luminescence spectroscopy. Luminescence is the process of relaxing electrons after excitation to a higher state (1). If the source of excitation is chemical—that is, fireflies, glow-worms—then the process is called *chemiluminescence.*

Chemiluminescence is used for numerous trace-detection techniques, not the least of which is high-performance liquid chromatography (HPLC) detection. However, the mode I am interested in is the excitation by light, or photons. This is called, fittingly, *photoluminescence.* (Aren't we chemists a poetic group?) This phenomenon has been noted for centuries and was identified by G. Stokes in 1852 as a potential analytical technique (2).

I do not intend to go into multiplicities or Jablonski energy-level diagrams here. I would like to highlight several theoretical considerations, mention some applications, and discuss the hardware involved.

The energies involved in various spectrophotometric techniques differ significantly. Rotational energies are in the 0.4 kJ/mol range, vibrational energies in the 20 kJ/mol area, and electronic transitions are about 400 kJ/mol. Clearly, the hardware and techniques vary greatly.

In all these processes, a photon of the correct energy strikes the molecule, and energy is transferred to the atom or bond. In one case—rotational—the atom(s) spin along a single covalent bond between two atoms. Because this is the lowest-energy transition, the changes are easiest to make, and the energy levels the closest. The subtle differences are best read in the condensed state, preferably the solid state. This is the backbone of the far infrared and is a subject for future consideration.

The next level of energy causes a bond (single through triple) to stretch or bend. This motion is the basis for infrared spectroscopy, and I have milked that topic sufficiently for now. I would be out of character if I failed to mention that the same types of absorptions give rise to near-IR spectra, as well. Right now, we are concerned with the highest-energy absorptions, or the UV/vis region of the spectrum.

Here, the bonds are left alone and the electrons are affected. A photon of sufficient energy strikes an atom or bond, and an electron is promoted to a higher-energy state. In absorbence UV/vis spectroscopy, the energetic electron then relaxes or is de-excited through an internal conversion such as collision with solvent molecules or other sample molecules. The absorbed energy is released, most often, as thermal energy. That serves to warm the solvent slightly, but no photons are re-emitted. In some cases, a photon is emitted.

In fluorescence, some electrons return to the ground state, or become de-excited, rather slowly (10^{-9}–10^{-7} vs. 10^{-12} s for an absorbance de-excitation). In this case, a photon having lower energy than the exciting photon may be emitted.

While in this slowed state, the excited electron may reverse its spin. In this case, it may no longer merely relax to its original position for quantum considerations. (Now, two electrons have the same four primary quantum numbers.) It must first "re-reverse" its spin. This longer relaxation time span is called *phosphorescence*. It is a truly different process, as the fluorescence and phosphorescence spectra differ (singlet vs. triplet for purists).

How does all this affect the price of butter in Wisconsin? Or, more to the point, how may we use these differences? The answer, simply put, is: How sensitive is your work? In absorbence spectroscopy, a relatively large number of molecules absorb a large proportion of the light. Thus the linear relationship of source, grating, cuvette, and detector is convenient. A chopper alternates the light transversing the sample cell with the light through the reference cell. The differences are usually sufficient for an accurate rationing for Beer's law work. UV/vis absorbence spectroscopy is used in the mg/ml range.

The fluorescence instrument setup is slightly different. Because photomultipliers are relatively dumb, they cannot distinguish between the light transmitted through a sample and the fluorescent glow given off by a sample. In this case, the detector is placed at right angles to the incident beam. Barring any light scattering, the only light impinging on the detector is from the fluorescence radiation from the sample.

In absorbence spectroscopy, the detector determines the ratio between a high-intensity light and a slightly lesser amount of light. In fluorescence spectroscopy, the detector measures the ratio between a small signal and no signal. This results in quiet, sensitive measurements of very small signals.

One drawback of fluorescence is a phenomenon called self-absorption. In this process, the emitted light from one molecule is absorbed by a second molecule and subsequently re-emitted as heat. The quenching process limits the concentrations used in the method to micrograms per milliliter. The result is a higher sensitivity, which is good because of concentrations limited by reabsorption phenomena.

Because phosphorescence is a quantum-forbidden process—where an electron flips, then flips again, before relaxing—it produces an even smaller number of photons than fluorescence does. The samples herein are routinely in the nanogram range. Because of this extremely small signal, interferences are more important. Higher-intensity light sources are used to populate the excited state, allowing a higher number of phosphorescing molecules. This high-intensity light (or laser) may lead to Raman radiation, scattering, or some Rayleigh scattering. Not much stray light is needed to disturb the incredibly small signal produced.

To obviate the interferences, the time delay peculiar to phosphorescence is utilized. Incident light may be pulsed with the detector only reading the light during these dark periods. The more sensitive instruments essentially count photons. Because of the low quantum yield, or actual number of phosphorescing molecules compared with the total number present, much work has been performed at lower temperatures to lessen the internal quenching of the excited electron.

Recent instruments provide excellent results at room temperature.

APPLICATIONS

To fluoresce, a molecule must (usually) be highly conjugated. Some examples include substituted and unsubstituted olefins, aromatic amines, amino acids, heterocycles, phenols, barbiturates, and aromatic acids. Sample molecules may also contain easily excited f electrons. Some inorganic species that fluoresce include several lanthanide ions, cerium (III) ions, and chelates of group IIIA ions.

The ability to quantify trace amounts has been applied to thin-layer chromatography (TLC) and liquid chromatography (LC) applications for some time. Luminescence is used to detect trace elements in pollution, and it is an excellent tool for detecting trace amounts of licit and illicit drugs in blood, plasma, and urine.

HARDWARE

Fluorescence equipment may be either single wavelength (filter or grating) or a scanning type. Because distinct spectra are produced whether the incident light or the emitted light is scanned, a dual-scanning instrument is often used for qualitative purposes. That is, the incident light may be scanned while a single wavelength is selected for the detector. Conversely, the sample may be excited by a single wavelength while the emission spectrum is scanned.

The above-described instrument is good for qualitative or quantitative work. It is often a shame that luminescence measurements are ignored for ancient reasons. It is assumed that fluorescence or phosphorescence techniques are difficult or time consuming. Current instruments are simple to run and provide quite excellent results. Try luminescence—it could be an enlightening experience.

REFERENCES

1. G. D. Christian and J. E. O'Reilly, eds., *Instrumental Analysis, 2nd ed.* (Allyn and Bacon, Boston).
2. G. G. Stokes *Phil. Trans.* **142**, 463 (1852).

CHAPTER II.C.4

FILE TRANSFER SOFTWARE FOR UV/VIS SPECTROSCOPY—AN OS/2-BASED PROGRAM FOR DYNAMIC DATA EXCHANGE

John Sanders

This chapter was written by John Sanders (Varian Australia Pty Ltd, Victoria, Australia 3170). Although I do not endorse commercial software, I do like to make readers aware of various programs. In the case of this chapter, I am pleased that (1) a major manufacturer of analytical equipment acknowledges the power of a spreadsheet approach with spectroscopic data and convoluted chromatographic peaks, such as those found in gel permeation chromatography, and (2) the software is written to facilitate the use of third-party software with Varian equipment. That second part makes this one of the rare instances I've encountered of a name manufacturer accepting the need for generic programs. [Another interesting detail about the Dynamic Data Exchange (DDE) interface software described here is that it is written in OS/2.]

The program, while specific for Cary equipment, is offered in enough detail that it may be adapted to older equipment as well. The program, as presented, may be used with DDE programs such as Microsoft Word for Windows, Excel for Windows, Lotus 1-2-3 for Windows, OS/2, or any compatible spreadsheet. The OS/2 software presented here allows the spectroscopist to export spectra without extensive manipulation prior to the exchange.

The increasing complexity of spectrophotometer data and the complex requirements of some Good Laboratory Practices/Quality Control Practices (GLP/QCP) protocols has meant that chemists are making extensive use of spreadsheets for data manipulation and reporting. This chapter describes the use of dynamic data exchange to automate some of the procedures required to transfer data between applications so that this analysis can be simplified and automated. For extensive manipulation of spectrophotometer data, users normally must export the results at the end of an analysis using the instrument's software, then reimport the data into a separate reporting program. DDE enables spectroscopists to import data directly into another program without these lengthy procedures.

The DDE capability in the software described in this chapter (Cary OS/2 software, Varian Associates, Palo Alto, CA) allows users to export data from the UV/vis instrument directly into a Windows-compatible spreadsheet or word processor while the data are being collected or displayed. In this case, the Cary OS/2 software is referred to as the *source application* or *server*. Many common word processing and spreadsheet programs feature DDE capabilities, including Microsoft Word for Windows, Excel for Windows, and Lotus 1-2-3 for Windows, or OS/2. These programs are referred to as the *destination application* or *client program*.

PROCEDURES

The spectrophotometer system referred to throughout this chapter is a Cary OS/2 UV/vis–near-IR instrument (Varian). To receive data from the spectrophotometer it is necessary to initiate communication through a series of commands, for example the string

CARY! DATA ! TYPE

creates a link to the host program to read data from the instrument's program. The data "TYPES" available to users are:

- STATUS: This provides the current status messages from the spectrophotometer. These messages are normally seen in the top line of the information bars in the Cary Base System.
- ERROR: This transfers any errors that are displayed in the Cary Base System. These are normally displayed in the second line of the information bar.
- IDLE: This displays the current readings in the spectrophotometer Cary Base window abscissa and ordinate display, for example, "550.000000, 0.0123456789."
- PLOT: This provides the abscissa, ordinate, and box identification generated by the instrument during the collection of a continuum. For example, during a wavelength scan, the wavelength measured in nanometers, the

Table I: Cary DDE PLOT data.

Item	Format
File name	Current user\Filename,X units,scan number-box number
Full path name	Drive\Path\Filename,Y units,scan number-box number
Ordinate range	Max-Min Scaling of PLOT area,blank,scan number-box number
Abscissa range	Max-Min Abscissa Range,blank,scan number-box number
Data	Ordinate,Abscissa,scan number-box number
Continuing data	Ordinate,Abscissa,scan number-box number
End of data	"End of scan",,scan number-box number

absorbance, and the continuum number and box number are sent from the instrument's software in the form of a comma-separated string. For example, 447.59000, 0.951832,0104 is interpreted as wavelength = 447.590000, absorbance = 0.951832, and continuum number 01 is in box 04.

The PLOT data are read into the DDE link whenever a "RETRIEVE," "SHOW" continua, or a "COLLECT" or "READ" operation occur. Because of the nature of the DDE function in Windows, the computer may be totally occupied during the collect period. When the PLOT data are sent to the client, the data stream consists of the components shown in Table I.

If multiple cuvettes are being used with the multicell holder, then the data stream consists of the first four lines in Table I for each cell, then the data stream for each cell reads as illustrated in Table II.

Table II: Cary DDE PLOT data for multicell operation.

Item	Format
File name	Current user\Filename,,scan number 01 box number 01
Full path name	Drive\Path\Filename,,scan number 01 box number 01
Ordinate range	Max-Min Scaling of PLOT area,,scan number 01 box number 01
Abscissa range	Max-Min Abscissa Range,,scan number 01 box number 01
File name	Current user\Filename,,scan number 01 box number 02
Full path name	Drive\Path\Filename,,scan number 01 box number 02
Ordinate range	Max-Min Scaling of PLOT area,,scan number 01 box number 02
Abscissa range	Max-Min Abscissa Range,,scan number 01 box number 02
Remaining cuvettes	
Data	Ordinate,Abscissa,0101
Data	Ordinate,Abscissa,0102
Remaining data for cuvettes	
End of data	"End of scan",,scan number-box number

SAMPLE PROGRAMS

The following examples illustrate using DDE in various spreadsheets. The data handling is then limited by the capability of the spreadsheet. These formulated or processed data, diagrams, and graphs can then be linked by DDE or Object Linking and Embedding (OLE) to other compatible programs.

Microsoft Excel for Windows v.4 or 5. The cell entries shown in Table III will display data from the Cary OS/2 program. It is always necessary to start the spectrophotometer software before attempting to start a link. To start a DDE link, highlight the cell where you require the information to be input and type the commands shown in Table III. When you press ENTER, the system will immediately attempt to access the spectrophotometer system. If the system is present and active on the desktop, the relevant information will be displayed. If no information is currently available, "N/A" will be shown. If the instrument's program is inactive, "#REF!" will be shown. These are Excel error messages, not text transferred from the instrument. A series of DDE-related messages will also be seen in the instrument's status line. Changes in the instrument will be immediately reflected in the relevant cell.

As the data from the Cary instrument will be received in a stream, it is necessary to place them in an array of cells. This is done by parsing the data stream as it is received. It is necessary to set a macro that will activate a parsing macro when data are received in the application. After the links are set, then changes in the links can be used to copy the data, parse the data stream, and write it to an array of cells.

Macro for Excel 4 and 5. The Excel macro commands shown in Table IV can be used to process the data supplied by the spectrophotometer. The macros show a typical application in gathering data from a plot into a three-column table that can then be automatically incorporated into a graph. This graph can then be linked using OLE2 into Word for Windows v. 6.

The function "ON.DATA("document_text", "macro_text")" is used to detect

Table III: Cary DDE data types.

Identification (optional)*	Actual cell entry†	Typical response
Status Messages	=Cary|Data!Status	Scan 500–400 nm
Error Messages	=Cary|Data!Error	N/A
Abs,Ord	=Cary|Data!Idle	553.6,0.1234
Plotted Abscissa,Ordinate	=Cary|Data!Plot	553.6,0.1234,0101

*Plain text to identify the data that will be displayed in the adjacent cell.
†Macro information that will collect data from the spectrophotometer.

Table IV: Excel 4 macros for Cary DDE.

Range or Macro names*	Formulas	Comments
Auto_Open_DDE	=ON.DATA("CARY|DATA!PLOT","PLOT")	Change in "CARY PLOT" data causes branch to PLOT macro
	=SELECT(SHEET1!A30)	GOTO cell A30
	=RETURN()	EXIT
PLOT		
PlotAbsOrd	=Cary|Data!Plot	DDE Link
CommaPos	=SEARCH(",",*PlotAbsOrd*)	Find first comma
CommaPos2	=Search(CommaPos + 1,",",*PlotAbsOrd*)	Find second comma
XVal	=MID(*PlotAbsOrd*,1,*CommaPos*-1)	Find Abscissa
YVal	=MID(*PlotAbsOrd,CommaPos*+1,CommaPos2)	Find Ordinate
ScanBox	=Right(PlotAbsOrd,4)	Find scan number and box number
	=FORMULA(XVal)	Put Abscissa value in first cell
	=SELECT(OFFSET(ACTIVE.CELL(),0,1))	Move pointer to next column right
	=FORMULA(YVal)	Put Ordinate value in cell
	=SELECT(OFFSET(ACTIVE.CELL(),0,1))	Move pointer to next column right
	=FORMULA(Scanbox)	
	=SELECT(OFFSET(ACTIVE.CELL(),1,-2))	Shift to next row first column
	=RETURN()	Macro return
Auto_Close_DDE	=ON.DATA(*PLOTDDE.XLM*")	Stop parsing routine
	=RETURN()	

*Range names indicated in italics.

185

changes in the range "document_text." When this occurs as information is received, then the parsing macro "macro_text" is run, and the data are translated into rows of cells.

Excel 5 Visual Basic. Excel 5 for Windows enables users to create macros in Visual Basic. This provides an increase in capability of data presentation and versatility. It should be noted that Microsoft will only support Visual Basic in further releases of its products.

Table V illustrates a simple procedure to gather spectrophotometer data into an array of cells.

The parsing subroutine assumes there is a range called "PAO" in the active worksheet that contains the "CARYIDATA!PLOT" formula, and a cell named "FILE_NAME_START" to act as references.

This visual basic macro is a simplified routine to provide a three-column layout of the data coming from the spectrophotometer. Now the data are translated into rows and columns, and further macros can be written to analyze or graph the data.

A number of macros allow data to be translated, graphed, and saved, creating extensive data arrays when multicell measurements are taken. These macros include:

- VBXL5.XLS. This program is a universal routine where data are collected, parsed, and displayed. It allows you to enter the number of cuvettes together with the SAT and DWELL times to ensure correct layout and ordering of the data.

- VBKIN.XLS. This program is designed specifically for the spectrophotometer's kinetics package. The incoming data stream is parsed and set out in rows and columns according to the individual cuvettes used in the multicell holder.

At the end of the analysis the data are automatically graphed and stored to an Excel file with the current sample name. Figure 1 illustrates the presentation of data from the kinetics program into Excel 5.

- VBTHERM.XLS. This program is designed for use with the instrument's thermal application package. The data are collected, parsed, and set out in rows and columns according to the number of cuvettes being used in the multicell holder. On completion, the data are automatically plotted.

- VBCONC.XLS. In this program, the incoming data are parsed into rows and columns, then at the end a calibration graph is constructed from the data and a regression line is fitted to the data. The equation and fit parameters are then displayed.

Table V. Visual Basic macros for Cary DDE.

Visual Basic codes	Comments
Sub DDE_START()	Macro subroutine name
Range("File_Name_Start").Select	Position cursor on active sheet
ActiveSheet.SetLinkOnData("CARYIDATA!PLOT","CODES")	On change in data, do "Codes" macro
End Sub	
Sub Codes()	Macro subroutine name
On Error GoTo handle_errors	Error handling
cp = InStr(1, Range("PAO"), ",")	Get first comma location
cp2 = InStr(cp + 1, Range("PAO"), ",")	Get second comma location
xv = Left(Range("PAO"),cp-1)	Get Abscissa value
yv = Mid(Range("PAO"),cp + 1,cp2-cp-1)	Get Ordinate value
scannum = Right(Range("PAO"),4)	Get cell and scan number
Selection.Formula = xv	Paste Abscissa in active cell
Selection.Offset(0, 1).Activate	Move to next column
Selection.Formula = yv	Paste Ordinate in active cell
Selection.Offset(0, 1).Activate	Move to next column
Selection.Formula = scannum	Paste cell and scan number
Selection.Offset(1,-2).Activate	Reposition to first column and next row
Exit Sub	
handle_errors:	Error handling routine
Range("File_Name_Start").Select	
Endrun:	
End Sub	

187

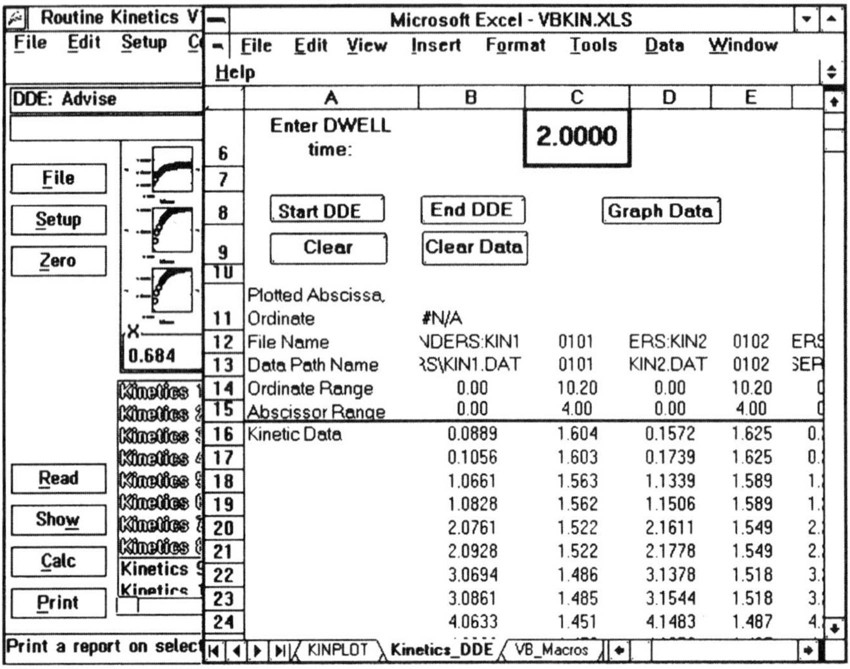

Figure 1.   Data layout in Excel VBKIN.XLS.

- VBSIMPLE.XLS. This program parses the incoming data and creates a plot of the data.

Lotus 1-2-3 for Windows Release 5.   The Lotus macros follow similar operation to those in Excel, allowing for the differences in nomenclature and protocols. To create the Link in Lotus use the EDIT LINK commands listed in Table VI.

Table VII illustrates a simple set of macros to create a data array in Lotus 1-2-3.

Table VI. Lotus 1-2-3 DDE Link command for Cary.

Link name	Cary Plot (option)
Application	CARY
Topic	DATA
Item	PLOT or IDLE or ERROR or STATUS (as appropriate)
Format	TEXT
Update	AUTOMATIC

Table VII. Lotus 1-2-3 macros for Cary DDE.

Lotus 1-2-3 macros	Comments
\A	
{GOTO}F1~	Position active cell
{DDE-OPEN "CARY","DATA",*NUMPLOT*}~	Open Communications place value of link in range NUMPLOT
{DDE-ADVISE *PLOTING*,"PLOT"}~	Do macro PLOTING if PLOT value changes
{DDE-CLOSE}	
\PLOTING	
{DDE-USE *NUMPLOT*}	
{DDE-REQUEST *ABSORD*,"PLOT"}~	
{LET *COMMAPOS*,@FIND(",",*ABSORD*,0)}~	Locate first comma separator in data string and insert in range COMMAPOS
{LET *PLOTLENGTH*,@LENGTH(*ABSORD*)}~	Find length of data string and insert in range PLOTLENGTH
{LET *COMMAPOS2*,@FIND(",",*ABSORD*,COMMAPOS+1)}~	Locate second comma separator in data string and insert in range COMMAPOS2
{LET *ABSCISSOR*,@VALUE(@LEFT(*ABSORD,COMMAPOS*))}~	Determine abscissa value and place in range ABSCISSA
{LET @CELLPOINTER("ADDRESS"),*ABSCISSA*}~	Insert ABSCISSA value in column
{RIGHT}~	Move curser right one column
{LET *ORDINATE*,@VALUE(@MID(*ABSORD*,(COMMAPOS+1), (COMMAPOS2-COMMAPOS)))}~	Determine ordinate value and place in range ORDINATE
{LET @CELLPOINTER("ADDRESS"),*ORDINATE*}~	Insert ORDINATE value in column
{DOWN}{LEFT}~	Move cursor to previous column, next row
{DDE-UNADVISE "PLOT"}~	Exit routine
{DDE-CLOSE}~	Close link

CONCLUSIONS

This chapter has described a tool for the simplified export of complex spectroscopic data from the UV/vis instrument to other programs for reporting and word processing. This approach has potential to ease the task of complying with regulatory protocols requiring spreadsheets for data manipulation and reporting.

IONIZATION METHODS

For the remainder of this column, we'll consider an instrument in which the electron ionization source is followed by a single focusing magnetic sector mass spectrometer. A commercial instrument of this type might use an acceleration potential of 5000 V. Confirm that ions 1^+, 1000^+, and $10,000^+$ reach velocities of 9.82×10^5, 3.11×10^4, and 9.82×10^3 m/s, respectively. If we manage to put several charges on the ion, the velocity changes. For example, we find that an ion $10,000^{10+}$ exhibits a velocity equal to that of an ion 1000^+. We'll then pass these ions through a source slit, focus them with small potentials applied to lens elements (note that the more highly charged ions will be more susceptible to focusing potentials than will singly charged ions), and then we will pass the ions into the magnetic sector mass analyzer.

The interaction of an ion with a magnetic field depends on the magnitude of the charge on the ion. Balancing angular momentum and centrifugal forces within the curved magnetic sector leads to the classic equation $m/z = B^2 r^2 / 2V$, where B is the magnetic field strength and r is the radius of the ion path passing through the curved magnetic sector. Ions of greater mass will follow paths of larger radius than ions of lesser mass. An ion path of finite width intersects the exit slit of the sector so that at any given field strength and instrument geometry, ions of only a small range of m/z values will pass through the instrument to the detector. We design the instrument and manipulate the slits to obtain the desired mass resolution by controlling the range of m/z values that can pass to the detector. How do we scan across the mass range of interest? We can change the acceleration potential V to change the m/z of ions reaching the detector, but this is not normally done because such a change will adversely affect ion focusing.

The usual procedure is to change B to bring ions of different m/z to the detector. An ion of higher charge state will interact more strongly with the magnetic field. It will be bent more strongly for a given B. A doubly charged ion will pass through the magnetic sector at a field strength equivalent to that required to pass an ion with half its mass. (How then can we determine the charge state of an ion? This is significant and will be detailed in a later column.)

Increasing the charge on an ion is an efficient and straightforward means of extending the accessible mass range of a magnetic-sector mass spectrometer. This idea was put forward by Gordon Wood (1), using doubly charged ions formed in field-desorption mass spectrometry as specific examples. Now, with electrospray ionization, we have the ability to generate ions bearing 30, 40, and even 50 charges, and we enjoy the benefits of a 30-, 40-, or 50-fold extension of mass range of our instruments. When we deal with standard electron or chemical ionization sources, the doubly charged ions (if any) are present only with low relative intensities. Triply charged ions in electron ionization are rare indeed. As a result, our procedures for dealing with the data, and for interpretation of mass spectrum, assume the predominance of ions of a single charge state in the mass spectrum.

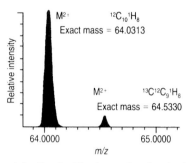

Figure 1. The isotope peak for the doubly charged molecular ion M^{2+} appears at a non-integral mass of 64.5330 Da.

Most often the ions are singly charged, but mass spectra consisting solely of doubly charged ions can be generated with an additional filtering element used to prevent singly charged ions from reaching the detector (2). In MS/MS experiments, charge-changing experiments probe the interconversion of singly and doubly charged ions (3). Otherwise, doubly charged ions are identified by the presence of ion signals at "half-mass" values; although the ion signals from lower mass organic molecules tend to fall very near integral mass values, isotopic peaks from doubly charged ions can appear nearly halfway between adjacent integer masses (see Fig. 1). In this manner, we can identify the presence of two different ion charge states in the mass spectrum.

Now, what about more highly charged ions, and the complications that ensue when ions of many different charge states are simultaneously present? If we generate an ion with an average of 30 charges, the electrospray process also gives us substantial relative intensities of ions with other charge states. A distribution of ions with charges from 20 to 40 follows a modified Gaussian-like curve that reflects the ability of a large molecule to accommodate and stabilize these charges. A typical distribution of multiply charged ions for a high mass biomolecule is shown in Fig. 2. Each of the signals in this mass spectrum repre-

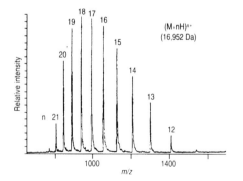

Figure 2. ESI-MS provides a distribution of multiply charged forms of the molecular ion for highmass biomolecules such as myoglobin.

sents the molecular ion, albeit in a different charged state. This ion charge-state distribution can be convoluted with the distribution of masses (molecular ions, cluster ions, and fragment ions) in the mass spectrum.

By fortunate coincidence to date, electrospray ionization methods tend to generate molecular ions that undergo little dissociation to fragment ions, and the mass spectrum is correspondingly simplified. This characteristic behavior holds most strongly for very high mass molecules. For lower-mass molecules examined by electrospray ionization, the mixture of charge states of molecular and fragment ions in the mass spectrum can present a formidable interpretive dilemma.

ION DETECTION

After the ions have been separated by m/z ratio in the mass analyzer, such as the single focusing magnetic sector above, the ions encounter the detector of the mass spectrometer. An early detection method was photographic film, held in an image plane. The amount of "fogging" of the film was proportional to the number of incident ions, and the position of the image on the plate could be related to the mass (the m/z ratio) of the ion. These detectors have since been replaced by electronic charge-coupled devices or photodiode-array detectors in which the position of the trace is still related to mass, but in which the ion impact generates a current flow into the appropriate array element. The number of electrons induced to flow is proportional in the array element to the number of incident ions.

Do we require a charge on the ion for registration by the photographic film or the array detector? If the neutral particle is sufficiently energetic, it can generate the appropriate chemical reactions when it strikes the photographic film, or it can stimulate the flow of current when it strikes the front element of an array detector. How much energy is enough? In the case of the electronic array detector, this depends on the number and wavelength of photons emitted on neutral particle impact (photodiode array) or the ability to generate electron–hole pairs in the semiconductor material first encountered.

The most widely used detector in mass spectrometry is the electron multiplier, and here we will explore the effect of the charge of the ion on ion detection by the electron multiplier. Briefly, the process of detection involves impact of an energetic particle on the front dynode surface of an electron multiplier, with release of several electrons. These electrons are accelerated to a surface held at a more positive potential, where each subsequent impact releases several more electrons. The process is repeated in a manner such that the initial ion impact generates a large number of electrons. A gain of 106 to 108 can be achieved in an electron multiplier.

A neutral particle that impacts the front dynode surface with enough energy to release electrons will produce a detectable signal, and it is for this reason that electron multipliers in line-of-sight instruments (quadrupole mass filters, for example) are located off the axis of ion movement through the instrument. Ions will change their trajectory to hit the dynode, but neutrals and photons cannot do so. The velocity with which the ions impact the dynode surface is established by their initial acceleration potential and the potential maintained at the front of the multiplier.

In modern electron multiplier detection systems, the ions are accelerated to a separate metal target maintained at a very high attractive potential (8000–30,000 V), and the electrons released upon impact move to the dynode surfaces of the electron multiplier.

The postacceleration detector (as this system is called) was developed in response to the expectation that ions must impact a surface with more than the threshold velocity required to release electrons from that surface. Ion velocities (at usual values of the acceleration potential) for high mass, singly charged ions are very near the experimentally determined threshold velocity for electron release. For multiply charged ions, the velocity imparted by the acceleration potential is increased linearly with respect to ion charge, as is any additional velocity gained by acceleration into the metal target. As a result, the sputtered electron yield remains acceptable even for higher-mass ions. In addition, a collective effect is observed for very high-mass ions, so that the impact of a large ion generates an electron yield equivalent to several smaller ion impacts.

We have summarized the effect of ion charge on mass analysis and detection. We've mentioned briefly some of the ionization methods that create singly charged ions (electron ionization and chemical ionization) and multiply charged ions (electrospray ionization). Once we create ions in an ionization source, are we stuck with what we make, or can charge be manipulated in a mass spectrometer? In fact, charge can be altered as an ion passes through a mass spectrometer. Collisions with gas molecules or surfaces can result in charge stripping ($M^+ \rightarrow M^{2+}$) or charge inversion ($M \rightarrow M^+$) reactions. Ions that neutralize, or neutral fragments that form in dissociation reactions, can be detected in certain instrument configurations. Neutralization–reionization mass spectrometry (4) is a means of ion structural elucidation developed extensively over the past decade. In each of these cases, the charge on the ion makes it amenable to selection, analysis, and reaction.

REFERENCES

1. G. W. Wood and W. F. Sun, *Biomed. Mass Spectrom.* **7**, 399–400 (1980).
2. J. R. Regalado, W. M. Holbrook, D. E. Bostwick, and T. F. Moran, *Org. Mass Spectrom.* **25**, 174–80 (1990).
3. R. G. Cooks, J. H. Beynon, and T. Ast, *J. Am. Chem. Soc.* **94**, 1004 (1972).
4. C. Wesdemiotis and F. W. McLafferty, *Org. Mass Spectrom.* **24**, 663–68 (1989).

CHAPTER II.D.2

PHOTON RESONANCE MASS SPECTROMETRY

John Wronka

This chapter deals with a method—photon resonance mass spectrometry—that isn't strictly molecular spectroscopy in the traditional sense. However, mass spectrometry certainly has a place in this volume because it has become so useful for the characterization of large macromolecules. In addition, the infrared CO_2 laser used in the instrument described is what I would call an "honest" spectroscopic tool. John Wronka, a senior applications scientist for mass spectrometry at Bruker instruments, has put together a nice description, with some applications, of Bruker's laser desorption mass spectrometer. I was particularly impressed with the system's ability to selectively desorb molecules from complex mixtures.

Time-of-flight mass spectrometry (TOF-MS) offers an extremely high rate of data acquisition. TOF-MS is ideal for studying fast reactions and detecting reaction intermediates, and is the mass analyzer of choice for pulsed ionization methods or any study requiring rapid sampling. In fact, because a TOF analyzer can acquire a mass spectrum in 100 μs, it was the first mass analyzer to be used successfully for gas chromatography-mass spectrometry (GC-MS). Because TOF-MS requires pulsed ionization sources and provides a complete mass spectrum for each ionization pulse, it is particularly useful for pulsed laser systems.

The past few years have shown a resurgence in the use of TOF-MS. This renewal is the result of new developments in the main components of a TOF spectrometer and new applications. The introduction of the gridless reflection offers both high mass resolution (10,000 fwhh) and efficient ion transmission (1–3). Improvements in transient digitizers and data systems now allow full use of the rapid sampling representative of the TOF technique. The application of standard tools of the molecular spectroscopist, such as lasers and supersonic jets, helped drive these developments and have provided some unique applications.

One such example of applying lasers and supersonic jets to TOF-MS is the photon resonance ion source and laser desorption unit originally introduced by Schlag and co-workers (4). This source is standard equipment on the Bruker TOF-1 mass spectrometer. In this source, volatile samples are introduced into a supersonic jet by seeding the cooling gas while thermally labile samples are desorbed into the jet using a low-power CO_2 laser. Resonant laser light ionizes the molecules cooled in the supersonic jet.

Supersonic cooling, laser desorption, and laser ionization have properties that are important to the operation of TOF-MS and its specific application to some unique problems. For example, mass selection in a TOF spectrometer depends on the flight time of ions between the ion source and detector. The flight time t is proportional to the square root of the mass m:

$$t = s/v = s(m/2E)^{1/2} \qquad (1)$$

where s is the effective path length in the field-free region; v, the ion velocity; and E, the ion kinetic energy. Effective TOF mass analysis requires short ionization pulses and a spatially resolved ionization location. Lasers provide extremely short pulses (5 ns or less) and are easily focused on a tight spot. With pulsed-laser ionization both the location and time of the ionization event are sharply defined.

Even with these short ionization pulses and sharply defined starting points, the presence of thermal energy can give a distribution of velocities that will limit the resolution. For a typical ion with accelerating potential energy of 700 eV and a flight time of 60 μs, an initial kinetic energy distribution of 0.025 eV (300 K) would add 5 ns to the peak width a molecule with molecular weight (MW) of 106. If the same molecule is cooled in a supersonic jet to 10 K, the contribution from the initial energy distribution will drop to 1 ns.

Cooling in a supersonic jet also greatly enhances the selectivity of the ionization process. Ionization with laser light requires that the sample molecule absorb more than one photon. Ionization efficiency is drastically increased if the laser wavelength is tuned to excited states in the sample molecule. Under these conditions of resonance-enhanced multiphoton ionization (REMPI), almost all molecules within the region of focus may be ionized.

Furthermore, the cooling of a neutral molecule will narrow its absorption bands and improve the selectivity of the REMPI as a function of laser wavelength. For example, Fig. 1 compares the wavelength scans of benzene (MW = 78) and the ^{13}C isotope of benzene (MW = 79) at room temperature (300 K) (Fig. 1a) to the theoretical curve for benzene cooled to 2 K (Fig. 1b). Figure 1c gives the actual wavelength-dependent mass spectra of benzene cooled to 2 K in a supersonic jet. The wavelength-dependent mass spectra at 2 K show that it is possible to selectively ionize isotopic species.

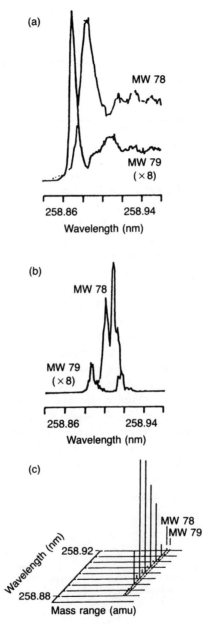

Figure 1. A comparison of (a) the wavelength scans of benzene (MW = 78) and the ^{13}C isotope of benzene (MW = 79) at room temperature (300 K), and (b) the theoretical curve for benzene cooled to 2 K. (c) The actual wavelength-dependent mass spectra of benzene cooled to 2 K in a supersonic jet.

Nonvolatile substances require a method to desorb neutral intact molecules into the supersonic jet. This is accomplished with the multistage ion source (4) including laser desorption, supersonic jet cooling, and REMPI, as shown in Fig. 2. Cooling of the desorbed neutral molecules is an important step because it greatly reduces the amount of thermal degradation in the desorbed molecules and also results in more structured absorption spectra that will allow for selective ionization.

As the laser power is varied, the amount of energy deposited into vibrational states is also altered. Likewise, as the laser wavelength is varied, the actual vibrational band that is excited will change. In this manner, both the amount and type of fragmentation present in the mass spectrum can easily be adjusted to yield either molecular weight or structural information. For example, the absorption spectra of Rhodamine 6G in Fig. 3 show two distinct bands centered at 250 and 275 nm, respectively. Figure 3 contains the mass spectra of Rhodamine 6G desorbed into the supersonic jet and ionized at wavelengths of 250, 260, and 275 nm. All three spectra were acquired using equal laser intensities. Desorption into the supersonic jet yields neutral Rhodamine 6G without C1. This neutral corresponds to the ion at mass 443.

Examination of these spectra shows that ionization wavelength will radically change the mass spectra. The peaks below mass 87 indicate that at wavelengths < 261 nm (4.75 eV) there is extensive fragmentation of the benzene rings. The fragmentation of the benzene rings is greatest at 250 nm. At 260 nm there is less fragmentation of the benzene rings, and ions at 414, 369, and 355 Da correspond to the loss of specific side chains of the Rhodamine 6G structure. Finally at 275 nm, the benzene rings are not broken at all and the major peak at 414 Da is due to the loss of C_2H_5. Likewise, in Fig. 4 a wavelength of 260 nm is selected and the laser power is adjusted. As power is increased and more photons are pumped into the vibrational bands, the relative amount of low-mass ions increases while the fragment ratios in the higher mass range remain constant.

Grotemyer and co-workers (5) used the ability to selectively ionize and control the amount of fragmentation to analyze complex mixtures of biomolecules

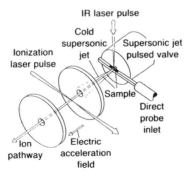

Figure 2. Schematic diagram of the multistage ion source, including laser desorption, supersonic jet cooling, and resonance-enhanced multiphoton ionization.

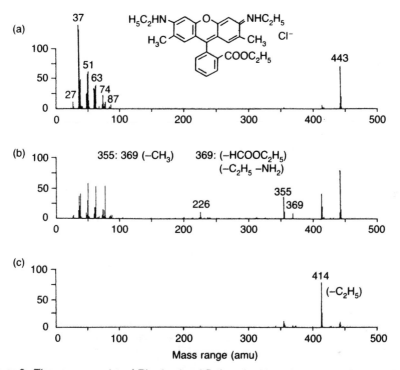

Figure 3. The mass spectra of Rhodamine 6G desorbed into the supersonic jet and ionized at wavelengths of (a) 250 nm, (b) 260 nm, and (c) 275 nm. All three spectra were acquired using equal laser intensities.

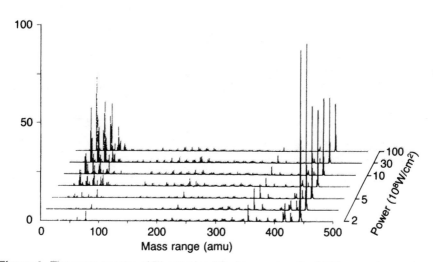

Figure 4. The mass spectra of Rhodamine 6G at a wavelength of 260 nm as laser power is adjusted.

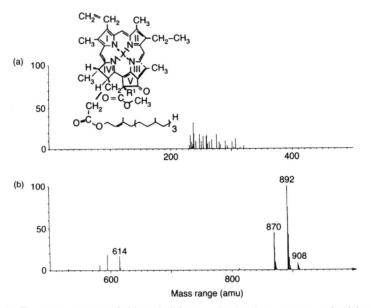

Figure 5. The mass spectra of chlorophyll from a simple algae extract under (a) soft ionization conditions and (b) hard ionization conditions. Chlorophyll is selectively ionized even though the mixture contains three similar structures: chlorophyll where X = Mg and R^1 = H; 10-hydroxy-chlorophyll where X = Mg and R^1 = CH; and phaeophytin where X = H,H and R^1 = H.

without extensive purification. Figure 5 shows the mass spectra of chlorophyll from a simple algae extract under soft (Fig. 5a) and hard (Fig. 5b) ionization conditions. Even with a simple sample cleanup, it is possible to get both molecular-weight information (soft ionization) and structural information (hard ionization). This technique promises to provide a unique method for the sequential analysis of samples such as polypeptides.

Instruments such as the TOF-1, which has a multistage ion source, are beginning to enhance the selectivity of spectroscopy by adding both the selectivity and sensitivity of mass spectrometry to provide new solutions for a broad range of analytical problems.

REFERENCES

1. B. A. Mamyrin, V. I. Karateev, D. V. Shmikk, and V. A. Zagulin, *Sov. Phys. JETP* **37**, 45 (1973).
2. R. Frey, G. Weiss, H. Kaminski, and E. W. Schlag, *Z. Naturforsch.* **40a**, 1349 (1985).
3. K. Walter, U. Boesi, and E. W. Schlag, *J. Mass Spectrom. Ion Proc.* **71**, 309 (1986).
4. H. V. Weissenhof, H. Seize, and E. W. Schlag, *Z. Naturforsch.* **40a**, 674 (1985).
5. J. Grotemyer, U. Bosel, K. Walter, and E. W. Schlag, *J. Am. Chem. Soc.* **108**, 4239 (1986).

CHAPTER II.E.1

A NOVEL SPECTRO-PHOTOMETER SYSTEM BASED ON AN ACOUSTO-OPTICAL TUNABLE FILTER

Since this chapter was written, a number of improvements have been introduced. The company cited now produces a hand-held device that works with a laptop holding all the necessary hardware and software needed for operation. Brimrose, Bran+Leubbe, and Rosemont should also be contacted for any new improvements.

I recently met a man named George Hooley who is involved in near-IR spectroscopy and the manufacture of IR fiber optics. Because fiber optics are one of the best tools for spectroscopists since Beer's law, I chatted at some length with him. I had the opportunity to visit his facility (Infrared Fiber Systems, Silver Spring, Maryland) and to look, listen, and touch. I had never actually seen fiber-optic material being manufactured, and I was impressed by the clever mechanisms used to pull and coat the fibers. Hooley described the types of fibers and end products made from them and showed me some of the instrumentation that his company is developing. The toy that impressed me most was the company's acousto-optic tunable filter (AOTF) spectrophotometer. Because this novel instrument is the only one of its kind under production [*Oops! I missed Brimrose, Bran + Leubbe, and Rosemont, but, hey, who's perfect?*], and because of its potential implications for industrial spectroscopic applications, it merits discussion in some detail.

The instrument is based on a pulsed AOTF manufactured by Westinghouse (Orrville, Ohio). The original AOTFs are manufactured from thallium arsenic selenide (TAS) or tellurium oxide (TeO_2), depending on the wavelength range desired, as a single-crystal birefringent optical material. The crystals used by Infrared Fiber Systems are entirely TeO_2, allowing continuous radiation throughput. In either case, an acoustic shear wave is produced by a transducer and is propagated across the optical path of the filter. This diffractive acoustic wave modulates the refractive index of the crystal, in essence creating a grating in the crystal. By tuning the transducer radio frequency, any wavelength of interest may be diffracted.

Three wavelength-range (TeO_2) crystals are now in production: 350–900, 1000–3000, and 2000–4600 nm. The speed of the electronic tuning allows for real-time optical monitoring. The large-filter internal aperture and light-acceptance angle allow a large throughput of light. Using a frequency synthesizer under microprocessor control, the transducer may be driven so that the tuned wavelength can be selected in microseconds. This frequency jumping allows for high-speed data acquisition. Continuous broadband scanning is done with incremental frequency steps or a sweep generator. A typical 1000-point scan takes ~50 ms. The speed of the instrument is a strong point. Of course, one cannot expect the same resolution that can be achieved with a research-grade FT-IR spectrometer, but it is typically better than 20 cm^{-1} and may be improved in future models.

The AOTF diffraction results in spatial separation of the tuned wavelength and a 90° rotation of polarization from the incident radiation. Both of these features are available for signal separation.

The manufacturer intends this instrument to be used as a rugged industrial on-line measuring device. Because the wavelength regulator is a single crystal sealed in a chemically resistant case, it appears to be more rugged than a grating device for high-vibration or dust situations. Although it is comparable to a filter instrument, its wavelength tunability makes it more versatile.

The models currently available are driven by IBM PC/AT or PS/2 30 computers, working through an IBM PC/AT plugin system [*actually, they use a 486 or Pentium-based laptop now!*]. The output is a parallel 16-bit bus and is compatible with Lotus, WordPerfect, and other software. The instrument is controlled through a menu-driven program. Currently, the data analysis modes are normalization, signal averaging, reciprocals, logs, differentials, peak detection, and calibration, and it uses either transmission or absorbance modes. For those of us who like color, it supports an enhanced graphics adapter (EGA) display and can be controlled with the keyboard or a mouse.

Optics can be optimized for conventional sensing or fiber-optic coupling (with either silica or ZrF_4 IR fibers). These fiber optics can also be multiplexed through an eight-channel system. The source may be a hot wire or Nernst glower, and the detector is usually thermoelectrically cooled with optional N_2 cooling.

Some of the uses demonstrated were analyses of gases such as methane, CO, CO_2, and water vapor; various medical applications; industrial applications in cases where space for the measuring apparatus is limited; and surface oil contamination applications. Obviously, I could not see any chemical plant in-process work at the facility; however, I think this will become a primary application of this type of instrumentation. Several novel multiple-pathlength cells were demonstrated for gases and some aqueous–alcoholic mixtures were run.

Some of the applications are proprietary, but I was shown enough to believe that the technique is viable. It won't replace FT-IR, near-IR, or UV/vis systems in the normal laboratory, but there will indeed be niches for the approach.

CHAPTER II.E.2

AN AOTF-ILLUMINATED MICROSCOPE

As a direct replacement for laboratory bench-top spectrometers, AOTF instruments offer few advantages. Used in industrial or peripheral situations such as in providing the illumination source for a microscope, however, AOTFs' speed and extremely small size are clear advantages.

Forensics and pharmaceutical laboratories have used midrange infrared (IR) microscopes for some time. They are quite useful in life sciences as well—the ability to explore the chemistry of minute portions of small species, such as cells, has been a godsend for life scientists. Attaching a microscope to an IR (usually FT-IR) instrument, however, can be cumbersome and expensive. Although the results are almost always spectacular and worth the effort, a more user- (and benchspace-) friendly system would be appreciated.

Ta-da! Brimrose Corporation (Baltimore, MD) has produced a teeny-weeny AOTF source, smaller than the keyboard of most computers, and attached it to a customer's microscope (it works with most high-end microscopes). The concept was developed by workers at the National Institutes of Health (NIH) in Bethesda, Maryland (2–4), and was recently commercialized by Brimrose.

I will briefly explain how an AOTF works. An anisotropic crystal of, say, tellurium oxide (TeO_2) is polished into a roughly rectangular shape, approximately one-quarter the size of a cigarette pack (remember those?). Along one of the narrow faces, an acoustic transducer is bonded, often by high-temperature metal bonding. Along the opposing face an acoustic absorber is bonded. When a high-frequency rf signal is applied to the transducer, a series of acoustic waves is produced. The distance between these waves determines the wavelength of light diffracted, according to Bragg's law. If the rf is changed, or swept, throughout a given range, the standing wavefronts vary in distance accordingly. The speed of sound and the changes taking place are orders of magnitude less than the speed of light; so even rapid scanning produces a standing wave, from the viewpoint of the light. The effect is that of an incredibly rapid scanning monochromator with a resolution of −2.5 nm, tunability of 0.1 nm, and reproducibility of 0.01 nm.

Because we all know that there's no such thing as a free lunch, certain precautions must be taken. First, the crystal must be temperature stabilized. Second, three beams emerge (normal, containing most of the incident light; ordinary, one wavelength with parallel polarization; and extraordinary, the same wavelength with perpendicular polarization), and because only one beam is usually used, the other two must be blocked or absorbed in some manner. Third, the beam is usually square, requiring additional optics such as beam condensers. Once these have been implemented, the spectroscopist has an instrument capable of working from the UV (~190 nm) to the faaaar-IR (~22,000 nm).

The application that caught my attention was the use of this microscope for Raman imaging. The schematic in Fig. 1 diagrams the instrument's ability to perform either transmission or reflectance measurements. Individual sources may be used or a single source directed through the chosen path. The charge-coupled device (CCD) detector is attached to an infinity corrected microscope.

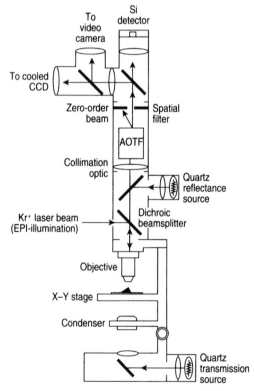

Figure 1. Schematic diagram of an AOTF microscope

Using partially mirrored reflectors, other detectors may also be placed in line: silicon detectors, video cameras, and, of course, the investigator's eye.

The AOTF in the figure is used as an after-the-fact wavelength sorter. Using this approach, a spectrum can be recorded or one or more wavelengths may be viewed using the AOTF as a bandpass filter. By configuring the system in this way, NIH scientists created "an all-solid-state two-dimensional imaging spectrograph" (3). Because the CCD is used as a true focal-plane array, no image transformation is needed. The system, as constructed, is capable of recording as many as 200,000 separate Raman spectra.

Investigation into this instrument's uses have only begun. The NIH group has published primarily on visible absorption of human epithelial cells and (dare I say the words?) the near-IR absorption images of a hydrated phospholipid suspension. Reference 3 shows the Raman image of polystyrene microspheres $\sim$45 μm in diameter. The detail observed in all photographs is quite remarkable. I highly recommend the original articles to anyone remotely interested in the spectra of small samples. I am happy that yet another tool is available to the spectroscopists of the world. (Not to mention the fact that I wasn't wrong about AOTFs!)

REFERENCES

1. E. W. Ciurczak, *Spectroscopy* **5**(1), 10–12 (1990).
2. P. J. Treado, I. W. Levin, and N. E. Lewis, *Appl. Spectrosc.* **46**(4), 553 (1992).
3. Y. Soos, *Photonics Spectra*, August 1992, p. 91.
4. C. D. Tran, *Analyt. Chem.* **64**(20), 971A–81A (1992).

CHAPTER II.F.1

SHEDDING LIGHT ON FIBER OPTICS

Scott Wohlstein

This month's guest author, Scott Wohlstein (SD Laboratories, Inc., Franklin, PA), is a photonics specialist, which is to say he uses light like a chemist uses chemicals. I asked him to tell us about fiber optics, but not necessarily how they are used. He obliged with this piece on their theory and production and added a bit of a glossary to help us understand the jargon. With the inclusion of fiber optics as an option in nearly every spectrometer sold today and their growing use in process control, I believe that this explanation is timely.

The field of fiber optics has enjoyed explosive growth since its humble beginnings more than 100 years ago, thanks to John Tyndall, a British physicist. AT&T was awarded a patent in the early 1930s for the "light pipe," the first glass fibers developed to transmit light over short distances. They were engineered by American Optical Corporation, and Corning Glass Works developed the technology to produce fibers capable of transmitting light over long distances.

The basic theory of fiber optics is straightforward and is primarily based on total internal reflection (TIR). This phenomenon describes the path taken by the light as it travels through the fiber and the way the light should encounter a higher refractive index in the buffer or cladding material than in that of the fiber core material (more on these materials further on). The ideal fiber-optic element would present 100% transmission, 100% reflectivity (off cladding material), and 0% absorption of light to be launched down the fiber (Fig. 1).

Fiber optics show many advantages over copper and electrical conductors, including low signal radiation; immunity to radio frequency interference (RFI), electromagnetic interference (EMI), and lightning; absence of groundloop prob-

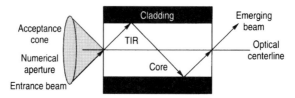

Figure 1. Light propagation through fiber optics.

lems (and associated hazards); low attenuation and high bandwidth; and a smaller size per relative volume.

Construction. Fiber optics are made up of core materials surrounded by cladding materials. The cores are made through a pulling or drawing process in which a molten glass boule is drawn (much like taffy) and then processed though various heating, forming, and cooling processes. Core materials typically used (followed by their respective wavelength bandwidths) include chalcogenide (5–15 μm), zirconium (1–6 μm), glass and glass-enhanced materials (0.4–6 μm), and silica (0.17–2 μm).

The final step in the process is to "clad" the bare fiber with a protective coating then wind it onto a spool for storage and handling (Fig. 2). Cladding materi-

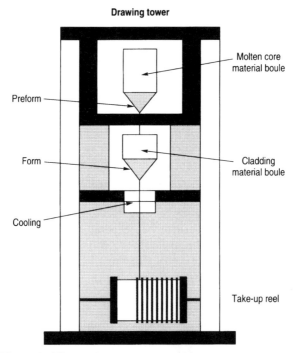

Figure 2. Fiber-optics construction process and equipment.

als typically used include vinyl, neoprene, Hypalon, polyethylene, polyurethane, thermoplastics, nylon, Kynar, Teflon FEP, Tefzel, and polyolefins.

Configurations. Fiber-optic systems are configured either as single fibers or as fiber bundles, in which groups of fibers are variously arranged (Fig. 3). Depending on its function, the single-fiber configuration may consist of step-index fibers, in which the core and cladding materials have distinctly different indexes of refraction; graded-index fibers, in which the core and cladding materials exhibit a decreasing index of refraction radially outward; and *polarization-preserving* fibers, which are manufactured with birefringent cores to maintain a specific level of linear polarization.

Fiber bundles can be configured randomly or in an arranged pattern. In a random configuration, groups of fibers are used primarily for raw power delivery when image- or signal-restoration is not a requirement. Arranged patterns are designed to keep some aspect of the detection and analysis section the same throughout the system or provide a unique function. For example, a random bundle may provide image gathering, light gathering, or light emission. An arranged bundle may provide image and illumination (in the same bundle) or photo-chemometrics (chemical reaction in the presence of light).

Design. Fiber optics are designed with some important characteristics in mind. The *numerical aperture* is the "light-gathering" number; the larger it is, the higher the light-gathering capability.

The *normalized frequency parameter*, or V number, is a dimensionless quantity that describes fiber parameters including the core diameter, wavelength, and refractive index.

The highest rate for which information can be transmitted with acceptable resolution is limited by the *bandwidth*, which can be further separated into three types of limitations caused by dispersion. *Modal* dispersion is caused when multiple modes in a fiber-optic system interfere with a signal or even modify its

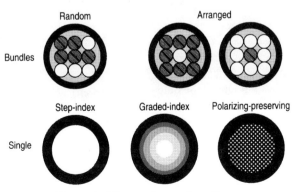

Figure 3. Fiber-optics configurations.

characteristics (for example, pulse lengthening). *Material* dispersion is caused by the material of which the fiber is manufactured (especially core material), which creates dynamic changes such as shifting interfaces and static changes due to impurities in the core.

Attenuation, which describes loss, is the most important factor to consider in designing a system because it can occur in so many different ways, including the following:

- *Absorption.* If materials, interfaces, and so on are not properly packaged, the surrounding environment can affect absorptive losses by changing the material. For example, when exposed to humid conditions, some optic, electro-optic, and acousto-optic materials will exhibit slight hygroscopic characteristics.
- *Scattering.* Loss by scattering occurs when an interface surface is broken by an abrupt discontinuity such as a crack, scratch, or reflective contaminant.
- *Microbending.* Losses of this type occur when a normally propagating beam follows a path that is not entirely linear, trapping light.
- *Mode scrambling.* Multiple modes can create interferences that limit power.
- *Polarization scrambling.* The light signal may lose its original polarization orientation, again limiting power.
- *Environmental.* Performance can be adversely affected by temperature, pressure, mechanical stress, radioactivity, light irradiation, or chemicals.
- *Nonlinear effects.* New frequencies, "hot spots," and varying spatial and temporal problems can occur from putting too much power (overdriving) into the fiber-optic system.
- *Surface preparation.* Losses can occur if proper cleaning, polishing, and splicing procedures are not followed.
- *Feedback.* This is caused when an unexpected abrupt interface, such as microbending, appears in the optical path.

A properly designed fiber-optic system—one that addresses the factors above—can reduce space, cost, weight, and need for operator attention, and at the same time increase performance, repeatability, reliability, and applicability for any system to which it is applied.

Fiber optics has influenced the field of spectroscopy in many ways, from helping make finicky instruments easier to operate, to introducing entirely new applications in new fields. It has brought us from the use of fragile optical breadboards to the development of spectroscopic "engines" able to work in remote locations.

Today we have not only accepted fiber optics as the technology of the future in most areas of society, but are spending a great deal of effort developing material and systems to take us well into the next century.

CHAPTER II.F.2

FIBER-OPTIC
MULTIPLEXERS

Bruce R. Buchanan and Patrick E. O'Rourke

Bruce Buchanan (Merck & Co., Inc., P.O. Box 4, West Point, PA 19486) collaborated with his former co-worker Pat O'Rourke (Westinghouse Savannah River Company, P.O. Box 616, Aikin, SC 29802) to share some work that they did at the Savannah River facility.

For those of you who do not recognize the name of the place, let me say two words: nuclear facility. You can see why they might not want to take samples up to the lab for testing. As we are wont to say, "Necessity is a mother!" In situ testing was more than a good idea for them. The use of minimum instrumentation for maximum measurements is a concept whose time has come in every industry (unless you sell the maximum instrumentation!)

For years the use of optical fibers in spectroscopy has been touted as the means to shield instruments from the rigors of the process environment. An additional advantage promised (but not always delivered) by most process analytical chemists is the use of one instrument to monitor several locations. The tool that allows this to happen is the fiber-optic multiplexer. This chapter will address the problems of and solutions to using fiber-optic multiplexers to link one instrument to multiple sample locations.

BACKGROUND

For the sake of this chapter, the term, *fiber-optic multiplexing* denotes (*my personal apologies to Bill Fateley*) the coupling of light from a single fiber sequentially into a set of other fibers. The light can be coupled either by physically moving the fibers or by moving a mirror. Either way, the fiber-optic multiplexer allows a single instrument to monitor several sampling points.

Optical fibers for the visible and near-infrared (near-IR) region have redefined on-line process analytical chemistry. Optical fibers allow instruments to run in clean, vibration-free environments while monitoring several locations. Because one instrument can monitor several sample sites, the cost per site is reduced. Reducing cost is always an excellent selling point with management. The fibers themselves are relatively inexpensive and easily replaced. Optical fibers can be pulled through conduits in a manner similar to copper wire for easy installation. However, to benefit from these advantages, a mechanism to couple the multiple sample points to the instrument must be used. In general, these devices are known as fiber-optic multiplexers.

TYPICAL SETUP

A typical setup is shown in Fig. 1. Fiber-optic multiplexers with as many as 50 positions have been developed; however, more practical units have 10–11 positions. A good practice is to use one position for a reference, one for a verification standard, and the remaining positions for samples. As shown in Fig. 1, the sample positions can be Swagelok assemblies (Swagelok Co., Solon, OH), and the standard and reference positions are usually made up of cuvette holders. Swagelok assemblies were developed to couple lenses with a process using established connectors. With this configuration, the sample cell can be flushed with the reference solution less often.

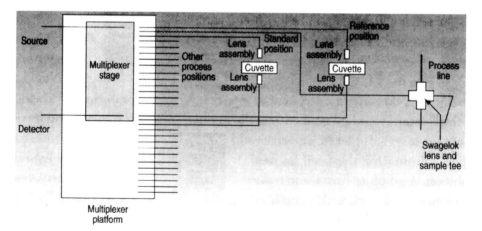

Figure 1. Example of a fiber-optic multiplexer setup.

ADVANTAGES

The most obvious advantage of using a fiber-optic multiplexer in process measurements is the ability to measure several positions using a single instrument. This addresses the problem in process analytical spectroscopy of transferring a calibration from one instrument to another—many times this translates to building instrument-specific models. This constraint, coupled with the cost of many of the instruments. makes it sensible to measure as many sites as possible with a single instrument.

One of the biggest advantages of using a fiber-optic multiplexer with instrumentation is the ability to have a check standard on line to verify instrument operation. Although the standard is usually similar to the process samples, it is not a necessity. Any standard can be used to verify that the instrument is operational as long as the standard is well characterized. Because the standard is connected through fiber optics. it is possible to isolate the standard and control the environment to ensure the standard' s stability.

PROBLEMS

The main problem in using a multiplexer to couple fiber optics to the process is that each fiber has an absorbance spectrum. Unless a model is developed on each position—an odious task—the individual fiber spectrum will interfere with quantitative analysis. Removal is accomplished by collecting a zero scan of each position. The zero spectrum consists of a spectrum of the fiber using solvent in the sample position versus the reference position. This zero spectrum can then be subtracted from each measurement. This assumes that the spectrum does not change over time, which is not a good assumption under many conditions. Thus the zero spectrum should be checked periodically. A new zero spectrum must be obtained if any major changes are made to the system, including changes to the instrument, fibers, or sample interfaces.

An experiment using holmium glass as a standard was performed to test the stability of the multiplexer. The experiment involved moving the stage to the reference, homing the unit, moving the stage to the sample, and then homing again. Scans were obtained when the stage was in the reference and sample positions. Figure 2 shows the correlation coefficient R versus time using the first scan for the comparison. An ideal value of 1.00 would indicate no difference between the first scan and the subsequent scans. The difference over a 3-h period is quite small, but an obvious downward trend appears. This means the multiplexer needs to be realigned periodically and a new zero spectrum obtained for the position.

Another problem is the fiber-to-fiber interface at the multiplexer, where there is a loss of signal at each window in the setup. This loss can be minimized by adjusting the distance between fibers using antireflection coatings, and ensuring

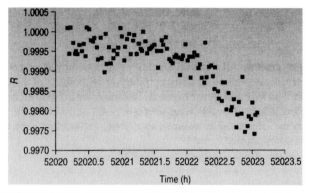

Figure 2. Experiment demonstrating the stability of the fiber-optic multiplexer. Scan 1 was used as the comparison scan.

a good polish. However, we have found that for many applications the losses at the interface do not prevent good spectroscopy.

Alignment of the fibers at the multiplexer interface also affects the signal. This effect is greater in the near IR than in the visible region. The main effect of misalignment is shown in Fig. 3, in which a baseline offset in the visible region can be seen. In other words, misalignment has the same effect as a neutral density filter. Thus the multiplexer can be used to ensure that the light intensity level remains similar at all positions. Unfortunately, this works only in the visible region. In the near-IR, the effects of the mechanism of transmission—modal effects—are pronounced, resulting in spectral changes when the multiplexer is misaligned.

Another factor to be considered is the switching time between positions, which limits the amount of time that can be used between measurements. The switching time is also affected by the number of positions on the multiplexer.

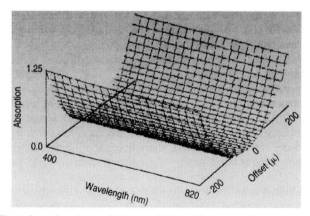

Figure 3. Effect of moving the multiplexer off the optimum position in the visible region.

For the standard 10–11-position unit, the time needed to move from position 1 to position 11 is ~45 s. This time limit, coupled with the time needed to collect the spectra and perform the data processing, allows ~1 scan every 2 min. This is usually adequate for most process systems.

Finally, the problem of reference stability arises. Figure 1 shows that there is a separate reference position where the reference intensity, 10, is found. The fiber-optic multiplexer is switched to the sample position, and the sample intensity is found. This allows the calculation of absorbance. Because the spectrum in Fig. 1 still contains absorbances of the fiber, the zero scan must be subtracted. This process allows the sample to be measured on line without needing to flush the sample cell with the reference solution as often. This procedure assumes that the fiber's zero spectrum has not changed and that no coatings have built up on the reference or sample cell windows. These assumptions are both process and time dependent. They hold quite well for many applications in the visible region, but less well in the near-IR. Each user must determine whether the assumptions hold well enough to implement a reference position. However, even if the assumptions prove false, it does not preclude the use of a fiber-optic multiplexer. It is still advantageous to measure multiple positions even if the cell must be periodically flushed for a reference scan.

CONCLUSION

Whenever an additional component is added to an instrument configuration, it adds to the complexity of the measurement. Even though the addition of a fiber-optic multiplexer to a spectrometer makes the collection of data more difficult, the advantages offset the disadvantages. For example, the use of a multiplexer allows one instrument to monitor a number of locations. In weighing the advantages versus the disadvantages, the use of a fiber-optic multiplexer usually comes out on top as an effective means of increasing an instrument's usefulness.

CHAPTER II.G.1

WHAT IS FT-RAMAN SPECTROSCOPY, AND WHERE CAN I USE IT?

Mary Ann Finch

The author of this chapter, Mary Ann Finch, was the national marketing manager for FT-IR products at Bruker Instruments Inc. When I first asked Mary Ann to explain FT-Raman spectroscopy, she said, "Sure, when I have time." Most editors automatically assume this means "go away" in polite terms; not so in this case. Being a chemist before all else, she was true to her word and prepared this overview of the oft-misunderstood technique. With the advent of laser systems utilizing Fourier transforms, Raman has become a very important addition to the molecular spectroscopist's toolbox.

This chapter has two main objectives. First, it is meant to provide current nonusers of Raman spectroscopy with a brief introduction to the principles and advantages of the little-practiced technique. Second, it is intended to demonstrate for both users and nonusers the analytical capabilities made possible by recent advances in the instrumentation of Fourier transform (FT)-Raman spectroscopy. FT-Raman's advent into the commercial instrumentation arena has in a relatively brief time brightened the chances that Raman spectroscopy will finally gain wide acceptance and application as a routine analytical problem solver.

It is well documented that a major advantage of FT-Raman over conventional dispersive Raman spectroscopy is its ability to render spectra that are generally free of fluorescence interference. The vast majority of samples either fluoresce or contain impurities that do when excited by the visible lasers used in conventional Raman spectroscopy. This is because many electronic transitions are active at visible laser frequencies, and the fluorescence is often many times

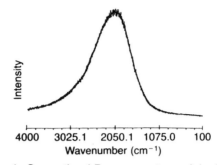

Figure 1. Conventional Raman spectrum of rhodamine.

stronger than the weak Raman scattering. This effect is illustrated in Figs. 1 and 2, which show the spectra of a rhodamine sample obtained on a conventional Raman spectrometer and on an FT-Raman system, respectively.

It has been long understood that this fluorescence could be generally avoided by going to lower-frequency near-IR laser sources. (*No! I did NOT ask her to write because she would refer to near-IR!*) However, the intensity of the Raman scattering is proportional to the fourth power of the frequency of excitation; so there would be a drastic decrease in sensitivity. Moreover, different spectrometer gratings would be required when switching from visible to near-IR sources.

Because of these difficulties, practicing Raman spectroscopists have been searching for new instrumental approaches that would make Raman's benefits

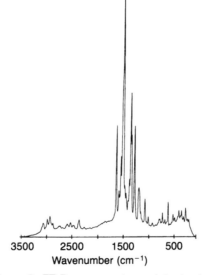

Figure 2. FT-Raman spectrum of rhodamine.

available to researchers in need of an easy-to-use, routine analytical tool. Several research groups in recent years have reported great success with FT-Raman experiments using a near-IR laser for sample excitation. Scientists at IBM, Du Pont, the National Institutes of Health, and Georgia Institute of Technology, to name but a few (1), all have active FT-Raman projects underway. Over the past three years, nearly every major producer of FT-IR instrumentation has introduced accessories that add FT-Raman capabilities to their existing instruments.

All these commercial systems were developed around Nd : YAG near-IR laser sources due to the obvious advantage arising from nearly eliminating fluorescence. True, the near-IR laser produces weaker Raman scattering than a visible laser would, but this disadvantage is offset by a new breed of sensitive near-IR detectors capable of measuring these signals with very high quantum efficiencies. In addition, FT-Raman enjoys the wavelength precision of FT-IR so that spectra may be co-added, resulting in a rapid improvement in signal-to-noise (S/N) performance. This precision also enables the spectroscopist to perform accurate spectral subtractions. Spectral collection times are reduced in FT-Raman spectroscopy because all the frequencies are measured simultaneously rather than sequentially as in dispersive spectroscopy. One can envision FT-Raman following the lead of FTIR into applications that utilize signal averaging to measure very weak samples or that take advantage of rapid spectral collection to monitor transient events.

Why bother? Nonusers of Raman spectroscopy are probably wondering why some of the best and busiest minds in analytical chemistry have invested so much effort into making Raman a routine method. If it is such a troublesome technique, why not put it on the back burner and concentrate on further improving more user-friendly methods that have already become established routine analytical tools?

Raman long ago attracted a loyal following of scientists—in academic, government, and industrial labs—who appreciated its many analytical advantages. As an alternative to IR spectroscopy, Raman can, quite simply, be easier to use in some cases. Water and glass, in particular, are strong IR absorbers. Thus special techniques must be employed to obtain IR spectra of aqueous samples, and glass is not a satisfactory material for holding liquid samples when using these IR techniques. Happily for the Raman spectroscopist, both water and glass are weak Raman scatterers; so it is easy to produce a good-quality Raman spectrum of an aqueous sample in a glass container.

IR spectroscopists are quite familiar with tedious procedures for preparing samples for IR analysis. These include pressing KBr pellets, casting thin films, preparing solutions with appropriate solvents, and using special sample accessories requiring careful alignment. In many cases, FT-Raman spectra can be obtained on samples "as is" by simply positioning them in the laser beam. A small cup may be used to hold powders and pastes. Glass tubes are ideal for liquids.

Attractive, too, is Raman's ability to glean useful information from very small samples. The laser excitation source is usually focused to a 100-μm-diameter spot size. Furthermore, a commercial FT-Raman microscope has

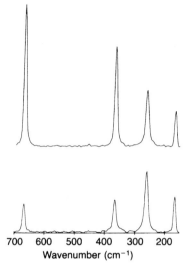

Figure 3. Results of an FT-Raman polarization study of chloroform. The upper trace shows the parallel and the lower trace shows the perpendicular polarized spectra.

recently been introduced. Alternatively, the laser can be defocused for samples where lower laser intensity is desirable, such as measurements on samples contained in rare gas matrices.

Raman has proved useful in polarization studies. These studies can be helpful to researchers wishing to assign vibrational modes to observed transitions. To illustrate, Fig. 3 shows the parallel and perpendicular polarized spectra of chloroform, in which the bands at 667 and 336 cm^{-1} are easily assigned to symmetric C–C13 stretching and bending modes. In the remainder of this chapter, I will present and briefly describe spectra selected to demonstrate these advantages of the Raman method in the analysis of a variety of sample types. With FT-Raman, these inherent advantages are complemented by factors of added instrumental convenience and general lack of fluorescence.

EXPERIMENTAL

FT-Raman spectra of the samples presented in this chapter were measured with an IFS 66 FT-IR spectrometer coupled to an FRA 106 FT-Raman accessory (both offered by Bruker Instruments, Billerica, Massachusetts). The IFS 66 FT-IR instrument, which can be hydrogen purged, incorporates a Michelson interferometer with a flexible design that allows the user to switch easily between FT-IR and FT-Raman modes. Data collection and processing were carried out with an Aspect 1000 data system (Bruker). The FT-Raman accessory, used with a near-IR beam splitter, collected Stokes (over a broad range) and anti-Stokes data simultaneously, without changing the Rayleigh filter or warming up the liquid nitrogen–cooled detector. Raman scattering was excited by a 1.06-μm continuous-wave Nd : YAG laser using multimode operation. Although the accessory easily accommodates both 90° and 180° sampling geometries, all spectra were obtained using 180° backscattering. Double-sided interferograms were acquired in both directions that the interferometer's moving mirror traveled. These were transformed into their corresponding power (magnitude) spectra to eliminate phase errors (2).

RESULTS AND DISCUSSION

FT-Raman spectroscopy is applicable to a wide variety of samples. Previous investigations have demonstrated the feasibility of the technique, as well as presented specific applications (3–8). These analyses have ranged from forest products (7) to illicit drugs (8). Following are additional examples illustrating the variety of samples amenable to FT-Raman investigations.

Specialty chemicals. The analysis of organic and inorganic compounds is of primary importance to a wide array of scientists. Such studies often involve functional group identification, as well as molecular structure and conformational elucidation. FT-Raman spectroscopy is a helpful tool in such investigations. Figure 4 shows the results of FT-Raman measurements of 1,4-di-(2-methylstyryl)-benzene and anthracene at 4-cm^{-1} resolution using laser powers

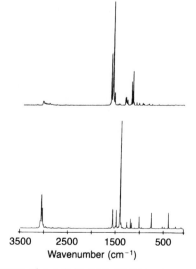

Figure 4. FT-Raman spectra of 1,4-di-(2-methylstyrl)-benzene with 10-mW laser power (upper trace) and anthracene with 200-mW laser power (lower trace), both with a 2-s collection time.

of 10 and 200 mW, respectively, and 2-s collection times. The spectra clearly demonstrate the high S/N and the wealth of spectral information that can be obtained with FT-Raman experiments. The data are even more significant when one considers that these samples are very difficult to analyze with conventional Raman spectroscopy due to fluorescence.

Polymers. FT-Raman spectroscopy can be used in the identification of polymers and as a tool in investigating conformational preferences. For example, FT-Raman can be used to measure *cis/trans* ratios and monomer content of polymers.

To illustrate, a commercial overhead projector transparency was analyzed. The spectrum obtained is shown in Fig. 5. In order to identify this sample, a number of common polymer fibers were studied, including nylon, Orlon, Acrilan, and Dacron. Comparison of the transparency's spectrum to that of Dacron (Fig. 5) clearly indicates that the transparency consists of a polyethylene terephthalate polymer.

Forest products. The ability of FT-Raman spectroscopy to differentiate between hard and soft woods as well as the ability to measure coatings on paper has been demonstrated (7). Additional studies suggest that FT-Raman can also be used to determine the amount of processing that a paper sample has undergone. Figure 6 shows the spectrum of a piece of yellow paper and that of a manila envelope. For the yellow paper, the band at 1601 cm^{-1}—attributed to lignin—is much less intense (if present at all) than it is for the manila envelope.

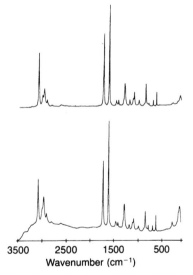

Figure 5. FT-Raman spectra of a commercial transparency (upper spectrum) and of a Dacron fiber (lower spectrum).

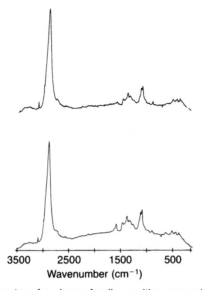

Figure 6. FT-Raman spectra of a piece of yellow writing paper (upper) and of a manilla envelope (lower).

The spectral differences clearly indicate that the more highly processed paper sample contains less lignin, which is not surprising, since manufacturers tend to remove lignin from paper products that are to be used for correspondence. Failure to do so leads to the eventual discoloration of the paper. As an additional verification, Fig. 7 details the FT-Raman spectrum of a wood matchstick, which

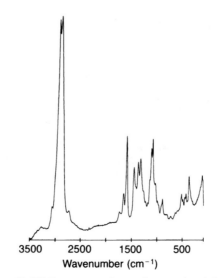

Figure 7. FT-Raman spectrum of a wood matchstick.

is not subjected to all the processes used in making paper. Examination of the bands produced by lignin demonstrates that FT-Raman can be used to monitor the amount of lignin remaining in processed forest products.

Food products. A common myth often espoused by detractors of FT-Raman spectroscopy is the inability of the technique to analyze colored or, in particular, black samples. The spectrum of milk chocolate in Fig. 8 demonstrates that useful information can be obtained from dark samples using FT-Raman spectroscopy.

The sensitivity of Raman spectroscopy to carbon–carbon double-bond stretches is illustrated in an FT-Raman study of a series of vegetable oils (Fig. 9). By observing the intensity of the band at 1656 cm^{-1}, the relative abundance of the *cis*-carbon–carbon double bonds in these oils, and hence the extent to which the oils are unsaturated, can be determined. To emphasize the complementary nature of Raman spectroscopy, an IR spectrum of soybean oil is also shown in Fig. 9. Clearly, the unsaturated carbon–carbon stretching vibration is detected much more easily in the Raman spectra.

Biochemicals/pharmaceuticals. The analysis of proteins is important in the study of biochemical systems. Figure 10 shows the resulting spectrum of an FT-Raman investigation of crystalline egg albumin. The spectral information found in the amide I, II, and III regions can be used to determine the secondary structure(s) of the protein.

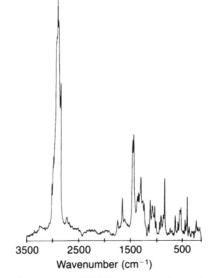

Figure 8. Baseline-corrected FT-Raman spectrum of milk chocolate.

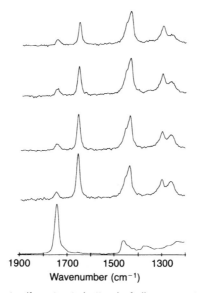

Figure 9. FT-Raman spectra (from top to bottom) of olive, peanut, soybean, and safflower oils and an FT-IR spectrum of soybean oil.

The study of drugs in tablet form is also important for the chemist involved in pharmaceutical applications. Figure 11 shows survey and expanded spectra collected for an aspirin tablet. The data exhibit high S/N and were collected in ~30 s. Even more important, the spectra clearly indicate that useful information can be obtained within 50 cm^{-1} of the Rayleigh line.

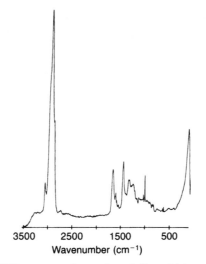

Figure 10. FT-Raman spectrum of crystalline chicken egg albumin.

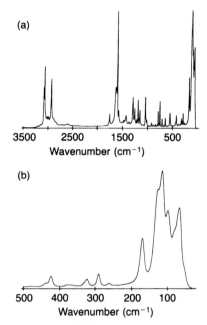

Figure 11. FT-Raman (a) survey and (b) expanded spectra of an aspirin tablet.

CONCLUSIONS

FT-Raman spectroscopy is a technique suitable for the analysis of many types of samples. In addition to the specific examples described in this chapter, the technique has been used in many fields, including petroleum, agriculture, and forensics. FT-Raman has an important role to play in analytical chemistry. When coupled with commercial FT-IR instrumentation, FT-Raman allows scientists to obtain a vast amount of information quickly and easily.

ACKNOWLEDGMENT

Spectra and technical assistance were provided by Dr. Roger C. Kenton and Dr. Ron Rubinovitz of the Bruker applications laboratory in Billerica, Massachusetts.

REFERENCES

1. Proceedings of the 7th International Conference on Fourier Transform Spectroscopy, David G. Cameron, ed., *Proc. SPIE* **Vol. 1145** (1989).

2. P. R. Griffiths and J. A. de Haseth, *Fourier Transform Infrared Spectrometry* (Wiley Interscience, New York, 1986), p. 93.

3. B. Chase, *Anal. Chem.* **59**, 881A (1987).

4. H. Buijs, *Spectroscopy* **1**(8), 14 (1986).

5. F. J. Bergin and H. F. Shurvell, *Appl. Spectrosc.* **43**, 516 (1989).

6. F. J. Purcell, *Spectroscopy* **4**(2), 24 (1989).

7. R. C. Kenton and R. L. Rubinovitz, *Appl. Spectrosc.*, in press.

8. C. M. Hodges, P. J. Hendra, H. A. Willis, and T. Farley, *J. Raman Spectrosc.* **20**, 745 (1989).

TECHNICAL BACKGROUND: THE RAMAN EFFECT

The Raman effect occurs when a sample is irradiated by monochromatic light, causing a small fraction of the scattered radiation to exhibit shifted frequencies that correspond to the sample's vibrational transitions. Lines shifted to energies lower than the source are produced by ground-state molecules, while lines at higher frequency are due to molecules in excited vibrational states. These lines, the result of the inelastic scattering of light by the sample, are called Stokes and anti-Stokes lines, respectively. Elastic collisions result in Rayleigh scattering and appear as the much more intense, unshifted component of the scattered light. The ratio of the intensities of the Stokes and anti-Stokes lines can be used to determine the temperature of the sample. Figure 12(a) of the spectra at the right shows the spectrum obtained from a lump of sulfur, which illustrates Stokes and anti-Stokes scattering as well as the suppression of the Rayleigh scattering by the instrument's filters.

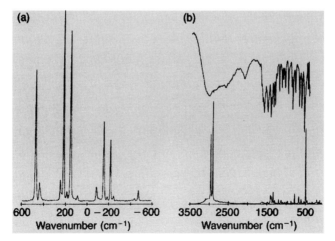

Figure 12. (a) FT-Raman spectrum of sulfur, and (b) FT-IR (upper trace) and FT-Raman (lower trace) spectra of l-cystine.

Raman, like infrared (IR) spectroscopy, is used for qualitative analysis, molecular structure elucidation, studies of molecular interaction, and determination of thermodynamic properties. There are several important reasons why routine collection of Raman spectra is desirable. Raman and IR transitions have different selection rules; so some modes are allowed in one technique and forbidden in the other. Both types of spectra are needed to completely characterize a molecule. Linkages such as C—C, C=C, C=S, C—S, S—S, N—N, and S—H produce bands that are generally stronger in Raman than in IR. This is illustrated in Fig. 12(b) of the spectra at the right for the amino acid I–cystine where the S—S stretch near 500 cm^{-1} is easier to see in the Raman spectrum than in the IR spectrum.

CHAPTER II.G.2

A COMMERCIAL NEAR-IR, DIODE-ARRAY SPECTROPHOTOMETER

I was recently invited by Isaac Landa, president of LT Industries (Rockville, MD) to play with his company's near-IR diode-array instrument, the Powerscan Process Analyzer. Diode arrays have been a common detector for UV/vis instruments for more than a decade now. While prototypes of near-IR instruments have been described in the literature for several years, I am aware of only two true commercial models. My definition of "commercial" includes the requisite that at least one instrument is in the hands of a customer and is actually in use, with a second model ready for demonstration and delivery. (Perkin–Elmer, Norwalk, CT, began shipping its PIONIR 1024 diode-array instrument in September.)

I was impressed enough during my visit that I returned with pharmaceutical samples to test. The crux of the present chapter is to describe the instrument itself.

Any diode-array detector consists of a line of light-sensitive diodes (100×30 μm) arranged in a tight, thin line. In the case of the 256 diodes in this instrument, the array was roughly 0.1 mm wide $\times$ 2 cm long. The actual material from which the diodes are manufactured depends on the wavelength range desired, which is based on the material to be analyzed. For example, silica diodes are best suited for the UV and visible regions of the spectrum. For the near-IR, InGaAs (indium–gallium arsenide, pronounced "ingas") or PbS (lead sulfide) are often the compounds of choice.

A carefully controlled bias is induced across the diodes, and the change in resistance that is induced by light impinging on the diode is measured. The resistance is correlated to intensity, and then spectroscopic measurements may be made. The difference between a "normal" scanning spectrophotometer and a diode array is that the latter has no moving pans.

The grating in the diode-array system is fixed while the resistance's of the diodes are read sequentially, often several hundred times per second. As with a rapid scanning system (grating or interferometer), a substantial number of scans may be co-added to reduce noise levels. The instrument's specification is listed as less than 20 μAU (or 20×10^{-6} AU). Although I didn't measure the actual noise levels for this particular instrument, I noticed that the spectrum smoothness was limited only by the pixels of the screen when the spectra were expanded. This has become a trademark of diode-array instruments at any wavelength.

The wavelength range of the instrument in question is 800–1760 nm. To purists, this corresponds to the second and third overtone region of the near-IR. (Another diode array has recently been introduced that consists of 512 diodes, which expands the wavelength region to 2200 nm.) Because existing commercial grating instruments extend to 2400 or 2500 nm, this will allow for comparable spectral information.

The strength of a diode array is that there is virtually no reproducibility factor. The better near-IR gratings have reproducibilities on the order of 0.1 nm from scan to scan. This is based on careful control of the speed and beginning and end points of the grating's motion. Much of the (minuscule) noise present in these systems is a result of scanning motion. In addition, there might always be a question of validating the actual wavelengths. The reproducibility of diode arrays may be their greatest advantage.

The weakest point in any diode-array system is that the resolution is limited by the physical presence of the diodes and their spacers. In UV or visible systems, this has been a criticism by competitors of the systems' manufacturers. The criticism hasn't stopped diode-array systems from becoming a mainstay for use in high performance liquid chromatography as well as in performing routine UV/vis analyses. Resolution is virtually a nonissue in near-IR because the peaks are rarely sharp and can be as wide as 100 nm. Also, most near-IR analyses fall into the category of "routine analysis," not "research." For routine work, a single wavelength, or at most, two or three wavelengths, are all that are necessary (note the popularity of filter instruments over the years).

The instrument that I inspected routinely comes equipped with the ability to read seven channels sequentially (for the beloved DOM, Bill Fateley, I won't say "multiplexed"). The interesting design feature of this particular "multiple-measuring device" is that the optics are fixed. In many manufacturers' earlier versions, the so-called multiplexer rotated some part of the optical train from channel to channel. This made alignment difficult (and sometimes impossible) to maintain during continuous operation. In this new design, all the channels are focused on the fixed grating, with a slit shield rotating to expose each optical path in turn.

With only one moving part, this is the simplest design of a production-level near-IR spectrophotometer that I have seen yet. I can envision several different reactions, processes, or operations monitored simultaneously. The software changes needed for that type of monitoring would be minimal. I cannot personally attest to this instrument' s workability over the long haul, because I have worked with it only through one of its channels. My job [*Well, job is usually something for which one gets paid. Let's just say I do this for fun*] is merely to present the latest trends in spectroscopic toys.

CHAPTER II.G.3

AN LED-BASED, NEAR-IR PROCESS SPECTROMETER

On-line analysis can speed the production of goods and lower the percentage of rejected materials, thus making industry more competitive.

Under traditional manufacturing conditions, a raw material is sampled, bagged, labeled, taken to a quality control lab, logged in, assigned to an analyst, assayed, the results logged in, the data checked by a supervisor, and then reported to manufacturing. This is a 1–2-week process just for raw materials—a time frame not conducive to just-in-time manufacturing.

The finished product or in-process materials suffer a similar fate. The assays for many products arrive well after the process has been completed and the line broken down for another product. If the process is controlled by these assays, there are often lags of as long as an hour between sampling and reporting. The product must either wait in limbo or be over cooked while results are "in the mail."

All of which leads me to this chapter's topic: a look at another of these rapid analysis instruments. This time, Katrina Inc. (Hagerstown, MD) caught my attention. Founded in 1991 by David Honigs, Katrina specializes in light-emitting diode (LED)-based systems covering the spectral range from 900 to 1050 nm (with 400 to 1050 nm available).

The heart of the company's two systems (Protronics models I12 and 412) is a 12-diode source. Each of the LEDs is faced with a bandpass interference filter, allowing specific and reproducible wavelengths to be emitted. These diodes fire sequentially through a fiber bundle to the sample. The light is transmitted through the sample, collected by a second bundle, and delivered to a precision silicon detector.

The nominal scan speed of 50 linear ft/min allows for averaging a large num-

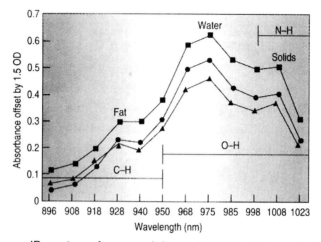

Figure 1. A near-IR spectrum of processed cheese that distinguishes peaks for fat, water, and solids. ■ = 45.6% water, ▲ = 43.9% water, and ● = 41.8% water.

ber of readings or rapidly scanning a large number of samples. The instruments may be used with liquids, powders, or discrete solids.

As you should know by now, I am always intrigued by clever toys. Honigs has found ways of measuring moisture and fat in frozen French fries. "So what!" you say? Well, how about while the French fries are moving along a conveyor belt and dropping in front of the near-IR beam at a rate of 50 linear ft/min? The spectra are averaged over a period of time and the thickness of the potatoes is then averaged over that time frame of several seconds.

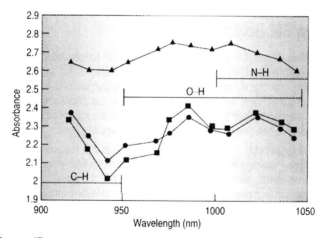

Figure 2. A near-IR spectrum of cereal flakes showing the sugar and water content. ■ = sugared flakes, ▲ = wet/dried flakes, and ● = granulated sugar.

Materials running along a belt are rapidly measured in the same fashion: thickness variations arc averaged out rather than being smoothed. Processed cheese is analyzed for fat, water, and solids. The spectra produced for this assay are seen in Fig. 1. Even using 12 wavelengths, the materials may be assayed. The optical density of absorbance is <2.2 in this case, attesting to the strength of the light sources.

Another example is the analysis of cereal flakes in which the light is transmitted through 1 in. of product. Again, the optical density of 2.4 implies that a 2-in. bed could be assayed (Karl Norris, after all, read spectra through whole eggs more than 30 years ago!) Figure 2 displays the regions of the spectra and some rationale for the assignments.

Filter near-IR instruments have been in use in the food industry for decades and are still the most abundant form of near-IR spectrometers. The updating of these approaches, along with the cost savings, is the logical next step. The elegance of this type of equipment is twofold: It is simple (thus inexpensive), and it works.

CHAPTER II.G.4

PICTURES FROM NUMBERS AND THE LITTLEST SPECTROMETER

In this chapter, I'd like to cover two very different topics: CCDs, or charge-coupled devices, and the little spectrometers we call HPLC detectors.

CHARGE-COUPLED DEVICES

Having worked at Goddard Spaceflight Center, I had many occasions to hear about the Hubble telescope. Imagine my surprise when I learned that the Hubble used CCDs for imaging (1). In reality, it meant that I'd have to learn a little more about how they worked or be "out of the loop." More researchers are using CCDs for such things as IR microscopes, and manufacturers are using them for quality control.

In a very simple example, assume that a solid-state detector is used in a spectrometer. In, for instance, a near-IR instrument, the detector could be a 1-cm-square lead sulfide (PbS) chip with a carefully selected voltage across it. When light strikes the detector, a resistance change proportional to the intensity of the light occurs. The intensity of light that strikes the detector after imping-ing upon (and being either reflected from or transmitted through) a sample is recorded. This intensity is compared with the intensity of light from a reference, and the magic of Beer's law is performed.

The normal mode of operation is for light to first be submitted to a wave-length selector (grating, prism, acousto-optic tunable filter), and the individual wavelengths monitored for intensity. An alternative is to have an array of these detectors, much smaller in size, as the mode of intensity measurement. In this case, the light is split into its component wavelengths, and the full spectrum is measured simultaneously as a set number (254, 512, and so forth) of discrete signals.

In both cases, the sample is considered a point source and the spectral signal is representative of the entire sample. In recent years, light has been focused down to smaller and smaller points, and the spectrum of incredibly small portions of a sample (for example, the nucleus of a cell) scanned. With CCDs, entire snapshots are taken.

Instead of the reflected light from a sample being focused to a small point, the light creates an image on the surface of a larger surface. This surface resembles, somewhat, a diode array, but in a square, not linear, pattern (2). It consists of a series of picture elements or, more commonly named, "pixels." Each pixel registers the light intensity striking it as an electric current. Signals from the array of pixels are translated into a picture, often with the aid of a computer.

Individual signals are digitized and given a value as to their brightness or intensity (3). Values range from 0 for the lowest intensity (or black for an image) to 255 for the brightest (or white as a color).

Strangely, this 256 value just happens to equal a byte on a computer. This technology is commonly used in camcorders that nearly every family now owns, as well as in the Hubble. The biggest difference is that the Hubble has an array of 64,000 pixels!

This breakthrough now allows a researcher to scan the sample with light from a monochromator and "see" the spectral image of many points across the material simultaneously. If the spectral images are recorded (as with a camcorder) for each wavelength, a spectrum of each point may be generated. The strength of this technique for medical research is yet to be tapped, but promises to be phenomenal.

For the record, color images in the Hubble are obtained by using three filters: blue, red, and green. The incoming light is shone through the filters onto three sets of pixels. The intensities are then compared with the amounts of light from each filter and reconstructed into a color portrait. This approach could even be used with IR, UV, vis, or near-IR filters instead of a scanning device for industrial uses. Oh my, I'm giving free hints—bad form for a consultant!

HPLC DETECTORS

Chemists often forget that UV/vis detectors are little spectrometers (our friends). They tend to neglect the simplest care and feeding of the little guys. I'd just like to slip in a few hints and experiences with them.

I cannot count the number of times that lamps have been replaced when there was no need to. The wavering baseline, interpreted as the noise associated with a failing deuterium lamp, often has a different cause. One reason is a changeover of an aqueous-based solvent system to nonaqueous and back without proper cleaning of the system. Droplets of, say, methylene chloride adhere to the face of the lightpath window and "wiggle" as the water-based mobile phase flows past. Conversely, water droplets can do their cha-cha while nonaqueous music flows over them.

I always rinse the system with isopropyl alcohol between changeovers. The solvent is completely miscible in almost any mobile phase (barring heavy salts) and will eliminate "wiggle" problems. I really don't want to get into a heavy HPLC discussion, but just one or two more, if you will bear with me.

Most people know that attaching a back-pressure regulator after using the detector will prevent dissolved gases from coming out of solution. What some forget is how delicate the little thing is. If the back pressure is too strong, the pulses present in any piston-type instrument may cause the windows of the flow cell to expand and contract. This changing pathlength, resulting in a sinusoidal baseline, may also be interpreted as a failing lamp.

One last LC problem and I'll stop. Adjustable, single-wavelength detectors are susceptible to improper wavelengths. Many manufacturers do not provide instructions about how to check, much less repair, this problem. I recommend (for the UV region) a methanolic solution of caffeine placed into the detector using a syringe. Because virtually all detectors are single-beam, the absolute intensities of the peaks may be used to gauge the strength of the light. The ratio of "known" peaks may be used to determine wavelength accuracy. (Hint: Always work in the same direction; start from ~360 nm and go down to 220 nm. Any "play" in the dial will, at least, be repeatable then.)

REFERENCES

1. Stargazer Series, NASA Information Services, 1993 production date: *Pictures from Numbers.*"
2. J. D. Ingle, Jr., and S. R. Crouch, *Spectro-Chemical Analysis* (Prentice Hall, Englewood Cliffs, NJ, 1988).
3. H. V. Malmstadt, C. G. Enke, and S. R. Crouch, *Electronics and Instrumentation for Scientists*, (Benjamin/Cummings Publishing Co., Reading, MA, 1981).

CHAPTER II.G.5

HADAMARD TRANSFORM SPECTROSCOPY: TEACHING OLD MONOCHROMATORS NEW TRICKS

Robert M. Hammaker, William G. Fateley, and David C. Tilotta

This chapter features the work of William G. (Bill) Fateley, of the Department of Chemistry, Kansas State University, and David C. Tilotta, who is now with the chemistry department at Baylor University, Waco, Texas 76798. The Hadamard transform technique has the potential to become as important as Fourier transform spectroscopy. As I listened to Professor Fateley's talk on the Hadamard transform at the 1988 Chambersburg Diffuse Reflectance Conference, I was quite impressed with the technique's possibilities. The research for this article was performed while David Tilotta was a Phillips Petroleum Fellow at KSU. David's analytical interests include infrared, Raman, and magnetic resonance spectroscopies. Fateley, as many readers know, is the former Editor-in-Chief of Applied Spectroscopy, the journal of the Society of Applied Spectroscopy. His current research is directed towards the areas of Hadamard transform spectrometry, mobile FT-IR spectrometry of volatile organic molecules, vibrational correlation spectroscopy, and surface-enhanced Raman spectroscopy. One of the reasons I like Dr. Fateley is that he clearly enjoys teaching as well as researching. His talks are always crystal clear and exciting.

It is quite enjoyable to periodically examine old techniques to see if new technology can be applied. Methods that were once considered old-fashioned, difficult, or simply "impossible" are often successfully resurrected when brought into the light of modern advances. Everything from the dirigible to Fourier transform Raman spectroscopy has benefited by advances and improvements brought about by the technological war. Recently, this has also been the case in dispersive spectroscopy.

As teachers, we enjoy describing the dispersive spectrometer to our students: It's simple! We start by describing a "hot" body of radiation as a source for the spectrometer and the focusing optics to collect and focus the radiation onto the entrance slit. Next, we show the collimating optics, which collimate the source radiation from the slit onto a grating (or prism) for dispersion into short-wavelength intervals of radiation, termed *spectral resolution elements*. The dispersed radiation is collected and focused onto a plane where, to the astonishment of the student, an exit slit is used to collect less than 1% of this radiation. Finally, the selected small portion of radiation is allowed to impinge upon a detector. By scanning the grating, the spectrometer records the entire spectrum of the source.

Upon mental reflection, the process of collecting a small portion of the radiation with the exit slit seems wasteful, and scanning the grating to record the entire spectrum appears tedious. The obvious question to ask is: Is there a way to multiplex the dispersed radiation (measure more than one spectral resolution element of the radiation at a time) in a dispersive instrument that would increase the efficiency of the data acquisition and eliminate or minimize the movement of the grating (1)?

The literature reports that nearly four decades ago (in 1949), Marcel Golay described and built an instrument—the static, multislit spectrometer—that partially answered the question raised in the previous paragraph (2). This anachronistic instrument utilized two sets of multislit arrays, with one set of arrays located at the entrance slit plane and the other set of arrays located at the exit slit plane, to selectively modulate the dispersed radiation in a dispersive polychromator. The multislit arrays, referred to as *encoding masks*, were based on complementary (orthogonal) binary code sequences and allowed a larger cone of radiation to be collected than would normally be accepted by a conventional dispersive instrument. An individual spectral resolution element was selected for detection by selective modulation via a scanning Littrow mirror and the use of a dual-detector detection system. The remainder of the spectral resolution elements were left unmodulated as a tuned amplifier demodulated the signal. Although Golay's spectrometer was not a multiplexing spectrometer, his early work showed the utility of using multislit masking devices both to minimize the movement of the grating (in this case, a mirror was scanned rather than the dispersive device) as well as to increase the throughput of a conventional monochromator.

In 1969, Decker and Harwit (3) constructed a multiplexing dispersive instrument—the Hadamard transform spectrometer (HTS)—that used a multislit array at the exit focal plane of a conventional dispersive monochromator. While sharing some similarities with Golay's static multislit spectrometer, Decker and Harwit's HTS possessed a single encoding mask that was sequentially stepped in the exit focal plane. This linear multislit encoding mask allowed more than one spectral resolution element at a time to impinge upon the detector; hence, the encoding mask multiplexed the dispersed radiation, The encoding patterns used on the encoding mask (that is, the various sequences of opened and closed slits) were derived from a binary simplex code. The matrix representation of their simplex code was related to the matrices first developed by the French mathematician, Jacques Hadamard (4). In contrast with the complementary binary codes used by Golay, the Hadamard simplex matrices, typically abbreviated as S matrices, are the most accurate choice of matrices to be used in minimizing the error if a single detector is to be used in the instrument (5). These S matrices specified which spectral resolution elements were to be exposed to the detector (the open slits in the array) and which spectral resolution elements were to be blocked and not exposed to the detector (the closed slits in the array). The simplex matrices also specified the total number of separate measurements to be made for a given number of spectral elements to be measured.

The operation of the encoding mask in the HTS was straightforward: A pattern of open and closed slits was placed in the exit focal plane of the dispersed radiation, which allowed only certain spectral resolution elements to impinge upon the detector. Once the encoding mask was accurately positioned, a detector reading was made. The encoding mask was then physically moved to allow a different combination of spectral resolution elements to impinge upon the detector and a detector reading was again performed. This procedure was continued until all the mask patterns had been cycled through the dispersed radiation. The spectrum was obtained from the recorded detector measurements via inverse Hadamard transformation. Although Decker and Harwit's HTS did not have a throughput advantage, their instrument multiplexed the dispersed radiation onto a single detector. (In subsequent instruments, a Hadamard encoding mask was developed that could be placed in the entrance focal plane to increase the throughput of the spectrometer.) Decker and Harwit successfully applied Hadamard mathematics to optical spectrometry in order to increase the total amount of light impinging upon a detector. This increase resulted in improved signal-to-noise ratios in comparison with those found in dispersive instruments possessing a single exit slit.

This chapter is a report on the revival of Hadamard transform spectrometry as a useful tool for spectral analysis. Several technological advances have resulted in the development of new and relatively fast electro-optic switching arrays that can be fashioned into Hadamard transform encoding masks. These new electro-optic arrays can remove some of the previous disadvantages of the Hadamard transform technique and make the method much more useful. It is our goal to briefly review the fundamentals of Hadamard transform optics and to present the interested reader with the current status of Hadamard transform spectrometry as applied to optical spectral analysis.

MULTISLIT OPTICAL CODING WITH HADAMARD MATRICES: A TUTORIAL

For a binary multiplexing spectrometer possessing a single detector, encoding mask slit patterns fabricated from S matrices are the most accurate choice for minimizing the errors incurred in the detection of the intensity values of a set of spectral resolution elements (6). In other words, if one wants to accurately estimate the intensity values of a set of spectral resolution elements via a binary multiplexing method, the combinations of spectral resolution elements allowed to be multiplexed onto the detector should be based on Hadamard matrices in order to minimize the error (the error in this case is the variance, σ^2). But how does one apply Hadamard mathematics to the design of an optical instrument? Before proceeding to the instrumental considerations, let us briefly reexamine some of the nuances of Hadamard mathematics.

Hadamard transform optical coding can trace its roots back to a branch of statistics that deals with weighing designs. Yates (7) was the first to point out that by weighing a collection of objects in groups rather than individually, it would be possible to determine the weights of the individual objects more accurately. If it is assumed that the error in a given weighing arises only from the weighing device and is independent of the total amount of weight being measured, the value of the error incurred by a single weighing of a collection of objects together will be distributed over all the objects in that collection.

Although Yates was literally referring to weighing the objects on a single- or double-pan balance, the principle can be applied in a straightforward way to an optical spectrometer. Sloane and Harwit (8) have shown that one need only replace the weighing balance with an optical detector (strictly speaking, the detector noise should be independent of the total intensity impinging upon it) and replace the collection of objects to be weighed with dispersed spectral resolution elements. We will now examine how the Hadamard transform coding procedure is applied to optical spectrometry.

Before generalizing the Hadamard encodement procedure, we shall first examine a simple example. Let us use the selected spectral elements, a_i, with 10 cm^{-1} bandwidths each, as shown in Table I. These spectral resolution elements arise in a conventional dispersive polychromator via dispersion and are

Table I. Selected spectral elements, a_i, with 10 cm^{-1} bandwidths.

Spectral element, a_i	Region of spectrum	Intensity*
a_1	1000 cm^{-1}	$\left[\begin{array}{c} I_1 \right.$
a_2	1010 cm^{-1}	I_2
a_3	1020 cm^{-1}	I_3
a_4	1030 cm^{-1}	I_4
a_5	1040 cm^{-1}	I_5
a_6	1050 cm^{-1}	I_6
a_7	1060 cm^{-1}	$\left. I_7 \right]$

*Although irradiance would be the more proper word, we shall use the term intensity to denote what the detector measures.

focused onto the exit focal plane of the instrument. In this example, the assumption is that each of these spectral resolution elements has a bandwidth of 10 cm^{-1}, although grating instruments disperse radiation approximately linearly with respect to wavelength, not wavenumber. It should be noted here that the intensity values, I_i, have been arranged in a column matrix form. The reason for choosing seven resolution elements will be discussed shortly.

To perform the Hadamard encodement, certain sets of spectral resolution elements need to be selected to impinge upon the detector. This example, as well as the rest of the discussion in this chapter, will assume the use of a single-detector Hadamard transform spectrometer, although Hadamard transform spectrometers can be constructed that utilize two detectors. As mentioned previously, the choice of sets of spectral elements reaching the detector provides the multiplexing. Which spectral elements are to be selected to impinge upon the detector, and which combination of spectral elements should we select? The answers lie in the Hadamard matrices.

The Hadamard S matrices are representations of the encoding mask patterns. The encoding sequences of the slits in the mask—or the patterns of opened and closed slits for the measurements—are represented by the rows in the Hadamard S matrix. Now consider an example of a seven-"slot" mask. One particular Hadamard S matrix that could be used to determine the slit patterns is given in Table II, where i is a row marker number and j is a column marker number. A zero (0) means that the slit on which the spectral resolution element is focused is closed (that is, the slit does not pass the radiation through it onto the detector); a one (1) means it is open (the radiation is allowed to impinge upon the detector). The E_i (where $i = 1, \ldots, N = 7$) are the slit pattern numbers, referred to as "encodements," and the S_{ij} (where $i = j = 1, \ldots, N = 7$) are matrix elements that will be used to designate a specific element in the two-dimensional encoding S matrix.

The S matrix is used to select the spectral resolution elements that will impinge upon the detector by multiplying the intensity column matrix of the spectral resolution elements by the two-dimensional S matrix. This is accomplished via conventional matrix multiplication—that is, by multiplying each of

Table II. A Hadamard S-matrix that can be used for a seven "slot" mask.

Slit pattern number (or measurement number)		Slit (or "slot") number $S_{i,1}$	$S_{i,2}$	$S_{i,3}$	$S_{i,4}$	$S_{i,5}$	$S_{i,6}$	$S_{i,7}$
E_1 =	$S_{1,j}$	0	0	1	0	1	1	1
E_2 =	$S_{2,j}$	0	1	0	1	1	1	0
E_3 =	$S_{3,j}$	1	0	1	1	1	0	0
E_4 =	$S_{4,j}$	0	1	1	1	0	0	1
E_5 =	$S_{5,j}$	1	1	1	0	0	1	0
E_6 =	$S_{6,j}$	1	1	0	0	1	0	1
E_7 =	$S_{7,j}$	1	0	0	1	0	1	1

the spectral resolution elements by each of the entries in a given row of the S matrix (the ones and the zeros) and then summing the result. This procedure is performed for all the rows of the S matrix and generates the measurement column matrix, M_i. For example, the first measurement of the intensity impinging upon the detector, M_i, would be:

$$M_1 = 0 \times I_1 + 0 \times I_2 + 1 \times I_3 + 0 \times I_4 + 1 \times I_5 + 1 \times I_6 + 1 \times I_7 + e_1$$
$$= I_3 + I_5 + I_6 + I_7 + e_1 \tag{1}$$

where the I terms are from the intensity column matrix and e_1 represents an error in the measurement resulting from random noise.

For this example, the total number of measurements that need to be made would be equal to seven; the Hadamard S matrix in this case would generate seven independent equations. We began this example with seven spectral resolution elements because we used a two-dimensional 7×7 Hadamard S matrix. It should be noted that the error, e_i, incurred in M_i for this example is distributed over four spectral resolution elements, I_i, and over seven total measurements. Of course, in a conventional scanning dispersive instrument, the error, e_1, would be distributed over a single spectral resolution element, I_1.

For the general case in which a Hadamard S matrix of order (size) N is used, the encodement procedure can be mathematically represented as:

$$M_i = \sum_{j=1}^{N} S_{i,j} I_j + e_i \qquad \text{for } I = 1, \dots, N \tag{2}$$

where the M_i are the N detector measurements in column matrix form (realize that for each mask encodement pattern, one detector reading must be made), I_j are the intensity values of the spectral resolution elements in column matrix form, the $S_{i,j}$ are the elements in the two-dimensional S matrix of order N, and the e_i are the detection measurement errors in column matrix form. Note that the subscript indexes are consistent with our example and that the integer, i, must be cycled from 1 to N to obtain all M_i. For the general case, the signal-to-noise ratio improvement induced by the distribution of the errors in the measurements is approximately $[(N)^{1/2}]/2$.

In the previous example, a two-dimensional 7×7 Hadamard S matrix was used for the encoding procedure, but Hadamard S matrices of other dimensions also exist. Some S matrices of dimensions other than seven are shown in Fig. 1. Several general properties of the Hadamard S matrices govern their generation and how they are applied to optical encoding spectrometers. The following section will briefly outline some of the more important properties of S matrices.

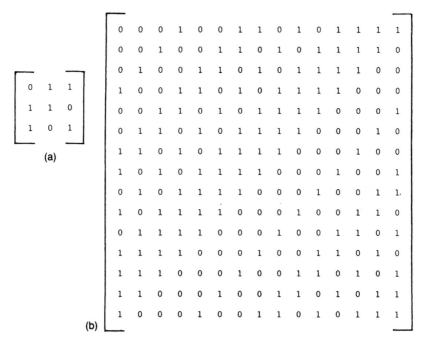

Figure 1. Two examples of **S**-matrices: (a) shows a 3 × 3 (N = 3) left-cyclic **S**-matrix; (b) shows a 15 × 15 (N = 15) left-cyclic **S**-matrix. It should be noted that the 15 × 15 **S**-matrix was not derived from the 3 × 3 **S**-matrix or the 7 × 7 **S**-matrix.

S MATRICES

Simply put, the S matrices dictate the mathematics of the HTS. The S matrices specify that the total number of detector measurements to be made must be equal to the number of unknown spectral resolution elements that are to be estimated. Furthermore, mainly for efficiency in computing the spectrum, the number of unknown spectral resolution elements should be equal to $2^k - 1$, where k can be any positive integer. In our example, the number of spectral elements was seven (with $k = 3$); therefore, seven independent measurements were necessary for describing the intensity of each spectral resolution element. The uniqueness of the Hadamard S matrices is derived from the pseudo-orthogonal properties that S matrices exhibit: A row of an S matrix correlated (vector multiplied) with itself will yield a different correlation constant than if that row had been correlated with any of the other rows.

For a given single measurement, the spectral resolution elements that are measured together are determined by a row in the Hadamard S matrix. For a given number of spectral resolution elements, N, each spectral resolution element is measured $(N + 1)/2$ times. This result is a consequence that for each encoding mask pattern, $(N + 1)/2$ slits are open and $(N - 1)/2$ slits are closed. There are three known methods of constructing S matrices, and each method will yield a different S matrix; although S matrices, by definition, must exhibit the aforementioned properties. As one can observe in Fig. 1, however, no obvious relationship exists between S matrices of different order N. The most useful type of S matrices are of the left-cyclic variety. Each successive row in the S matrix is obtained by shifting the column entries in the previous row one place to the left, with the overflow on the extreme left being transferred to the column on the far right. For our example with $N = 7$, a left-cyclic S matrix would be constructed as shown in Table III. We note that this encodement is cyclic because $S_{1,1} = S_{2,7}$; $S_{2,1} = S_{3,7}$; and so on.

After N detector measurements have been made, the data must be mathematically transformed to obtain the spectrum. The intensity, I_i, for each spectral element is what is desired. The relationship relating the measurements, M_i; the mask encodements, $S_{i,j}$; and the intensity of the spectral resolution elements, I_i, is given as:

$$M_i = \sum_{j=1}^{N} S_{i,j}I_j, \qquad \text{for } i = 1, \ldots, N \qquad (3)$$

where the terms in this equation have the same definition as before, and, for the moment, we will explicitly assume that the measurements are error free.

Recalling that the detector measurements are obtained by matrix multiplying the actual spectral resolution element intensities with the S matrix, it can be easily shown that the spectral resolution element intensities can be extracted from the detector measurements. This is accomplished by left multiplying the detector measurement column matrix with the inverse S matrix. The inverse S matrix is, by definition, the matrix that will yield the unit matrix when multiplied with the S matrix. The unit matrix is a two-dimensional, $N \times N$ matrix possessing ones along the main diagonal and zeros elsewhere. Therefore, the inverse Hadamard transform S matrix transforms the intensity, M_i, measured at the detector to yield the intensity, I_i, of each spectral element. However, recalling our assumption about the detection error, what the detector actually measures is a combination of the intensities of the spectral resolution elements,

Table III. The first three rows of a left-cyclic S-matrix.

	$S_{i,1}$	$S_{i,2}$	$S_{i,3}$	$S_{i,4}$	$S_{i,5}$	$S_{i,6}$	$S_{i,7}$
$S_{1,j}$	0	0	1	0	1	1	1
$S_{2,j}$	0	1	0	1	1	1	0
$S_{3,j}$	1	0	1	1	1	0	0

$S_{1,1}$ goes to $S_{2,7}$; all remaining elements move left.

$S_{2,1}$ goes to $S_{3,7}$; all remaining elements move left.

M_i, and a noise component, e_i. Because I_i can never be truly measured from an experimental standpoint, what is actually obtained in the transformation is I'_i, the estimated intensity of the spectral resolution elements. This can be generalized as:

$$I'_i = \sum_{j=1}^{N} (S_{i,j})^{-1} M_j, \qquad \text{for } I = 1, \ldots, N \tag{4}$$

$$I'_i = I_i + \sum_{j=1}^{N} (S_{i,j})^{-1} e_j, \qquad \text{for } I = 1, \ldots, N \tag{5}$$

where $(S_{i,j})^{-1}$ are the inverse S-matrix elements as defined previously.

The inversion of the data matrix, M_i, is performed with a fast Hadamard transform (9) rather than by directly multiplying it with the S matrix, although the latter can be performed.

HADAMARD TRANSFORM INSTRUMENTATION: THE NEXT GENERATION

So what's new? Up to this point in the discussion, we have simply summarized the method for obtaining spectra by the Hadamard transform technique. As stated in the introduction, the previous HTS methods used a movable mask for affecting the desired encoding. For each detector measurement that was to be made, a different mask pattern was placed in the exit focal plane of the dispersed radiation. The positioning of the previous multislit masking devices to afford the various encodings was performed mechanically by either rotating a disk mask (10) or piecewise stepping a linear mask (11). These moving masks, however, were subject to problems typically associated with moving parts; problems such as instabilities in field environments, jammed or misaligned parts, and the need for frequent realignments or precise tracking systems to follow their movements.

One possible solution to the problem of moving parts—and the solution currently being investigated in our laboratory—is to replace the previous moving encoding masks with a stationary encoding mask fabricated from an array of electro-optic switches. This electro-optic mask, which would be a one-dimensional "picket-fence" array mask, would be placed in the exit focal plane of the dispersive instrument to replace the moving encoding mask. The encoding elements of the electro-optic mask can be computer controlled either to turn some of the elements "on" and let the dispersed radiation be transmitted through, or to turn them "off" and let the dispersed radiation be blocked. The various mask encodements, as dictated by the Hadamard S matrix, can be programmed onto the mask directly. This removes the necessity of moving the mask. An excellent choice for an electro-optic masking array for use as a Hadamard encoding mask would be the readily available liquid crystal optical shutter array (LC-OSA). In several previous reports from this laboratory (12–14), we have demonstrated the feasibility of using LC-OSAs as HTS encoding masks in the spectral range of 300–2500 nm.

Briefly, LC-OSAs are either linear or two-dimensional arrays of liquid crystal cells fitted with a radiation polarizer before the array and a radiation analyzer behind the array. A schematic diagram of a liquid crystal electro-optic switch is shown in Fig. 2. Randomly polarized radiation, represented in Fig. 2 as an orthogonal combination of two linearly polarized radiation vectors (**E**) passing through the front polarizer (P$_1$), has the polarization component perpendicular to the transmission axis removed. Therefore, the randomly polarized radiation passing through the polarizer is linearly polarized. Linearly polarized radiation passing through an unenergized liquid crystal cell ($V = 0$ in Fig. 2a) will undergo a 90° "twist" (rotation) upon exit. If the polarization analyzer (P$_2$) behind the cell has its transmission axis oriented in a parallel manner with respect to the transmission axis of the radiation polarizer, the radiation exiting the cell will be absorbed by the polarization analyzer. Hence, that particular cell will be opaque to the radiation (Fig. 2a). If the liquid crystal cell is now energized (generally through the application of a low-potential ac electrical field, $V \neq 0$), the light

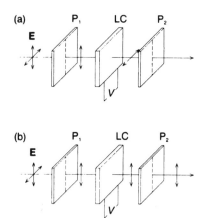

Figure 2. A schematic diagram of a liquid crystal electro-optic switch. Randomly polarized radiation (**E**) passing through the first polarizer (P$_1$) has the polarization component perpendicular to the polarizer transmission axis removed. In the upper diagram (a), linearly polarized light passing through an unenergized, $V \neq 0$, liquid crystal cell (LC) undergoes a 90° rotation upon exit. This rotated linearly polarized light vector, which is now polarized perpendicular to the transmission axis of the polarization analyzer (P$_2$), will be absorbed by the polarizer and, hence, will be opaque to the incident radiation. In the lower diagram (b), linearly polarized light passing through an energized, $V \neq 0$, liquid crystal cell (LC) will not undergo a 90° rotation upon exit. Therefore, the linearly polarized light will pass through the polarization analyzer, which is transmissive to the incident radiation.

passing through the cell will not undergo a 90° twist (Fig. 2b). Therefore, the linearly polarized radiation will be transmitted through the polarization analyzer. External computer control applies the appropriate ac electrical voltage and makes each slot either transparent or opaque to the selected radiation.

The details of the LC-OSA encoding mask and how it is utilized in the spectrometer are presented in another paper (12). A simplified block diagram of the optical bench is shown in Fig. 3. The optical dispersion system of this HTS is based on a modified, flat-field, coma-corrected Czerny–Turner double monochromator. The upper polychromator disperses the source radiation and focuses it in a flat plane. The 127-element encoding mask ($N = 127$), Ma in Figure 3, intercepts the dispersed radiation and Hadamard encodes it. Those spectral elements that are allowed to exit the encoding mask then need to impinge upon a single detector. Experimentally, this is accomplished by "dedispersing" the radiation. Dedispersion is performed by reversing the optical path of the radiation through the polychromator, which generates pseudowhite light of the source (the light is pseudowhite in the sense that some of the spectral resolution elements have been removed via the Hadamard encodement). In the instrument shown in Fig. 3, a second Czerny–Turner system is used to perform the dedispersion. Although an instrument with two complete polychromator sections is used in this case, methods exist that only make use of a single polychromator. The LC-OSA is placed in the focal plane of the spectral elements. The effective use of an LC-OSA encoding mask and an f/9 Czerny–Turner optics system allows one to focus all the chosen spectral elements onto one detector.

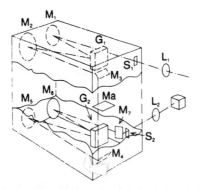

Figure 3. The optical design of the Hadamard transform spectrometer. The radiation follows the dot-dash line beginning with L_1. Lens L_1 focuses the light onto the entrance slit, S_1. The concave collimating mirror, M_1, collimates the radiation onto the plane grating, G_1. The dispersed radiation is collected and focused onto the encoding mask, Ma, by the concave camera mirror, M_2. Planar mirror M_3 deflects the radiation onto the mask. Radiation transmitted through the encoding mask is deflected by the planar mirror, M_4, onto the concave collimating mirror, M_5. Mirror M_5 directs the dispersed radiation onto the plane grating, G_2, for dedispersion. After being dedispersed, concave mirror M_6 collects the pseudo-white light and directs it toward the slit, S_2, and the focusing lens, L_2, by means of the planar mirror, M_7. The radiation is detected by the detector, D.

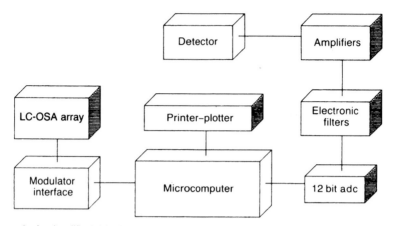

Figure 4. A simplified block diagram of the electronic components of the LC-OSA Hadamard transform spectrometer.

The electronic components of the HTS—as shown in Fig. 4—need only a brief comment. Because the LC-OSA Hadamard transform spectrometer has no moving parts, the electronic components are relatively simple (in comparison, for example, with those of a Fourier transform spectrometer). The detection and amplification system is based on dc measurements rather than modulated measurements as in the Fourier transform spectrometer. The preamplifier, the amplifier, and the electronic filters are readily available or easily constructed. The electronic filters lock the signal detection in a bandpass configuration, with the electronic bandpass being the switching frequency of the encoding mask. Because the dynamic range of the signal on the detection system in Hadamard transform spectrometry is generally less than one order of magnitude, the analog-to-digital converter need not be of high resolution. The LC-OSA control circuits are conventional, and details of their construction and operation are available (13).

APPLICATIONS

Before discussing the applications of LC-OSA Hadamard transform spectrometry, a word about signal-to-noise ratio improvements *via* quantum detectors is in order. As previously stated, the LC-SLM HTS operates in the spectral region of 300–2500 nm. Detectors that possess good sensitivities and high specific detectivities ($D*$) in the visible spectral region are generally quantum detectors. The fundamental postulate of the Hadamard transform technique, as with all multiplexing techniques, is that the noise output of the detector is independent of the total amount of radiation impinging upon it.

Quantum detectors do not obey this relationship; the noise output of a quantum detector can be proportional to the square root of the light intensity impinging upon it. Therefore, one would expect that by multiplexing spectral elements together, provided the system is not source-noise limited, one would defeat the purpose of multiplexing. One can, however, show that a multiplex advantage can exist in quantum detector LC-OSA Hadamard transform spectrometry (15) provided that the spectrum does not have a large number of strong emission lines. This condition can be true for flame atomic emission experiments, Raman spectrometry (provided the Rayleigh line is removed), as well as others.

The spectra in Fig. 5 show the green atomic emission region of a manganese hollow cathode lamp (because the hollow cathode lamp fill gas is neon, neon atomic emission lines are also present) and demonstrate the signal-to-noise ratio improvement when the LC-SLM is used as a Hadamard encoding mask rather than when it is used as a single scanning slit. (It should be noted that the spectra in Fig. 5 have been expanded by a factor of five and that the neon emission line at 585.2 nm is off scale.) The spectrum in Fig. 5a was obtained by sequentially opening successive elements of the encoding mask to simulate a single-slit scanning monochromator. When the encoding mask is programmed

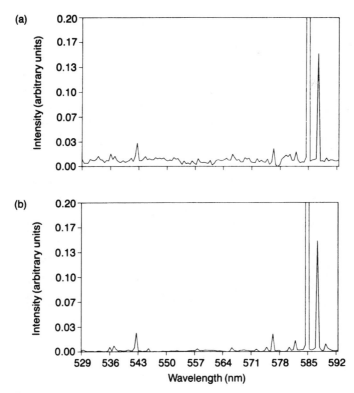

Figure 5. Spectra of the green region of a manganese lamp (with neon fill-gas). The upper spectrum (a) is a single slit mask scan; the lower spectrum (b) is a single Hadamard multiplex scan.

with the patterns given by the S matrix code, the result is the spectrum shown in Fig. 5b. Both spectra were acquired in 11 sec; however, the signal-to-noise ratio of the spectrum in Fig. 5b has been improved by a factor of 8.6 because of the multiplexing.

For visible absorption spectroscopy—where a large background intensity is present—the loss of the signal-to-noise ratio due to multiplexing can be recovered by the coaddition of multiple scans. In near-infrared absorption spectroscopy—where thermal detection is possible—the spectrometer benefits from the full multiplex advantage.

The applications of LC-SLM Hadamard transform spectroscopy are as broad as the applications of any focal-plane sensor array. Any spectroscopic application that involves either high resolution and a small spectral window or low resolution and a large spectral window can be coupled to the LC-SLM HTS. These applications can include Raman spectroscopy, atomic emission spectroscopy, industrial and clinical analysis, near-infrared absorption spectroscopy, photothermal beam deflection spectroscopy, and many others. The Hadamard transform technique can yield valuable information where other methods encounter difficulty.

Near-infrared analysis is an example of a potential use of HTS. Figure 6 shows a near-infrared transmittance spectrum of neat dibutyl phthalate in the spectral region 1145–1227 nm that is the result of eight co-added scans. This spectrum is of the C–H stretching second overtones of the methyl groups, and it has not been corrected for the 1-cm glass sample cell absorptions. The radiation source was a 1-W tungsten lamp, and the LC-SLM HTS detector was an indium–gallium arsenide (IGA) photodiode operated at zero bias. Although this LC-SLM HTS possesses a small spectral observation window for near-infrared analysis, the spectral resolution is 1.4 nm.

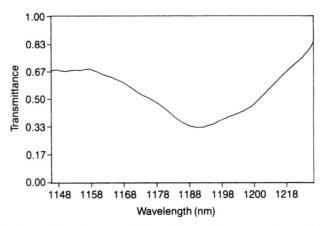

Figure 6. Hadamard transform near-infrared transmittance spectrum of neat dibutyl phthalate.

Another example of a potential use for an LC-SLM HTS is in Raman spectroscopy. In order to avoid the multiplex disadvantage resulting from the noise in the Rayleigh scattered radiation, the Rayleigh scattered radiation must not be multiplexed together with the Raman scattered radiation. In FTS, the Rayleigh line is removed with optical filters. Because the radiation is spatially presented on a focal plane in LC-SLM HTS, a convenient intraspectrometer knife-edge optical rejection filter can be formed by simply positioning the encoder mask in such a manner so as to selectively pass through it only Raman scattered radiation. Figure 7 is a Hadamard transform Raman spectrum of benzene that was collected in 11 sec with a room-temperature silicon photodiode detector, and it possesses a root-mean-square signal-to-noise ratio that is approximately equal to 100. The Raman exciting laser was an Ar^+ laser lasing at 514.5 nm with 300 mW of sample power that was utilized in a 90° Raman scattering geometry. This spectrum, which resulted from a single scan, clearly demonstrates the advantages of a multiplexing dispersive spectrometer or, in this case, of LC-SLM HTS.

As an aside for the interested reader, the benzene Raman's spectrum (Fig. 7) encodegram is also given (Fig. 8). The encodegram, which is a discrete function, is plotted with linear interpolation for ease of viewing. It is a graphical presentation of the detection measurement matrix, M_i, and it corresponds to a plot of detector intensity measurement versus the S matrix pattern number (row number). For the Hadamard transform Raman spectrum of benzene, a 127-element encoding mask was used: hence, 127 total detector measurements were taken. When the encodegram in Fig. 8 is transformed via the fast Hadamard transform, the spectrum in Fig. 7 results.

For any LC-SLM HTS experiment, a spectral resolution element impinges on the same encoding element for a given measurement. Therefore, as in Fourier transform spectrometry, spectral subtraction is possible. In Fig. 9, we apply

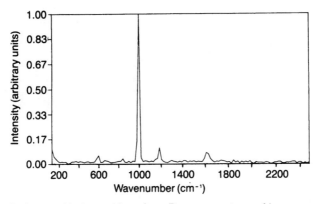

Figure 7. A single scan Hadamard transform Raman spectrum of benzene with a silicon photodiode detector.

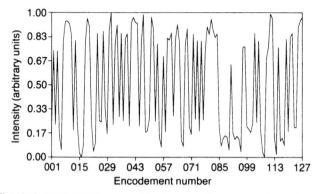

Figure 8. The Hadamard encodegram of the benzene Raman spectrum shown in Figure 7.

these subtraction techniques to Raman spectroscopy. For the spectrum in Fig. 9a, a 70 : 30 toluene : benzene (v/v) mixture was illuminated with 300 mW of the 514.5-nm line from the Ar$^+$ laser (with 90° sampling geometry) and the resulting Raman spectrum was recorded. A room-temperature silicon photodiode detector was used. When the toluene background Raman spectrum was subtracted from this mixture, shown in Fig. 9b, and was recorded under the same conditions, the result was a very identifiable benzene spectrum, as shown in Fig. 9d. For comparison, the Hadamard transform spectrum of pure benzene

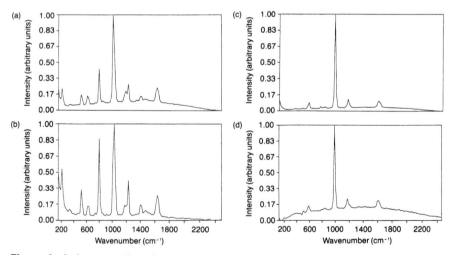

Figure 9. A demonstration of spectral subtraction: (a) is a Hadamard transform Raman spectrum of a 70% toluene and 30% benzene mixture; (b) is a Hadamard transform Raman spectrum of pure toluene; (c) is a Hadamard transform. Raman spectrum of pure benzene; (d) is the difference spectrum of benzene obtained by subtracting spectrum (b) from spectrum (a).

is shown in Fig. 9c. The complete experimental details of LC-SLM Hadamard transform visible Raman spectroscopy are available (14).

CONCLUSIONS

We have explored the basic principles and utilities of LC-SLM Hadamard transform spectroscopy in this chapter, but considering that entire books have been written on HTS, it is impossible to be comprehensive in these few pages. However, the LC-SLM Hadamard transform spectrometer can be a simple instrument to implement and offers advantages in a number of spectroscopic applications. In summary, we feel that Hadamard transform spectrometry defies the adage, "You can't teach an old monochromator new tricks."

ACKNOWLEDGMENTS

Support for this work was provided by the Phillips Petroleum Company in the form of a fellowship for D. C. Tilotta, and by the U.S. Department of Energy (DOE), Chemical Science Division, Grant No. DE-FG02–85ER13347. Any opinions, findings, conclusions, or recommendations expressed herein are those of the authors and do not necessarily reflect the view of the DOE.

REFERENCES

1. D. N. Waters, *Appl. Spectrosc.* **41,** 708–9 (1987).
2. M. J. E. Golay, *J. Opt. Soc. Am.* **39,** 437–44 (1949).
3. J. A. Decker. Jr., and M. Harwit, *Appl. Opt.* **8,** 2552–54 (1969).
4. J. Hadamard, *Bull. Des Sciences Math.* **17,** 240 (1893).
5. M. Harwit and N. J. A. Sloane, *Hadamard Transform Optics* (Academic Press, New York, 1979), pp. 44–95.
6. H. Hotelling, *Ann. Math. Stat.* **15,** 297–306 (1944).
7. F. Yates, *J. Toy. Stat. Coc. Supp.* **2,** 181 (1935).
8. N. J. A. Sloane and M. Harwit, *Appl. Opt.* **15,** 107–14 (1976).
9. E. D. Nelson and M. L. Fredman, *J. Opt. Soc. Am.* **60,** 1664 (1970).
10. P. Hansen and J. Strong, *Appl. Opt.* **11,** 502 (1972).
11. R. D. Swift, R. B Watson, J. A. Decker, Jr., R. Paganetti, and M. Harwit, *Appl. Opt.* **15,** 1596 (1972).
12. D. C. Tilotta, R. M. Hammaker, and W. G. Fateley, *Appl. Spectrosc.* **41,** 727–34 (1987).
13. D. C. Tilotta, R. M. Hammaker, and W. G. Fateley, *Appl. Opt.* **26,** 4285–94 (1987).
14. D. C. Tilotta, R. D. Freeman, and W. G. Fateley, *Appl. Spectrosc.* **41,** 1287–94 (1987).

CHAPTER II.G.6

LASERS: THE QUANTUM MONKEY WRENCH

Scott Wohlstein

I am blessed with the ability to find people who know more about things than I do. Before you say, "all you have to do is open the phone book," let me amplify that statement. Indeed, light amplification is the subject of this chapter. When l first asked Scott Wohlstein about lasers, I didn't know exactly what to ask about. Scott gave me a bit more than I bargained for. Scott is a consultant in the application of lasers and electro-optics and President of SD Laboratories Inc., P.O. Box 230, Convent Station, New Jersey 07961. In addition to a nice pithy explanation of laser types and some uses, he gave me a rather comprehensive listing of the different types of lasers, where they lase, their output, and some uses for each. Needless to say, I was impressed. The chapter can be read for purposes of instruction or used as a reference for future acquisitions.

During the past decade, the laser has been finding more applications and integrations than its inventors ever thought possible. The following is a discussion of many of the types of lasers available, along with characteristics and typical applications of each.

Reduction in overall size, increased resonator efficiency, and available emission wavelengths have contributed to the laser's emergence as a viable, low-cost tool for smaller businesses and research facilities. This chapter is a compilation of most of the available laser system types and a thumbnail description of each. Table I presents this information in a format for convenient reference. I shall start with a quick refresher in basic laser science.

A REFRESHER IN LASER SCIENCE

The word *laser* has seen a fair share of abuse since its conception. Technically the word is an acronym: Light Amplification by the Stimulated Emission of

Table 1. Laser systems that are currently available (sources 1–4)

Type	Name	Wavelength	Output[a]
Excimer	Argon fluoride (ArF)	193 nm	Pulsed (5–25 ns, ≤500 mJ/50 W)
	Krypton fluoride (KrF)	248 nm	Pulsed (2–50 ns, ≤1 J/100 W)
	Xenon cloride (XeCl)	308 nm	Pulsed (1–80 ns, ≤5 J/150 W)
	Xenon fluoride (XeF)	351 nm	Pulsed (1–30 ns, ≤500 mJj/30 W)
	Krypton chloride (KrCl)	222 nm	CW or pulsed (1–50 ns, ≤4 mJ/200 mW)
Dye	Excimer-, nitrogen-, Nd:YAG-, and Nd:glass-pumped	300–1000 nm (tunable)	Pulsed (3–50 ns, ≤200 mJ/15 W)
	Flash-lamp-pumped	340–940 nm (tunable)	CW or pulsed (0.2–4 μs, ≤50 J/50 W)
	Ion-Pumped	400–1000 nm (tunable)	CW (≤2 W) or pulsed (1–10 ps, ≤50 mJ)
Single gas	Neon (Ne)	1.152, 3.39 μm	CW (≤100 mW) or pulsed (0.1–10 ns, ≤10 mJ)
	Nitrogen (N)	337 nm	CW (≤330 mW) or pulsed (0.3–10 ns, ≤20 mJ)
Ion gas	Argon (Ar)	~351–528 nm (tunable); main lines 488, 514.5 nm	CW and can be mode-locked, ≤10 W
	Krypton (Kr)	~350–800 nm (tunable); main line 647.1 nm	CW and can be mode-locked, ≤20 W
	Argon/krypton (Ar/Kr)	~450–670 nm (tunable)	CW (≤10 W)
	Xenon (Xe)	540 nm	CW (≤10 W) or pulsed (0.1–20 ms, ≤10 mJ)
Helium based	Helium cadmium (HeCd)	442 or 325 nm	CW, ≤50 mW (442 nm) and ≤10 mW (325 nm)
	Helium neon (HeNe)	~540 nm to 3.4 μm by tuning or filtering; main line 632.8 nm	CW, ≤100 mW (632.8 nm)

Medium	Overall Efficiency	Applications
Gas mixing of Ar and F	Up to 1%	Spectroscopy, photochemistry
Gas mixing of Kr and F	Up to 2%	Spectroscopy, photochemistry
Gas mixing of Xe and Cl	Up to 2.5%	Spectroscopy, photochemistry, dye purification and pumping
Gas mixing Xe and F	Up to 2%	Spectroscopy, photochemistry, dye purification and pumping
Gas mixing of Kr and Cl	Up to 2%	Spectroscopy, photochemistry, range-finding
Dye in liquid solvent	5 to 25% (conversion of pump light)	Spectroscopy, photochemistry, fluorescence studies, entertainment/light shows
Dye in liquid solvent	0.2 to 1%	Spectroscopy, photochemistry, fluorescence studies, entertainment/light shows
Dye in liquid solvent	10 to 20% (conversion of pump light)	Spectroscopy, photochemistry, fluorescence studies, entertainment/light shows
Ne gas in sealed tube	Up to 0.5%	Spectroscopy, nonlinear spectroscopy, dye pumping, photochemistry, fluorescence studies, Raman-scattering studies
N gas in sealed tube	Up to 0.1%	Spectroscopy, nonlinear spectroscopy, dye pumping, photochemistry, fluorescence studies, Raman-scattering studies
Ar gas in sealed tube	0.01 to 0.1%	Recording, spectroscopy, photochemistry, dye pumping, reprographics, medicine, fluorescence studies, entertainment/ light shows, displays
Kr gas in sealed tube	0.001 to 0.05%	Recording, spectroscopy, photochemistry, dye pumping, reprographics, medicine, fluorescence studies, entertainment/ light shows, displays
Ar/Kr gas mixture in sealed tube	0.005 to 0.02%	Spectroscopy, photochemistry, reprographics, fluorescence studies, entertainment/light shows, displays
Xe gas mixture in sealed tube	0.01 to 0.1%	Spectroscopy, stroboscopy, solid-state pumping, photochemistry, reprographics, fluorescence studies, entertainment/light shows, displays
Ionized Cd vapor with He buffer gas in sealed tube	0.01 to 0.1%	Recording, spectroscopy, photochemistry, dye pumping, reprographics, medicine, fluorescence studies, entertainment/light shows, displays
Ionized Ne vapor with He buffer gas in sealed tube	0.01 to 0.1%	Construction, holography, measurement and control, recording, spectroscopy, photo-chemistry, dye pumping, reprographics, medicine, fluorescence studies, entertainment/light shows, displays

Table 1. (*Continued*)

Type	Name	Wavelength	Output[a]
Solid state	Ruby (Al_3O_2/Cr_3O_2)	694.3 nm	Pulsed (10 ns to 10 ms, ≤100 J/100 W)
	Nd:YAG	1.06 μm	CW (≤600 W) or pulsed (0.1–1.0 ms, ≤200 J)
	Nd:glass	1.06 μm	Pulsed (0.1–1.0 ms, ≤100 J/100 W)
	Erbium:YLF	850 nm, 1.73 μm	Pulsed (5 ns to 100 ms, ≤100 J/100 W)
	Erbium:glass	1.54 μm	CW, ≤5 mW
	Holmium:YLF	2.06 μm	CW, ≤100 mW
	Neodymium- and/or chromiumdoped: gallium scandium gadolinium garnet (GSGG)	745–835 nm	CW (≤500 W) or pulsed (0.1–1.0 ns, ≤150 J)
	Chromium-doped: potassium zinc fluoride (Crd:KZF)-perkovskite	780–850 nm	Pulsed (10 ns to 100 ms, ≤175 J/200 W)
Crystalline	F center	1.43, 1.58, 2.3–3.5 μm	CW (≤100 mW) or pulsed (1–10 ps, ≤5 mJ)
Semi-conductor diode	GaAlAs	750–905 nm (composition dependent)	CW (≤40 mW) or pulsed (≤5 mJ)
	InGaAsP	1100–1600 nm (composition dependent)	CW (≤40 mW) or pulsed (≤5 mJ)
	PbNaCl	3–30 μm (continuously tunable)	CW (≤100 mW)
Chemical	Hydrogen fluoride (HF)	2.6–3 μm	CW (≤150 W) or pulsed (50–200 ns, ≤600 mJ)
	Deuterium fluoride (DF)	3.6–4 μm	CW (≤100 W) or pulsed (50–200 ns, ≤600 mJ)

Medium	Overall Efficiency	Applications
Synthetic Al_3O_2 crystal dope with Cr_3O_2	0.1 to 0.5%	Holography, materials processing
Synthetic yttrium–alumina crystal doped with Nd	0.1 to 3%	Holography, materials processing, range-finding, medicine, dye pumping,
Synthetic glass crystal doped with Nd	1 to 5%	Materials processing, range-finding
Synthetic YLF crystal doped with Er	0.5 to 3%	Materials processing, range-finding, medicine, dye pumping
Synthetic glass crystal doped with Er	0.1 to 2%	Materials processing, range-finding, medicine, ophthalmology
Synthetic holmium crystal coped with YLF	0.1 to 2%	Materials processing, range-finding, medicine, ophthalmology
Synthetic GSG crystal doped with Nd and/or Cr	0.1 to 5%	Material processing, range-finding, medicine, dye pumping
Synthetic KZF crystal doped with Cr	0.1 to 0.8%	Materials processing, range-finding, communications
F-centers (impurities) in alkali-halide crystals	Up to 10% of conversion light	Spectroscopy, nonlinear spectroscopy, dye pumping, photochemistry, fluorescence studies, Raman-scattering studies
p–n junction in semi-conducting GaAs or GaAlAs	1 to 20%	Printing, recording, reading, optical communications, gauging
p–n junction in semi-conducting InGaAsP	1 to 20%	Communications
Several levels of diode arrays with PbNaCl as p–n "hole" material	Up to 30%	Communications, printing
Low-pressure gas containing chemically produced hydrogen fluoride	0.1 to 1%	Atmospheric research, military
Low-pressure gas containing chemically produced deuterium fluoride	0.1 to 1%	Atmospheric research, military

Table 1. (Continued)

Type	Name	Wavelength	Output[a]
Vapor	Copper	510, 578 nm	Pulsed (0.1–20 ns, ≤20 mJ)
	Gold	628 nm	Pulsed (0.1–20 ns, ≤20 mJ/400 mW)
	Iodine	1.3 μm	CW (≤100 W)
	Organic	30–1000 μm (continuously tunable)	CW (≤50 W)
Carbon/ oxygen based	Sealed	10.6 μm	CW (≤100 W)
	Pulsed-TEA	10.6 μm	Pulsed (50–100 ns, ≤150 J)
	Waveguide	10.6 μm	CW (≤50 W) or pulsed (5–50 ns, ≤5 mJ)

Radiation. As an acronym, the word should appear in all capital letters; however most journals, including this one, reject this usage. "Lazer" (ugh!), or any other convoluted or otherwise mutated version of the word, in my book, is unacceptable. A laser is a device that, by the process of "lasing," produces a very coherent, almost monochromatic, beam of highly directional light.

All typical laser devices consist of four main components:

- an excitation source (dc, ac, rf, flash lamp, or other type of laser);
- a lasing medium (gas, solid, liquid, or semiconductor);
- a geometrically designed resonator (for proper excitation of the medium);
- a feedback mechanism (typically mirrors).

The basic lasing process can be explained as follows. The excitation source applies its energy, which is called "pumping." A dc, ac, or rf excitation source is applied if the medium is gas or semiconductor, and a flash lamp or "seed" laser is applied if the medium is liquid (dye) or solid state. Once excitation has occurred, the medium absorbs the energy and simultaneously raises the outer-shell electrons to their higher states. When there are more electrons in the upper state than in the ground state, a population inversion is created. As the electrons "relax" back to ground state, photons are released that are characteristic of the fluorescing element in the medium.

These photons, by virtue of the design of the resonator assembly, travel along the optical axis, where they encounter one side of the feedback mechanism, or

Medium	Overall Efficiency	Applications
Prevaporized copper in a sealed tube	Up to 0.5%	Dye pumping, forensics, medicine
Prevaporized gold in a sealed tube	Up to 0.5%	Dye pumping, forensics, medicine, spectroscopy
Prevaporized iodine in a sealed tube	Up to 5%	Dye pumping, forensics, medicine, spectroscopy
Prevaporized organic gases in a sealed tube	Up to 10%	Dye pumping, forensics, medicine, spectroscopy
Mixture of CO_2 and other gases in sealed tube	Up to 15%	Low-power materials processing, medical, military
Mixture of CO_2 and other gases at near-atmospheric pressures	Up to 10%	Materials processing, medical, military, laser radar (LIDAR), photochemistry, remote sensing
Mixture of CO_2 and other gases in waveguide tube	Up to 5%	Materials processing, medical, military, laser radar (LIDAR), photochemistry, remote sensing

[a]Output ratings are given as the typical power output.

more specifically, a mirror. After reflecting from the first mirror (M1), the photons travel back along the optical axis, through the medium, where they excite other electrons to emit photons that are identical in every aspect. The photons emerge from the medium, and, after they are reflected from the second mirror (M2), the cycle is repeated. Because one of the mirrors, either M1 or M2, is 100% reflective and the other is ~92% to 99% reflective, the feedback of stimulating photons continues until the sustained energy level is higher than the losses. A truly coherent (both spatially and temporally) and nearly monochromatic beam (good news for spectroscopy fans) emerges through the partially reflective mirror.

MODES OF OPERATION

Lasers can have two main modes of operation. Continuous-wave (CW) laser devices have longer upper-state conditions than lower-state conditions: The beam appears to be produced continuously. Pulsed laser devices have longer lower-state (zero) conditions than upper—or in electronic terms "duty cycles." Pulsed systems can be either "single shot" or repetitively pulsed.

Both the pulsed and CW modes of operation can be enhanced by techniques such as Q switching, mode locking, and superpulsing.

Q switching is the introduction of "*Q*" loss in the resonator (typically between the medium and the 100% mirror). This *Q*-loss device can comprise several different devices and techniques. Mechanical techniques include the use of electricity to drive a mechanical device that disturbs the Q of the system. The timing of the disturbance is directly related to the amount of energy dumped out in the form of a powerful beam. Devices can include rotating mirrors and piezoelectric transducer (PZT) operated shutters.

Electro-optical (EO) techniques include the use of electricity to align the optical path in nonlinear crystals to divert or block the coherent radiation until a threshold limit is reached. The EO device aligns the path to the optical axis and "dumps" the energy in the form of a high-power beam.

Acousto-optical (AO) techniques include the use of acoustic standing waves in a nonlinear crystal to divert or block the coherent radiation until a threshold limit is reached. The AO device aligns the path to the optical axis and "dumps" the energy in a powerful beam. Chemical techniques include the use of saturable dyes to absorb the coherent radiation until a threshold limit is reached and then to "dump" the energy in the form of a powerful beam. This use constitutes one of the truly passive methods of Q switching.

Mode locking is the introduction of either constructive or destructive interference in the resonator, typically between the medium and the 100% mirror. Standard mode locking can be accomplished either by a specially designed resonator (superpulse) or by an intracavity device. The basis for mode locking is the compression of secondary or spurious emission/longitudinal modes to create more enhanced and oscillating modes and adopt equal frequency separation, while maintaining constant amplitudes and phase relationships. Mode locking is one of the more important operational techniques that spectroscopists can apply to a laser.

Materials that lase are being found in increasing abundance and exoticness, and so are the systems that facilitate the lasing. This increased specialization of available laser types is a mixed blessing. The field that was once dominated by the HeNe, ruby ($Al2O3/Cr3O2$), and occasionally ion (typically argon) lasers is now populated by exotic excimer, semiconductor, chemical, and tunable solid-state (for example, Ti : sapphire) systems. Users must be fairly knowledgeable about the requirements of an application in order to make the right system choice.

LASER SYSTEMS

Table I presents the major classes of lasers that are currently available.

Excimer systems. Excimer systems operate by breaking down a stable, noble gas with electrical or rf discharge and combining it with a reactive gas to create lasing.

Dye systems. Dye systems operate by irradiating a flowing stream of fluorescing dye with a laser or flash lamp to create lasing. Spectral tuning is accomplished by grating or prism. Because the standard configuration of a dye laser is fairly uniform throughout the industry, this section in Table I is categorized by pumping schemes.

Single-gas systems. Single-gas systems operate by exciting a single gas in a sealed tube to create lasing. Neon and nitrogen systems have been known to demonstrate superradiant characteristics. Superradiance is a condition in which the metastable lifetime of the related electrons in their higher-energy state is increased until the net losses of the system are relatively insignificant. This condition requires very little pumping energy to sustain the lasing process and, in most cases, eliminates the need for feedback (mirrors) in the resonator assembly.

Ion–gas systems. Ion–type gas systems operate by directly ionizing a noble gas in a sealed tube to create lasing. Ion systems are considered the workhorse lasers of the spectroscopy field, providing anything from reliable laser light sources to pumping dye systems.

Helium-based systems. Helium-based gas systems operate by directly ionizing a gas, which is buffered by helium (to increase energy conversion, coupling efficiency, and lifetime), in a sealed tube to create lasing. The HeNe laser is the most widely used laser ever devised because of its extremely rugged and time-proven design.

Solid-state systems. Solid-state systems operate by either laser or flash-lamp irradiation of a fluorescing element suspended in a crystal matrix to create lasing.

Crystalline systems. Crystalline systems operate by either laser or flash-lamp irradiation of a fluorescing element (for example, color F centers), typically an impurity suspended in a crystal matrix, to create lasing.

Semiconductor-diode systems. Semiconductor-diode systems operate by biasing a $p–n$ junction in a semiconducting material matrix to create lasing. The actual resonator and feedback mechanisms are formed out of the subsIrate material. These devices are similar in appearance to a typical transistor.

Chemical systems. Chemical systems operate by mixing two or more gases that are highly reactive to one another. This reaction, when given all the other necessary requirements of laser devices, will start to lase. The additional electrical discharge sustains and controls the reaction.

Vapor systems. Vapor systems excite prevaporized organic or inorganic gases to create lasing.

Carbon/oxygen-based systems. Carbon/c˙.ygen-based systems ionize gas (which is buffered by carbon to increase energy conversion, coupling efficiency, and lifetime) in a sealed, flowing, or "shot-filled" resonator system to create lasing. The rugged CO_2 laser is the most widely used industrial laser, followed closely by the Nd : YAG.

This listing contains most or all of the categories of lasers, but does not and cannot include all the actual laser devices developed into systems to date. Such a list could very well fill an entire textbook. Again, it should be stated that the above listing is for approximate system descriptions only. Readers should contact a manufacturer if they desire more detailed information (2–4).

REFERENCES

1. Heather W. Lafferty, "Lasers: Are They More Than Just Fancy Light Bulbs?" *Spectroscopy* **1**(6), 24 (1986).
2. "1989 Buying Guide," *Lasers & Optronics* **148** (1989).
3. "1989 LFBG—The Laser Focus World Buyer's Guide," *Laser Focus World* **48** (1989).
4. "1989 Spectroscopy Buyers' Guide," *Spectroscopy* **4**(5), 36–37 (1989).

CHAPTER II.G.7

DETECTORS IN INFRARED SPECTROSCOPY

I have been striving to keep the chapters relatively basic. I believe there is a need for primers on techniques, not just applications thereof. Recently, I got into a discussion (anytime shots are not exchanged, I refer to an encounter as a "discussion") about detectors used in vibrational spectroscopy. Based on that discussion, I can see a need for basic tutorials on the different parts of spectrometers, too. Because I've been working with detectors lately, I am choosing them as a topic for this chapter.

Because infrared radiation (4000–400 cm^{-1} or 2500–25,000 nm) is only heat, you might ask, "How hard can detection be?" That depends on the application you wish to perform. Is the work online and rapid, lab-bench and slow, microscopic in size, or a huge drum of material. Is the weather a factor? Do you want a rough estimate between 40% and 50%, or are you measuring part-per-million levels? Are you analyzing for a carbonyl group, or trying to judge the climate of the carbonyl group with 0.001 cm^{-1} precision?

If the answer to any of these questions is "Huh?" then pass this book to a friend. If you begin to get my point that there is no single IR system that can do everything, then pass "GO" and submit an expense report for $200 (good luck!) Let's look at some of the detectors that have evolved as instruments have become more sophisticated. For this information, I referred to several recent instrumental texts (1–3) and my sometimes questionable memory.

In essence, the two major types of IR detector are those that (1) measure some physical property of a material as it is heated (thermal effects) and (2) use quantum effects of impinging photons causing a change in the electrical properties of a semiconductor (photon detectors.)

THERMAL DETECTORS

Thermal detectors contain a sensitive element that is as small as possible and is universally black for maximum absorbance. The element radiates heat and returns to ambient when the source no longer shines upon it. This return to baseline follows a decay pattern that determines the speed of response. For this reason, detectors may be cooled for speed and sensitivity.

The response to heating within a thermal detector may cause an expansion of a solid, a gas, or a liquid (Golay detector); a change in resistance of a thermistor; an induced voltage at a thermocouple or thermopile junction; or electric polarization (pyrolytic effect).

The versatility of thermal detectors—that is, their ability to work at room temperature and over a large range of wavelengths—is offset by their slow response time and low relative sensitivity. The slow response leads to a limit to which the signal may be modulated (pulsed or chopped). With the older instruments, a "high-resolution" scan of (as long as) 22 min was a standard. This was a result of the time required for the heating and cooling cycle of the detector.

Golay detector. A Golay detector is essentially a gas thermometer in which xenon is placed into a small cylinder containing a blackened membrane. At one end of the detector is an infrared transparent window, while at the other end is a flexible diaphragm silvered on the outside. The silvered surface reflects a light beam off its surface onto a photodiode. Heating the blackened surface causes the gas to expand, distorting the silvered face. This is, in turn, measured by the amount of light intensity diminished as it strikes the diode. Its strength is in the far-Infrared (<200 cm^{-1}).

Thermocouple. Made from two dissimilar metals such as bismuth and antimony, a thermocouple produces a small voltage at the junction proportional to its temperature. The surface where the IR beam actually strikes is coated with gold oxide or bismuth black to enhance detection.

Thermopile. This consists of up to six thermocouples in series. Half the thermocouples are "hot," or sensing, while the other half are "cold." The assembly is mounted in a vacuum to minimize conductive heat losses. Response times are in the 30-ms range.

Thermistor. *Thermal resistors* are sintered oxides of manganese, cobalt, and nickel. These have a high-temperature coefficient or resistance on the order of 4%/degree. Two 10-μm flakes of the material are placed in the detector or pile. One is black and is "active," while the other is shielded and acts as a reference or compensating detector against changes in ambient temperature. Connected in a bridge circuit, a steady bias voltage is maintained across the bridge. The time constants are in the low milliseconds, but in general, speed is balanced against sensitivity.

Pyroelectric detectors. These contain a noncentrosymmetric crystal. When kept below its Curie temperature, it has an internal electric field along its polar axis (resulting from the alignment of electric dipole moments). Electrodes are attached to the crystal normal to this axis. These are connected to an external circuit so that a free charge is brought to the electrodes to balance the polarization charge, generating a current in the external circuit.

The detector is a thin plate of this material placed between two electrodes, forming a capacitor. Various materials are used for pyrolytic detectors: triglycine sulfate, deuterated triglycine sulfate, $LiTaO_3$, $LiNbO_3$, and assorted polymers. The glycine salts tend to be hygroscopic and have low Curie points. The lithium salts have higher Curie points and, with a 100 Mohm load resistor, have response times in the order of 1 ms with a responsivity of 100. With this speed and responsivity, pyrolytic detectors are quite suited to Fourier transform–infrared (FT-IR) instruments.

PHOTON DETECTORS

Photon detectors are quite sensitive and depend on the quantum interaction between the incident photons and a semiconductor. The resulting production of electrons and holes gives rise to an internal photoelectric effect. An energetic photon strikes an electron in the detector, raising it from a nonconducting to a conducting state. This energy limitation causes the detectors to fall off toward the far-IR region.

Photoconductive detector. This is a homogeneous semiconductor chip in which the presence of electrons in the conduction band lowers the resistance (for example, lead sulfide or InGaAs). When electrons are raised from valence bands to conduction bands, hole–electron pairs are created. This may also be accomplished by doping, called *extrinsic excitation*. Either way, the change is monitored through a bias current or voltage.

Photovoltaic detectors. These generate small voltages when exposed to light. Single crystal detectors, for example, InSb, use a diffused *p–n* junction. *p*-Type InSb is laid over the n-type material. Light strikes the *p* surface and energetic photons generate hole–electron pairs that are separated by the internal field existing at the *p–n* junction. The resultant voltage is amplified and used for detection. The detector must work at liquid nitrogen temperatures and is only good to 5.5 μm.

Lead–tin–tellurides (PbSnTe) are used to expand the range of detection and are optimum from 5 to 13 μm (under liquid nitrogen temperature) or, when liquid helium cooled, have a range from 6.8 to 18 μm. The most sensitive types are composed of mercury, cadmium, and telluride (MCT detectors). These are used with a current-mode amplifier and have speeds as high as 20 ns with comparable sensitivity to other members of this group.

SUMMARY

When rapid FT instruments are needed or sensitive measurements made, the photon-type detectors are required. Although liquid nitrogen or helium cooling is somewhat tedious, the spectacular results make it worthwhile. Besides, all the newer nuclear magnetic resonance instruments are also cooled, and inductively coupled plasma instruments are often fed by liquefied argon supplies. I can't see that liquid gases are the problem they were twenty years ago.

If speed is not important for an IR detector, then the simpler (read "cheaper") detectors are sufficient. Dispersive instruments have been used for decades with more than satisfactory results; so don't throw away your old system. Remember, the simpler the detector, the less maintenance!

REFERENCES

1. D. A. Skoog, *Principles of Instrumental Analysis*, 3rd Ed. (Saunders College Publishing Co., Philadelphia, PA, 1988).
2. H. H. Willard, L. L. Merritt, J. A. Dean, and F. A. Settle, *Instrumental Methods of Analysis*, 7th Ed. (Wadsworth Publishing, Belmont, CA, 1988).
3. J. D. Ingle, Jr., and S. R. Crouch, *Specrtrochemical Analysis* (Prentice Hall, Englewood Cliffs, NJ, 1988).

CHAPTER II.G.8

COMMON SOURCES AND SIGNAL PROCESSING TECHNIQUES USED IN VIBRATIONAL SPECTROSCOPY

The better topic for this column is light sources for vibrational spectroscopy. In this chapter I continue to describe components available to instrument manufacturers. Of course, there are more specialized or individually fabricated components available, but these are better relegated to later pages under the heading "Research."

In a subsequent chapter, we'll look at the puzzle and see how a manufacturer puts all the components into a box to solve your particular problem. I'll use case histories to show how "the right tool is chosen for the right job" (as Commander Scott stated in "Star Trek IV").

As I've said before, it is wasteful to build the most sensitive (read "expensive") research-grade instrument to measure water in methanol. It is equally silly to try to perform research with a twenty-year-old instrument that only cost $5000 new. Let's look at some more pieces of the puzzle. There aren't as many sources as there are means to crunch infrared (IR) radiation; so sources are a good starting point.

RADIATION SOURCES

While you may never be called on to actually choose the source type for your instrument, knowing the rationale behind a manufacturer's decision can be helpful. Sources for vibration, rotation, and bending are essentially pseudoblackbody radiators. They are heated to 1500–2000 K and emit light between 10,000 and 667 cm^{-1} (1000 to 15,000 nm). The maximum intensity at these operating

277

temperatures is between 5900 and 5000 cm^{-1} (1700 to 2000 nm). The shape of the intensity curve is somewhat asymmetric, with a gradual falloff toward the longer wavelengths to about 1% of maximum at 667 cm^{-1} (15,000 nm). A steeper falloff is seen toward shorter wavelengths, with the minimum occurring at about 10,000 cm^{-1} (1000 nm). The reduction is related approximately to the fourth power of the wavelength.

The main variations in the sources are based on cost, range of maximum intensity, and instrument lifetime. Ease of replacement and cost are often the two main factors in choosing a source. The monochromator will also dictate certain parameters for light sources: interferometer and diode-array-based instruments are single-beam spectrometers and demand extremely stable sources. Most dispersive instruments are dual beam, wherein small drifts may be tolerated.

Incandescent wire. A closely wound spiral of Nichrome wire around a ceramic core is a low-intensity, yet durable, source. At temperatures of 1100 K, a black oxide coating is formed and acts as a blackbody. That is, throughout the range of emitted radiation, a continuum of energy exists. No cooling is required other than radiation losses to ambient. This source is rugged and long-lived enough for classroom or quality control instruments and is decidedly simple to replace.

Because the wire is sturdy, albeit not very intense, it is recommended for in-process, filter-type, and nondispersive spectrometers. The initial low energy is further diminished if gratings and mirrors are used, thus the simpler applications. A rhodium wire heater sealed in a ceramic cylinder may be used in all applications recommended for Nichrome and will have similar characteristics. There is definitely a tradeoff in cost versus intensity for these two sources.

Nernst glower. The Nernst glower consists of a cylinder of rare earth oxides (zirconium, yttrium, and thorium) 1–3 mm wide and 2–5 cm long. The glower is heated through platinum leads sealed in the ends of the cylinder and is fairly fragile. The glower has a negative temperature coefficient of electrical resistance. That is, as it is heated, the resistance to current goes down. In fact, an external heater is required to lower the resistance before the material can conduct enough current to heat itself.

This necessitates a circuit with a current-limiting device, otherwise burnout will occur. The glower may be operated between 1200 and 2000 K and may be twice as intense as the other sources described here. The higher temperature necessitates ventilation, but the glower must be protected from drafts, because the poor conductivity of the oxides causes stress fractures with rapid cooling.

Glowbar. A glowbar consists of a silicon carbide rod ~5 cm long × 5 mm in diameter. The advantage of the glowbar is its positive coefficient of resistance, allowing simple control of intensity and temperature. Unlike the glower, it needs no booster. One drawback is that the electric contacts of the glowbar need water cooling to prevent arcing, adding to the complexity of any unit it fuels.

The spectral output of the glowbar is ~80% that of a theoretical blackbody radiator. It is a better choice than the glower below 2000 cm^{-1} (5000 nm) and beyond 667 cm^{-1} (15,000 nm). It has been used out to 200 cm^{-1} (50,000 nm).

Mercury arc. Simple blackbody sources lose enough intensity beyond 200 cm^{-1} that other types must be used in the far-IR range. High-pressure mercury arc lamps, jacketed with extra layers of quartz for thermal stability, produce intense energy in the far-IR. The overall output is similar to other blackbodies, but the plasma created by the arc through the mercury vapor emits fairly large amounts of long-wavelength radiation.

Incandescent lamps. Using a filament composed of tungsten, incandescent lamp sources are used primarily for near-IR work. The output is mostly between 12,800 and 4000 cm^{-1} (780 and 2500 nm). Most commercial near-IR instruments use this source, as it is reliable for 1000 to 2000 hours of continuous use and is quite inexpensive. Because this type is similar to car headlamps, high quality may be produced for about $100.

As with any other source, the intensity may be increased by increasing the voltage. The maximum wavelength is blue shifted with increasing temperatures, and lamp life is shortened in the process. Because most near-IR instruments are single-beam types, careful voltage regulation is required. This is mandated by the low noise inherent in near-IR detectors—they will see defects in lamp intensity immediately.

LASERS

Tunable laser diodes. A direct offspring of the fiber-optics communication age, tunable laser diodes are solid-state creations. They are mostly used in nondispersive spectrometers in process environments to monitor a small number of specific components. The laser diode emits radiation over a very narrow band of wavelengths at surprisingly high intensities.

Because common tunable diodes are used at cryogenic temperatures, the wavelength emitted may be selected by varying either the diode current or the operating temperature. The sensitivity demands careful temperature control and voltage stability.

Carbon dioxide lasers. Tunable CO_2 lasers have been used for measuring concentrations of atmospheric pollutants and various species in water. The CO_2 laser produces radiation in the 1100–900 cm^{-1} (9,000–11,000 nm) range, consisting of ~100 closely spaced discrete lines. These lines are extremely strong and pure and, at the same time, occur where many materials have absorption bands.

Some typical applications of a tunable laser source are for ammonia, butadiene, benzene, ethanol, nitrous oxide, and trichloroethylene. The power is amenable to the extremely long pathlengths used in environmental monitoring. The source and detector may be separated by many meters, allowing for incredibly low concentrations of materials to be quantified and qualified.

ASSORTED AND SUNDRY ELECTRONICS

Two things that most transducers (a transducer is a device that transforms a physical or chemical change into an electric signal) have in common are (1) a small signal output and (2) a continuous or analog output. One or both of these conditions must usually be changed before useful information is obtained from an analytical instrument. That is, the signal must be amplified and/or digitized for an integrator or computer to crunch the numbers.

We will examine these two steps as if they were independent, although anyone who has ever turned a radio louder because of static knows that the static also becomes louder. The last thing, then, that one needs from an amplifier is additional noise. This is another area where price versus performance comes into consideration. We'll look at some components used in the signal processing of instruments and then look at S/N enhancement techniques.

Operational amplifiers. Don't look now, but we don't use tubes in our equipment anymore—everything is solid-state electronics. We won't bother with the theory of transistors and other solid-state electronics (kids are taught these "basics" in middle school nowadays, anyhow) and jump right to integrated circuits. Integrated circuits and components such as operational amplifiers (op-amps) are used in place of discrete transistors, resistors, and capacitors.

A simple comparison circuit, seen in Fig. 1, uses a single op-amp for both detectors. The meter may easily be replaced by any measuring device. The point here is that any "noise" the detectors produce or any stray radiation they pick up is simply amplified without discrimination.

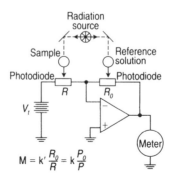

Figure 1. A simple comparison circuit with a single op-amp for both detectors.

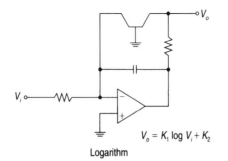

$$V_o = K_1 \log V_i + K_2$$

Logarithm

Figure 2.   An op-amp used for logarithms.

An op-amp used for mathematical operations (for example, logarithms, as seen in Fig. 2) also passes along noise without a filtering effect. To understand how a filtering device (passive or active) may be added to eliminate much of the noise present in a signal, the nature of noise must be reviewed. Without going into too much detail, the common noise(s) may be either low or high frequency and of long or short duration. In other words, noise varies all over the playing field. How, then, might it be avoided?

Let us assume that some of the line current at 60 Hz is being picked up by the detector. Compared with the typical frequencies observed in the IR, this is quite low. A simple capacitance/resistance (CR) circuit, as shown in Fig. 3, will passively and effectively block signals below a frequency determined by the values of R and C. It is called a *high-pass filter*.

High-frequency noise, possibly generated by a high-speed motor or rf source, may be blocked by a resistance/capacitance (RC) circuit, as seen in Fig. 4. Again, by choosing proper values for R and C, high-frequency signals are effectively blocked. This type is referred to as a *low-pass filter*. A combination of high- and low-frequency interferences may be treated with an impedance/resistance (LC) circuit, as seen in Fig. 5, and is called a *bandpass filter*.

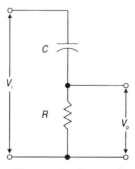

Figure 3.   A high-pass filter, or simple capacitance/resistance circuit.

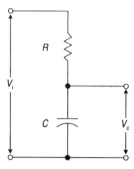

Figure 4. A low-pass filter, or simple resistance/capacitance circuit.

A simple, active filter may be affected by periodically sampling the same portion of a signal over a fixed interval. It then averages the samples using a low-pass filter. This device, a boxcar integrator, provides reasonably good S/N enhancement.

Digital electronics. All of the above manipulations are on the raw, or as-is, signal. The signal arising from most detectors is continuous, whether or not the beam is modulated or chopped. Computers and integrating devices work on discrete points of data. To turn a continuous or analog signal into a series of pulses or a digital signal, an analog-to-digital (A to D) converter is used.

This is simply a device that samples a signal at predetermined intervals for predetermined periods of time. The time intervals are usually chosen to provide the maximum amount of useful information using the minimum amount of computer space (*not as critical now with gigabit disks, but still a concern*) and time. If an interferometer is being used to generate a signal, it is obvious that a rapid sampling rate is required to avoid losing important portions of the spectrum. If a chemical process is being followed, often data points may be up to minutes apart. Once the sampling rate has been optimized for the measurement, further noise reduction may be used.

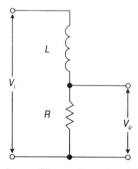

Figure 5. A band-pass filter, or impedance/resistance circuit.

Ensemble averaging. Used in all forms of science, the concept of ensemble averaging is based on the statistical assumption that noise, like assay errors, is random. The reason we perform titrations in triplicate is to assure random error averaging and to minimize its effect.

In ensemble averaging, the same signal (for example, a spectrum) is measured n times and co-added point by point. The summed signal is then divided by n and used. If the noise is truly random, the signal averaging will cause "destructive interference" and the signal will "constructively interfere," significantly enhancing the S/N ratio.

The limit to this technique is that the noise is diminished by roughly $n^{1/2}$ (barring chemical or physical interferences, carbon dioxide or water in the path of a single-beam instrument, or detector drift). Thus to bring to $\frac{1}{3}$ the noise generated by 10 co-added scans, the instrument must scan $100 \times$ and co-add the results.

Smoothing (weighted digital filtering). Assigning different levels of importance (weights) to points as a function of their position relative to the central point of a moving average may produce better filtering than mere co-adding or equally weighted moving averages.

The number of times smoothed and the relative weightings are open for discussion, but most statisticians agree that the procedure works. The smoothing usually occurs after the signal has been captured, mostly because the procedure take longer than scanning the data.

Boxcar averaging. For situations in which the signal changes relatively slowly with respect to, say, wavelength, this procedure is worth trying. In some IR spectra of liquids or near-IR spectra of almost everything, the peaks are wide and relatively featureless. Loss of resolution is balanced against loss of noise.

In this technique, also called *n-point averaging*, equally spaced grouped of data points or "boxcars" are averaged and replaced with a single data (*I know someone will scream "datum", but "data" flows here!*) point. The resultant spectrum is simplified by the number of points in each boxcar (for example, $\frac{1}{6}$ as large if the boxcar contains 6 points). The enhancement of the S/N is calculated by the equation

$$S/N = n^{1/2}(S/N)_0$$

where $(S/N)_0$ is the signal-to-noise ratio of the untreated data and n is the number of points averaged in the boxcar. The cost of S/N enhancements is loss of resolution. Only by using enhanced data in real situations to predict analyte values can the "correct" amount of averaging be determined.

Fourier transforms. The Fourier transform is heart of interferometer-based instruments, but may be used with "normally" generated data as well. One common application is in collecting spectrophotometric data at a much greater than normal rate. In essence, data are rapidly collected in the time domain and converted to the conventional frequency domain.

Because spectra are gathered so rapidly, smoothing is more easily performed on multiple scans. Conventional signals may be converted by a Fourier transform (FT) operation, the signal modified as needed, then an inverse FT converts it back to a usable signal.

This column is not the place for detailed math treatments, but an example of an FT pair is

$$F(\nu) = \sum_{-\infty}^{\infty} f(t)e^{-i(2\pi)\,dt}$$

(frequency-amplitude function)

$$f(t) = \sum_{-\infty}^{\infty} F(\nu)e^{i(2\pi)\nu t}2\pi\,d\nu$$

(time-amplitude function)

The most common functionality is based on cosines, for example,

$$f(t) = A\cos\,2\pi\nu_0 t$$

but may include square waves of most common repeating forms.

SUMMARY

Many types of detectors are available for the infrared spectrum. Manufacturers have to balance lifetimes with cost and ease of operation. In multiple-process monitors it is necessary to choose a durable instrument, but the cost of a large number of instruments cannot make the process nonprofitable. In a laboratory setting, an analyst might well choose the best and most versatile available instrument, because he or she will normally only have one instrument for method development and QC use. These facts must be taken into consideration in the initial engineering of equipment.

We also see that a number of commonly used data treatments are used to enhance the accuracy of the data. These will also depend on the type of work being performed. After we discuss monochromators in the next chapter, we'll tie all these components together.

REFERENCES

1. D. A. Skoog, *Principles of Instrumental Analysis*, 3rd Ed. (Saunders College Publishing Co., Philadelphia, PA, 1988).
2. H. H. Willard, L. L. Merritt, J. A. Dean, and F. A. Settle, *Instrumental Methods of Analysis*, 7th Ed. (Wadsworth Publishing, Belmont, CA, 1988).
3. J. D. Ingle, Jr., and S. R. Crouch, *Spectrochemical Analysis*, (Prentice Hall, Englewood Cliffs, NJ, 1988).

CHAPTER II.G.9

MONOCHROMATORS USED IN VIBRATIONAL SPECTROSCOPY

I seem to remember promising a yarn about monochomators. I will turn to a few good books (1–3) and the dusty caverns of my mind and see if I can do these important parts justice. Simply stated, my definition of a "monochromator" is any device used to segment a polychromatic beam into a single wavelength or narrow band thereof to enable spectroscopy to be effected. Without repeating myself, these devices are (in my opinion) the most critical pieces in any spectroscopic system. The entire system can be no better than the purity of light used.

Several distinct types of wavelength selectors are available, and each has strong and weak points. The first general grouping is based on a physical selection of wavelength and includes filters, prisms, and gratings.

FILTERS

Filters are used to pass a discrete band of wavelengths or to block wavelengths higher or lower in length than a particular value to be used for spectroscopic work. The most common are absorption and interference types.

Absorption filters. These are passive devices used to limit or totally absorb unwanted radiation. This is accomplished with colored glasses, crystals of various materials, solutions of inorganic and organic materials, sintered substances, and thin films. These are durable, inexpensive, and easiest to use.

Bandpass filters. These allow light transmission in a Gaussian pattern with the width of the band ranging from 30 to 250 nm at half-height. The filters made from an organic dye dispersed in a gelatin matrix, positioned between two glass plates, are called Wratten filters and may be purchased commercially.

Cut-off filters. These cause wavelengths higher or lower than the "chosen" wavelength to be totally absorbed. The cutoff is not a sharp line but a sharp "S" curve (looks like a titration curve!). These can be constructed from the same materials as the bandpass types.

Interference filters. The premise of these filters is that they are constructed such that the majority of wavelengths of light striking the filter are subject to destructive interference. A small range of wavelengths is subject to constructive interference and is passed. Two kinds of filters are most commonly used: a single layer and a multilayer type. The first type consists of two thin sheets of glass with a transparent dielectric spacer between them. The dielectric consists of quartz, CaF_2, MgF_2, ZnS, ThF_4, or sapphire. The center wavelength (peak of Gaussian or bell-shaped curve) depends on the refractive index of the spacer (and, of course, the angle of incidence). This type is named a Fabry–Perot (no relation to the Texas billionaire/presidential candidate) filter.

Varying the wavelength of some Fabry–Perot filters may be accomplished by changing the angle of incidence. This was the basis of the tilting-filter instrument produced by Pacific Scientific. While there is one particular band of interest in these filters, second- and third-order bands exist as well. They are usually eliminated with absorption filters referred to as "order-sorters."

Multilayer filters are constructed from alternating layers of high- and low-refractive-index materials. This "layer-cake" effect acts as a series of Fabry–Perot filters. The bandpass half-widths can be quite narrow for these filters and the percent light transmitted quite high. Wavelength ranges from the ultraviolet through the infrared with bandwidths between 1 and 100 nm are commercially available.

PRISMS

These were the first (practical) devices used to separate polychromatic light into discrete wavelengths (a spectrum), and there are still a fair number of these puppies around. The principle involved in prisms is simply that each wavelength of light has a different refractive index within any transmitting material. Each wavelength is "bent" to a different angle as the light beam is passed from the air (vacuum, argon, etc.) through the prism and back to the air again. The result is a fanlike dispersal of wavelengths in an increasing wavelength (decreasing energy) pattern.

Midrange infrared prisms are constructed of various salts, depending upon the wavelength range employed. Sodium chloride is used between 4000 and 650 cm^{-1} (2,500 to 15,400 nm). Potassium bromide is used down to 400 cm^{-1} (25,000 nm), while cesium iodide extends down to 270 cm^{-1} (37,000 nm).

Mirrors and lenses must be made of compatible materials so as not to absorb the infrared light. It may be inferred from the materials used that great care must be taken to exclude moisture from the workings of such instruments. The most excellent separating power of a prism is balanced by its delicate nature.

On the plus side, prisms give the purest spectra. No overtones or higher orders of light are generated by them. In fact, prisms from the earlier Cary spectrometers (models 11 through 15) were used by service reps to calibrate the earlier grating-type instruments. (A trick showed to me by a former P–E tech.)

While prisms designed for the near-infrared portion of the spectrum could be constructed from quartz-type glasses, no commercial instruments employ a prism. This is possibly a consequence of commercial (dedicated) NIR instruments not being constructed until after inexpensive and reliable gratings were readily available.

DIFFRACTION GRATINGS

A grating is either a planar or concave plate with precisely ruled grooves. These are spaced rather closely with up to several thousand lines per millimeter. Most current instrumentation use mirrored or reflection gratings. The current (relatively) low price is due to the production techniques now used.

Originally gratings were "ruled" mechanically with precise engraving tools. With the advent of the laser, gratings became simple to produce. A surface is coated with a photosensitive material. An interference pattern is projected on the surface using lasers. The surface is then developed, producing grooves that are subsequently etched into the surface for durability. Over these grooves, aluminum is vacuum deposited, creating the mirror effect. While slightly less efficient than ruled gratings, holographic gratings are orders of magnitude cheaper to construct.

The sawtooth pattern of grooves cause constructive interference between parallel photons. As the angle of incidence of the face of the grating to the incident light varies, different wavelengths are emitted toward a fixed slit. The distance between grooves and the angle of scoring (named the *blaze angle*; if there were no such angle, the reflected light would be (specular reflection only) determine the wavelength range separated.

As with interference filters, higher orders of light are reflected and must be removed with absorbance filters. Gratings are also used in conjunction with prisms or with a second grating. Often the first is a low-resolution type, used primarily to separate orders. The second is a higher-resolution type, used for narrow bandwidth resolution. Numerous mountings have been devised for gratings with some features in common: a grating, collimating mirrors, and slits.

Unlike prisms, gratings used in the infrared are not as easily destroyed by the atmosphere. This is not to say that they are truly rugged, just more stable and longer lived. When held in place, gratings are far more adapted for process work. The light of a stationary grating, for instance, may be impinged on a diode-array detector for long-term stability in a process environment.

Another similar approach is to have a stationary grating disperse the light and have selected wavelengths focus onto a single detector. The combinations of wavelengths are selected by a "mask" using Hadamard transforms. This approach was introduced commercially by D.O.M., Inc (founded by some minor spectroscopist from Kansas State University).

INTERFEROMETERS

First developed by Michelson in the 1800s to measure the speed of light, the interferometer named after him is the workhorse of Fourier-transform IRs. It is an example of a two-beam interferometer.

A focused beam of light strikes the beam splitter, where one-half of the intensity is directed to a movable mirror via a half-silvered mirror. The rest of the light continues through to a fixed mirror. At the "proper" lengths, the light will recombine and be directed to the sample, then to the detector.

When the movable mirror is moved to $\frac{1}{4}$ the wavelength away from this spot, the two beams are 90° out of phase and 180° out of phase when they reach the detector. If the mirror is moved at a constant rate, a sinusoidal pattern is achieved. When polychromatic light is used, this resultant interferogram may be used much as monochromatic light for developing spectra. The interferogram must first be deconvolved (in most cases) to yield meaningful information.

While the work of M. Fourier was known at the time of Michelson, computers weren't. Thus the rapid Fourier transform was not available as it is now for spectroscopists to utilize. The basic concept of FT-IR is that any signal (spectrum) is merely the sum of a series of sines and cosines. The computer-generated spectrum is indistinguishable from one generated by a prism or grating-based instrument.

Rugged instruments have been developed by several companies for routine work as well as in-process monitoring. While research-grade models are too sensitive for factory work, there are enough good instruments available for the job.

One big advantage of the interferometer is the throughput advantage. Without losing most light to mirrors, gratings, and lenses, an FT-IR produces a far stronger beam of incident light. The stronger signal produced also ensures a better signal-to-noise ratio. One minor problem is that the tolerances needed become more precise when the NIR portion of the spectrum is scanned. The movement must be smoother as the spectral lines are nearer and the length of sweep of the movable mirror is far shorter in this range.

ACOUSTO-OPTIC TUNABLE FILTERS

Covered in prior chapters (II.E.1 and II.E.2), this type of monochomator uses standing sound waves as diffractors for light waves. A transducer transforms radio frequencies into high-frequency sound waves within an anisotropic crystal, such as TeO_2. The device contains no moving parts and can scan 1000 nm in less than 50 msec.

While speed and durability are strong points here, few devices have been in use in industrial settings long enough to evaluate their long-term durability. Another minor problem is growing and polishing of the crystals, and successfully bonding the transducer material to the surface of the crystals. There are currently few commercial sources of the crystal in this country where reproducibility may be guaranteed.

CONCLUSIONS

The last set of pieces are now on the table for designing our "perfect" spectrometer. The wavelength specificity of the wavelength selectors varies from bands of 100–200 nm down to 0.1 cm^{-1}. We may have one, two, or several wavelengths or a full spectrum. Speed may vary from several minutes to several microseconds.

Of course, the price of these gems runs from several dollars to tens of thousands of dollars. In the next chapter, we will pick and choose among the pieces to make our own "supertoy."

REFERENCES

1. James D. Ingle, Jr., and Stanley R. Crouch, *Spectrochemical Analysis* (Prentice Hall, Englewood Cliffs, NJ), 1988.
2. Hobarth H. Wilard, Lynne L. Merrit, Jr., John A. Dean, and Frank A. Settle, Jr., *Instrumental Methods of Analysis* (Wadsworth Publishing Co., Belmont, CA), 1980.
3. Larry G. Hargis, *Analytical Chemistry—Principles and Techniques* (Prentice Hall, Englewood Cliffs, NJ), 1988.

CHAPTER II.G.10

BUILDING THE PERFECT SPECTROMETER OR PICK ONE FROM COLUMN A AND ONE FROM COLUMN B ...

I considered putting in "with two you get eggroll," but I don't know how many people reading the column have eaten in a big city Chinese restaurant or remember Buddy Hackett's classic routine; so I left the ending alone.

As promised (or threatened), here are the menus of components for the perfect spectrometer. Sometimes I even amaze myself. The general idea of this tome is quite simple: Decide on the application for your instrument, consult the tables, and piece together the ideal instrument for that application (based on function and financial risk). Then you take the description from booth to booth and find a company that makes said instrument. Simple? I didn't think so, either. However, work with me on this and we'll be fine.

The tables will consist of the components described over the preceding three chapters (1–3). The names of the components will be accompanied by various (simplified) parameters and, where needed, a word or three of descriptive information. A second table of the same components will contain "financial" considerations such as ease of repair, initial cost, and long-term durability. After the tables are presented, I'll give some examples of how to use them.

SOURCES

Along with this list, I feel that some terms used need be defined.

Mode of operation is simply a short description (reminder?) of the construction of the light source.

Table 1. Sources (Operational Parameters)

Type	Mode of operation	Intensity	Advantages	Caveats
Wire	nichrome wire around ceramic, heated to incandescence	low	rugged, inexpensive	
Nernst glower	Cylinder of rare earth oxide, heated	high		fragile, needs ventilation and secondary heater
Glowbar	silicon carbide rod	high	good to 200 cm^{-1}	needs water cooling
Mercury arc	high-pressure arc lamp	high	high amt of long-wave-length light	high heat, UV light
Incandescent lamp	heated filament	high	broad band	mostly vis and NIR light
Tunable laser diode	semiconductor source	med–high	selected wavelengths	lower number of wave-lengths
CO$_2$ laser	gas laser	very high	100 discrete lines available	only good in 1100 to 900 cm^{-1} region

Table 2. Sources (Financial Considerations)

Type	Cost	Ruggedness	Durability	Notes
Wire	low	high	high	best for student-type instruments
Nernst	med	low	medium	best in labs
Glowbar	high	low	high	only for labs
Mercury arc	high	med–high	medium	best in lab
Incandes. lamp	low	high	high	lab or plant
laser diode	low	high	high	good for process work
CO$_2$ laser	high	low	medium	excellent for lab; OK for process

Table 3. Monochromators (Operational Parameters)

Type	Quality	Speed	Bandwidth
Absorbance filter	band	v. fast	30–250 nm
Interference filter	band	v. fast	low
Prisms	discrete	slow to med.	excellent
Gratings	discrete	med to fast	v. good
Interferometer	full spectrum	v. fast	excellent
Acousto-optic tunable filter	discrete	v. fast	good

Intensity refers to its relative output of light. Something like a candle would be "very weak" while an arc lamp would be "very intense."

Advantages are just that. If a source is especially long-lived or resistant to vibration, that might be mentioned.

Caveats are simply areas of special concern such as a need for high power or ventilation.

Durability of any component is, in my definition, the lifetime *when used as directed.* In other words, a component may need to be in a dry atmosphere, but, if so protected, may last for many years. It would simultaneously be sensitive *and* durable.

Table 4. Monochromators (Financial Considerations)

Type	Cost	Ruggedness	Durability	Notes
Absorbance filter	low	high	good	hard to match for replacements
Interference filter	low/med	high	good	"
Prisms	med	low	good	need dry atm. for IR instrm.
Gratings	low/med	high	high	need order sorter
Interferometer	high	med/high	med	most cannot be used in process
AOTF	high	med/high	high	limited sources of supply

Table 5. Detectors (Operational Parameters)

Type	Mode	Response	Sensitivity	Notes
Thermal	absorbs heat	slow	low	universal detector
Golay	gas thermometer	slow	good	best for far-IR
Thermocouple	dissimilar metal junction	medium	medium	
Thermopile	series of thermocouples	fast	good	
Thermistor	oxides with high resist.	v. fast	low	
Pyroelectric	crystal capacitance	v. fast	v. high	
Photon	semiconductor	medium	medium	shorter wavelengths
Photoconductance	semiconductor	fast	good	
Photovoltaic	semiconductor	v. fast	v. good	needs liquid nitrogen temps

Table 6. Detectors (Financial Considerations)

Type	Cost	Ruggedness	Durability	Notes
Thermal	low	high	good	RT operation large range
Golay	med	med	good	lab instruments
Thermocouple	low	high	good	
Thermopile	med	high	good	good response time
Thermistor	med	medium	good	
Pyroelectric	high	medium	medium	highest speed
Photon	med	high	high	poor response in far-IR
Photo-conductance	low	high	high	
Photovoltaic	med	good	good	must work at liquid N_2 temps. and falls off after $5.5\mu m$

We may now address *monochromators*. For these, we need a few more definitions:

Quality: is in no way a judgment. In this case, I merely refer to whether one obtains a band, discrete wavelength, or a cluster of wavelengths.

Bandwidth: this could range from a single wavelength (ideal, not in any thing but lasers) to a full spectrum emerging from the device.

Speed: the speed of the resultant "spectrum" may range from 22 min to instantaneous. Thus slow to very fast, will be the indication.

When we talk about *detectors*, we also need to remember the ranges in which we wish to work. This will, of course, determine the source needed in some cases, but most sources are broad band in the IR and most monochromators are available in all wavelength ranges. Detectors are less forgiving.

Two more terms used herein must be defined (i.e., as *I* mean them.)

Response time: since we are dealing with heat, after all, the speed with which a detector may move from one wavelength to another will depend on how fast it returns to a base temperature or state. This ranges from nanoseconds to milliseconds for the detectors discussed in this chapter.

Sensitivity: there are more specific terms such as D^* to specify detection sensitivity, but for our purposes, we will use subjective terms such as low, medium, or high. All things are, after all, relative *n'est ce pas*?

"What in the world will I do with all this information?" you might well ask. Let's look at one or two examples of how to put the pieces together:

1. Assume that you are looking for a laboratory-based, research-grade instrument for your lab. We first need the best source money can buy ... for the range of operation you need. Suppose you are considering the standard 4000 to 400 cm^{-1} range of frequencies. The Nernst glower or Glowbar would work nicely as a source.

The monochromator would need to give very good discrimination of wavelengths as well as good speed. For this we might consider an interferometer or AOTF. If we consider size, the AOTF wins, **BUT** it can't compare with a good interferometer for bandwidth.

A detector must be very quick to match either of these monochromators. We would want to look at a pyroelectric or photovoltaic type for good speed. (Keep in mind that many good, fast detectors will need to be cooled ... a small inconvenience in the lab.) The odds are that you will use a Fourier transform algorithm if you choose an interferometer ... good guess since FT-IRs are the most popular lab instruments today.

To summarize, we put together a Nernst glower or Glowbar, an interferometer or AOTF, and a pyroelectric or photovoltaic detector. Just check the booths for the best performance and price!

2. Now we'll look for a process instrument to monitor the appearance of the product in a chemical reaction (I will ignore the NIR for now ... you've

heard enough from me on the subject in the past). A durable, broad-spectrum source might be the Nichrome wire. With it we might use a filter and add a plain vanilla thermal detector.

It will probably require a surface or internal reflectance cell or perhaps even a fiber-optic probe. The actual interface will be dictated by the temperature of the pipe or reactor wall and the vibration that might hit the instrument. Your friendly technical rep will be happy to spell out the options.

CONCLUSION

Quite simply, there are several ways to use the information in this column. You could piece an instrument together as I did or you might just use the information to "double-check" the information given to you by a salesman. Strange as it might seem, some of them aren't up to date on all the possible combinations available.

Several observations I have made over the years:

1. Seldom, if ever, is the best lab instrument of any value in the plant. The opposite is also quite true. Suspect any salesman who claims that one instrument can do everything.

2. While I don't believe that the big companies are always the best, they do have the greatest selection of equipment. The importance of technical support and repair personnel cannot be overstated.

3. Buy equipment to the level of personnel you have performing the analysis. Chemists are seldom insulted by simple equipment, but unsophisticated operators are often confused by complex equipment.

In general, I have found most companies basically honest. Where the problem comes in is that they don't eat if they don't sell. So, often, the merits of any model may be, er, uh, "slightly" exaggerated. Arm yourself with knowledge before shopping ... or bring me. I'll be the one with the sign saying "will consult for food!"

REFERENCES

1. E. W. Ciurczak, *Spectroscopy* **8**(9), 12 (1993).
2. E. W. Ciurczak, *Spectroscopy* **8**(10), 12 (1993).
3. E. W. Ciurczak, *Spectroscopy* **9**(1), 12 (1994).

CHAPTER II.G.11

TAKING THE SPECTROSCOPY OF FLUIDS BEYOND THE LIMITS
Advances in Sampling at High Temperatures and Pressures

Valentine J. Rossiter

In this chapter Valentine Rossiter describes a rather interesting cell for liquids that operates to extremes of temperature and pressure.

Loss of mechanical strength in optical window materials is a limiting problem under extreme conditions, but Rossiter describes solutions using the cell with samples such as jet fuel and lauric acid. He presents data from FT-IR investigations, but the cell has potential for use in all types of optical spectroscopy. I could see the range of sampling open up a little more and thought I would share this gem with you. Enjoy!

An earlier series of developments concerned with high-temperature spectro-scopic cells for solids had proved to be very successful (1, 2). Designs had evolved to provide a very wide temperature range—from – 170 to 700°C, now extended to 950°C (3)—combined with a very wide pressure range (high vac-uum to 2000 psig). The design also provided a variety of other techniques including transmission, specular reflectance, large-angle reflectance infrared (LARI), sample irradiation with UV, Raman, and emission spectroscopy (2).

These optical modes also were available over a very wide temperature and pressure range. The resulting cell designs have since found applications in such areas as polymer research, coatings, superconductors, and many other topics.

In the area of liquids research, designs were developed that solved many of the problems in the spectroscopy of liquids (4). These problems included the difficulty caused by the presence of interference fringes in cells with very short pathlengths, how to engineer such cells to be pressure tight, how to provide variable pathlength in cells of high structural integrity, variable temperature, variable pressure, and so forth. But it was still impossible to provide anything approaching the same high-temperature plus high-pressure experimental facilities for fluids as had been developed for solids with full wavelength range spectroscopy (1, 2).

The difficulty is caused by the fundamental difference in the design problem between solid sampling and fluid sampling. In the case of solids, it is possible to heat the solid in a region remote from the optical windows containing the sample environment in the experimental chamber. This still allows control of the atmosphere surrounding the sample. But in the case of a fluid, the retaining window must remain in contact with the sample material when it is at high temperature and high pressure so the window material and the sample material are then at the same temperature. Like all materials, the window material loses mechanical strength as the temperature is increased and its ability to withstand pressure is lost. This is a particularly severe problem with window materials that demonstrate good wide-range transmission in the mid-infrared spectral region. This region ($\sim$600 to 4000 cm^{-1}) is the one widely used in infrared studies of samples with window materials such as ZnS and ZnSe. Such optical materials have limited strength even at room temperature. Loss of strength with increasing temperature is a general problem to materials (2), but as with many optical materials, it becomes severely limiting. Even at relatively low elevated temperatures, the loss of strength is severe and unreliable.

Although the earlier development (4) of the liquid cell designs had achieved important technical advances, a critical review of the design was undertaken recently to see if it was possible to simplify the earlier format. This indicated that the fundamental optical design and the consequent complex engineering was necessary. Comparative alternative design studies concluded that simpler (and perhaps less complex) designs of liquid cells would not deliver the needed technical benefits. One change was made to integrate a simple optical unit into the design, rather than having to use an external optical system for beam diversion (4). The arrangement is shown schematically in Fig. 1. It was during this redesign exercise that the solution to the fundamental window problem became obvious (5).

DESIGN

As discussed, the retaining, or primary, window of fluid cells is in contact with the fluid and must be at the same temperature and pressure as the fluid. However, consider the situation if the primary window could be supported by exter-

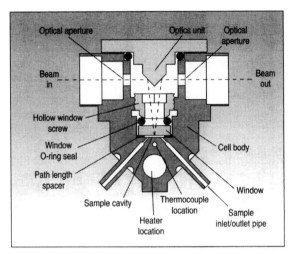

Figure 1. Simplified schematic cross section of a variable-pathlength, variable-temperature transflectance liquid flow cell with integral optics unit (RA4000, Aabspec Instrumentation).

nal gas pressure. This external pressure can be used to reduce the pressure differential across the primary window to zero. Under these circumstances, loss of mechanical strength in the primary window becomes irrelevant, Of course, optical access is still required, but this can be through windows that are not at high temperature and so have the mechanical strength to retain the gas pressure. The design is illustrated in Fig. 2. This shows how a liquid cell of the earlier

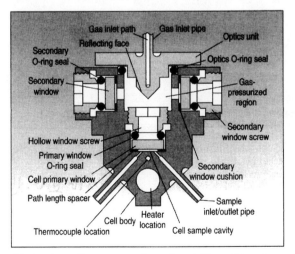

Figure 2. Simplified schematic cross section of variable-pathlength, variable-temperature, high-pressure transflectance fluids cell (RA4000-EXP) incorporating new window technology.

type (Fig. 1) can be modified to incorporate the deflecting optics unit and the new window technology. The liquid enters through either of the inlet ports into a cavity contained by the primary window. Pressurized gas can be admitted through the port formed in the optics unit. The pressurized gas is contained by an external chamber that incorporates secondary windows. Cooling the location of the secondary windows allows them to operate at their optimal mechanical strength. The gas pressure can be adjusted to match the pressure of the liquid in the cell. In principle, the primary window now needs no significant mechanical strength to retain the liquid that is at high pressure and high temperature.

Naturally, the gas-pressurized region should be of low volume, and other appropriate safety measures are required for the pressurized gas in the system in addition to the safety requirements for the pressurized sample fluid.

SPINOFF BENEFITS

This development in window technology has a number of spinoff benefits. From the spectroscopist's point of view, maintenance and reproducibility of exact cell pathlength is critical. As a conventional system is pressurized, the window pushes back against the high-pressure seal, increasing the pathlength in an uncontrolled and nonreproducible way. The exact pathlength change will depend on factors such as the resilience of the window seals (generally O rings), which may change with aging. Deformation of the seals or the window as functions of temperature and pressure are among the factors that can cause nonreproducible path lengths. As system pressure is increased using the new window technology, the external gas pressure can follow the change and the window does not move. Pathlength is accurately defined and reproducible.

Another benefit is that the method of sealing the primary window can be changed. Because there is no pressure differential, a conventional high-pressure seal is not required. So alternative material can be used, and the system can now operate at temperatures higher and lower than would be possible with conventional sealing methods.

Again, a wider choice of window materials is now available. We can consider using materials that would not be strong enough to be considered for high-pressure systems. We can select on the basis of chemical inertness, for example, irrespective of strength. Again, because of the removal of the strength constraint, windows of much reduced thickness can be used.

The technology can be incorporated into all types of optical cells. It is applicable to all types of optical elements, including windows, optical fibers, and multiple-internal-reflectance elements. The spinoff benefits are of such significance that their combined benefits can rival the importance of the discovery of the basic cell technology itself in some applications.

EXPERIMENTAL

Testing jet fuel. The first unit of this type (RA4000-EXP, Aabspec Instruments, Wantagh, NY) was provided to Henk Meuzelaar of the University of Utah's Center for Microanalysis and Reaction Chemistry. The system was supplied with a dual specification. In the standard operating range, the cell provides temperatures to 250°C at pressures as high as 4000 psig. In the experimental range, temperatures to 450°C at pressures to 4000 psig are available. Within the standard range, conventional sealing techniques are used for the primary window, while in the experimental range we are exploring new materials and methods with various user groups.

The cell's pathlength is stepwise variable from 0.05 to 2 mm within a single-piece, 316 stainless steel body. No welds, brazes, or joins are used in the body. Connections are by modified Swagelok connectors (Crawford Fitting Company, Solon, OH). Temperature control is provided by a dual-range digital temperature controller (DTC-2, Aabspec) or programmer (LTP-2A, Aabspec). The cell's dual cooling systems provide the necessary cooling for the secondary window system and have been developed to provide the necessary thermal distribution within the cell. This ensures that the sample-containment region of the cell is at high temperature while the secondary window region operates at near ambient.

The engineering required for optimal thermal distribution in this design of fluids cell is considerable. A temperature readout location is provided in the cell body, close to the sample cavity. This location accommodates a stainless steel–sheathed thermocouple that monitors real-time sample temperature. Readout thermocouples can also be introduced directly into the sample through the cell inlet/outlet ports. Figure 3 shows such a cell with the dual-range digital temperature programmer option, which provides a range of thermal programs, including controlled cooling rates and subambient temperature operation.

Meuzelaar's research group is interested in a variety of high-temperature and -pressure applications. This includes work on supercritical fluid extraction and the high-temperature/high-pressure oxidation of jet fuels. The high-tem-

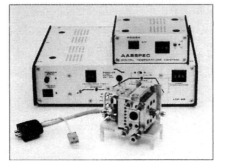

Figure 3.   The RA4000-EXP cell shown with the LTP-2A temperature programmer.

perature cell allows the research group to obtain full midrange infrared spectra using ZnS windows to greater than 400°C and 4000 psig. Such a temperature, pressure, and optical combination exceeds anything previously possible. Figure 4 shows preliminary spectra obtained for JP7 jet fuel showing progressive degradation as a function of time under high-temperature/-pressure conditions.

Testing fatty acid decomposition. Brian D. Kybett at the Energy Research Unit of the University of Regina (Alberta, Canada) has been using the new technology to examine the thermal decomposition of fatty acids at high pressure. Experiments are being conducted in a carbon tetrachloride medium using the high-temperature cell. The spectrum shown in Fig. 5 is that of lauric acid (1% in carbon tetrachloride) at 260°C and an initial pressure of 900 psig. This full-range spectrum was obtained at 4-wavenumber resolution with 64 scans on a Fourier transform infrared (FT-IR) spectrometer (FTS40, Bio-Rad, Digilab Division, Cambridge, MA) using the manufacturer's 3200 software. The pressure is allowed to increase during the experiment to 1200 psig, and the thermal decomposition can be observed as a function of time. The deconvolution spectra for the C=O region are shown in Fig. 6. The spectra can be interpreted as showing the thermal decomposition of the acid to ketone and possibly some anhydride; this is subject to further work. The intensities of the ketone C=O stretch at 1810 wavenumbers increase, while those of the acid dimers (1720 wavenumbers) and monomers (1760 wavenumbers) decrease.

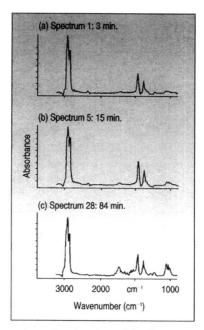

Figure 4. Time-based spectra showing oxidization processes in JP7 jet fuel.

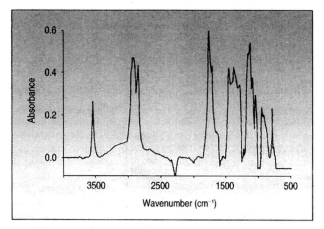

Figure 5. Initial full-range infrared spectrum of 1% lauric acid in carbon tetrachloride at 260°C and 900 psig. Total scans: 64; resolution: 4 wavenumbers.

DISCUSSION

The applications reported here are based on what is effectively transmission spectroscopy, but other forms of spectroscopy are also made possible. For example, it is now possible to provide a backscattered Raman cell (BS-RAM-10K, Aabspec) that can operate to 10,000 psig and to 250°C. This opens up further areas of study and will provide interesting new data.

In addition to the standard pathlength range available in the RA4000-EXP cell format (0.05 to 2.0 mm), it is also possible to work at high temperature and pressure with an extremely short pathlength, down to 5 μm (RA-SP-4K).

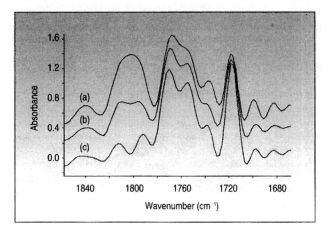

Figure 6. Deconvoluted spectra of lauric acid decomposition (1% solution in carbon tetrachloride) in the C=O region. Total scans: 64; resolution: 4 wavenumbers. Spectrum taken (a) initially; (b) after 10 min.; (c) after 35 min.

This will enable the system to handle fluids that, in spectroscopic terms, are very strongly absorbing.

Future work. Continuing work is aimed at exploring the most appropriate sealing techniques and materials for various applications and to raise the upper temperature limits even further. It is also possible to use the spinoff benefits of the sealing techniques in the low-temperature region to provide pressure-tight, low-temperature fluid cells. This has been a problem area for spectroscopists for some time.

Topics of immediate interest include FT-IR interfacing for both supercritical fluid chromatography and supercritical extraction. Polymer research is another interesting area, as it is now possible to obtain full-range spectra of molten polymers under previously inaccessible process conditions. Fuels and lubricants are among the many other areas in which the extension of the available experimental operating range is likely to be significant.

CONCLUSION

A new method has been developed to solve the problem of loss of strength in the retaining window of fluid cells. The method is general to all types of optical arrangements and all forms of optic elements, including fibers. The method provides significant spinoff benefits for optical spectroscopy and greatly extends the temperature range available. Applications in the areas of jet fuel decomposition and fatty acid decomposition have been shown.

ACKNOWLEDGMENTS

The author wishes to express his appreciation to Henk D. Meuzelaar and his research group at the University of Utah for providing the data on jet fuel samples and to Brian D. Kybett and his research group at the University of Regina for providing their data on fatty acids. The engineering assistance of Alan Wilkie and Brian O'Neill of Aabspec International (Dublin, Ireland) in implementing the cell designs discussed in this chapter is very much appreciated.

REFERENCES

1. V. Rossiter, *Research & Development* **26**(2), 162 (1984).
2. V. Rossiter, *Research & Development* **30**(2), 94 (1988).
3. V. Rossiter, *Research & Development* **34**(1), 180 (1992).
4. V. Rossiter, R. Milward, and D. Chenery, *Research & Development* **30**(11), 89 (1988).
5. V. J. Rossiter, U.S. Patent 5,223,716, (1993).

CHAPTER II.G.12

MORE PERSPECTIVES ON THE DECADE IN SPECTROSCOPY
Reader Input Covers Microscopic, Computational, and Chemometric Advances

Emil W. Ciurczak, Dan Higgins, and Claudio Puebla

In compiling the tenth anniversary issue of Spectroscopy *(October 1995), the editorial staff received a significant amount of input from scientists anxious to express their views about the preceding decade in spectroscopy. Inevitably (and to the benefit of this article) not all the contributed information fit into the commemorative issue. Several contributors had some highly relevant things to say on the subject of molecular spectroscopy, and we did not want to let this useful information go to waste.*

The result is this three-part chapter. In the first section, I am pleased to present the insights of Dan Higgins, who reports on recent work conducted in Paul Barbara's group using near-field scanning optical microscopy. This emerging microscopic method is highly compatible with a range of spectroscopic techniques and is making inroads in the field of single-cell analysis and in the characterization of materials used in detector and display devices. In part two, Claudio Puebla comments on the past 10 years in the rapidly expanding field of computational chemistry and its impact on UV/vis, infrared, and Raman spectroscopy. Finally, I will end the chapter by discussing the use of chemometrics in vibrational spectroscopy.

I recently decided that computers just might be around for awhile. Consequently, I sold some stock in my diamond mine (in South Florida) and invested in a new Pentium-based toy. In addition to the box, I received a modem, which led me right onto e-mail. So, while I listen to ZZ Top on my SoundBlaster, I can commune with fellow spectroscopists. (As an aside, anyone who has not yet visited the news group sci.techniques.spectroscopy *should definitely do so. It is very much worth perusing.)*

Dan Higgins, University of Minnesota, Minneapolis, MN

Researchers in Paul Barbara's group at the University of Minnesota chemistry department are involved in many projects combining spectroscopy and imaging with the emerging technique of near-field scanning optical microscopy (NSOM). Useful in semiconductor and materials research, medical research, DNA sequencing and gene mapping, and other fields, NSOM is applied by our group principally for the investigation of mesoscopic materials with applications in emerging display and detector technologies.

Developed primarily at AT&T Bell Laboratories (Murray Hill, NJ), NSOM effectively combines the advantages of optical microscopy with the high-resolution capabilities of a scanning probe microscope. It can overcome the diffraction barrier that otherwise limits the image resolution of a conventional microscope and allows imaging of features much smaller than the wavelength of light.

Single-molecule detection (SMD) is one of NSOM's most compelling applications and has been a major focus of numerous research groups. SMD was first reported by Betzig and co-workers, who employed an NSOM system built in-house (1). Our group has demonstrated SMD with the same compound as Betzig (DiI, a dye compound) on a modified commercial system (Aurora, TopoMetrix, Santa Clara, CA). The experiment involves spreading a submonolayer of dye molecules on a surface from a dilute solution so that single molecules can be isolated under the NSOM probe. Combined with spectroscopic sample investigation, SMD provides:

- A benchmark for the high spectroscopic sensitivity of the NSOM instrument;
- The capability of using a single chromophore (fluorescent molecule) as a probe of the local environment via the temporal and spectral characteristics of the single-molecule fluorescence;
- More specific sample information, because the detected signal is no longer an average of the signal collected from multiple molecules;
- Minimal external tagging of biological specimens for fluorescence imaging.

Figure 1 shows single molecules of DiI on a PMMA film imaged with the commercial NSOM system. The x, y axes are labeled in micrometers (the

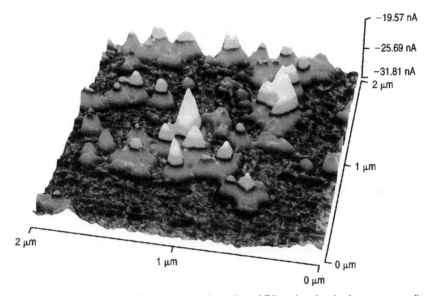

Figure 1. Single-molecule fluorescence detection of DiI molecules (a dye compound) on PMMA on a silica substrate acquired with a near-field scanning optical microscope (Topo-Metrix, Santa Clara, CA). Image courtesy Dan Higgins and Paul Barbara, Department of Chemistry, University of Minnesota.

scan dimensions) and the z axis in photon counts (the fluorescence from the molecules). It is believed that the spatial intensity profile for an individual molecule indicates its orientation with respect to the NSOM probe (1).

Verification that SMD has been achieved is accomplished first by comparing the observed single-molecule surface concentration to that expected for the spin-coating solution employed, and second by correlation of the number of photons counted by the system with those expected from a single molecule. According to Sunney Xie at Pacific Northwest Laboratories (Richland, WA), SMD "sensitivity results from the high photon flux delivered by the tapered single-mode fiber probe, efficient background rejection resulting from the small illumination area, and possibly, excitation by strong evanescent wave components near the tip end" (2).

Our group modified the commercial NSOM instrument by adding a silicon-based avalanche photodetector for photon counting and a high-numerical-aperture oil immersion transmission objective lens.

Investigating mesoscopic materials. Mesoscopic materials are of interest because of their important applications in novel display and detector technologies and also because of their interesting optical and electronic properties that vary over distance scales as small as a few tens of nanometers. Our group has extensively studied the J aggregates of pseudoisocyanine (PIC). Until the recent advent of NSOM, such systems could not be fully investigated because optical microscopes were not capable of subwavelength resolution. With NSOM, optical imaging of such samples is achieved via a number of different contrast mechanisms, including absorption, fluorescence, and scattering. In addition, the implementation of visible light as the imaging mechanism in NSOM provides for utilization of numerous well-known spectroscopic techniques in NSOM experiments. The static properties of local, nanometer-scale regions of mesoscopic materials can be directly probed with NSOM, as can important dynamic processes such as excited-state decay and charge carrier dynamics. These important photophysical properties can be directly probed by coupling NSOM with time-resolved spectroscopies such as time-correlated, single-photon counting, or ultrafast pump-probe methods.

In these experiments, fluorescence NSOM has been used to directly study the morphology of the highly structured films of *J* aggregates that form on quartz substrates when a PIC/polymer mixture is spin-coated in a thin film on a quartz substrate (3). In subsequent experiments, we used polarization-dependent fluorescence excitation and emission methods with NSOM to directly measure the local orientation of the excitonic transition dipoles in the aggregates (4). From these results, we determined the local molecular orientation and extent of orientational order within the aggregates. We measured the local fluorescence spectra of the aggregates by coupling the NSOM to a monochromator and CCD; the

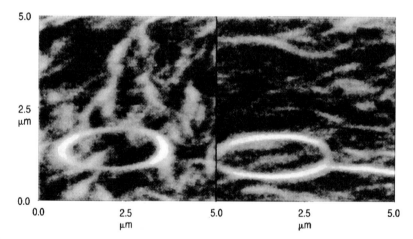

Figure 2. Polarization-dependent fluorescence image of pseudoisocyanine *J* aggregates grown in a poly(vinyl sulfate) film coated on fused quartz. The image demonstrates that the fluorescence is highly polarized along the long axis of the aggregates. The polarization effects are shown most dramatically by the circular aggregate in the image, which shows uniform intensity in unpolarized fluorescence images. The *J* aggregates were excited with 514-nm light from the NSOM tip; the fluorescence (λ_{max} = 574 nm) was collected in the far-field, passed through a polarizer, and detected with a photon-counting detector. Image courtesy of Josef Kerimo, Dan Higgins, and Paul Barbara, Department of Chemistry, University of Minnesota.

results of these local spectroscopic measurements yielded further information on structural order in the aggregates (3).

The important dynamic properties of the aggregates, including excited-state lifetimes in local regions, have been probed using time-correlated single-photon counting methods coupled with NSOM (4). We also measured excited-state migration distances in the aggregates by locally photobleaching the aggregates. In this experiment we illuminated one point on an aggregate with the NSOM probe, optically measured the size of the photobleached hole, and compared it to the size of the originally illuminated area (3). Figure 2 shows NSOM images representative of the work described above. In the future, we plan to develop these and other static and time-resolved near-field methods on the *J* aggregates *and* on new samples, including luminescent porous silicon and organic polymer-based light-emitting diodes. These new studies will focus on the effects of material structure on the static and time-resolved electronic properties of these interesting materials.

TEN YEARS OF COMPUTATIONAL CHEMISTRY: WHERE WE'VE BEEN AND WHERE WE'RE GOING

Claudio Puebla, Ciba–Geigy, Grenzach, Germany

One of the main aims of spectroscopy is the determination of molecular structure and properties. Much information can be gleaned from studying the interaction of electromagnetic radiation and matter, and this information can then be interpreted with adequate models. The task of theoretical work is to devise and refine such models that, when checked against experimental results, provide deep insight into the principles of the phenomena and, ultimately, nature itself. This field is currently based on the application of quantum-mechanical principles to study molecular phenomena and is known as quantum—or computational—chemistry.

The past 10 years have seen an extraordinary leap forward in how theoretical methods are applied to molecular spectroscopic problems in the real world. All three main paradigms of computational chemistry—molecular mechanics, semiempirical, and *ab initio* methods—have profited from the increasing and direct availability of computer power, especially *cheap* computer power. Although applied in different contexts, all three paradigms are strongly dependent on computer speed and storage to produce results within a reasonable time frame. Advances in parallel processing have had a great impact on the size of the molecules amenable to calculation; this trend has definitely characterized the past 10 years and will continue in the future. Although the past decade has seen the emergence (and what now appears to be the decline) of supercomputers, growth in the use of workstations has been steady—it will probably dominate the numerical world in the future. Massive parallelization of low-cost processors is an emerging area and will develop extensively in the years to come.

Ten years ago computational methods were severely limited by the size of the molecular problem. Large systems (≤ 1000 atoms) were the exclusive domain of molecular mechanics—the only paradigm that is based on classical rather than quantum mechanics. Semiempirical methods were applied for medium-sized (≤ 100 atoms), while *ab initio* methods were strictly limited to small systems (<10 atoms). While size is still a limiting factor, the order of magnitude has changed. This is particularly true with *ab initio* methods, where big clusters and fullerenes are now within reach. Recent development of *ab initio* methods based on density-functional formalism has definitely aided in the analysis of larger systems and in the calculation of molecules containing transition metals. Excited states and weakly bonded molecules have also become manageable; applications in spectroscopic and photochemical processes in matrix-isolated and jetcooled molecules are good examples of theory and experiment working together to produce better insight. Another exciting area is the examination of light from outer space. These data have proved to be rich in information about chemistry under extreme conditions and are thus a fruitful field for the evaluation of theoretical models.

Cheap computer power also had enabled the development of customer-friendly software. The past few years have introduced many computational packages that run on PCs right in the user's office. They incorporate semiempirical methods for calculations of molecules of respectable size and can calculate UV/vis, infrared, and Raman spectra with enough accuracy to obtain working parameters such as oscillator strengths, vibrational frequencies, and force-field constants. They also provide data for charge distributions and highest occupied molecular orbital/lowest occupied molecular orbital (HOMO–LUMO) analyses that allow insight into the path of chemical reactions.

Current challenges such as solvent effect and molecular interaction are being tackled by *ab initio* methods and Monte Carlo simulations. Much development is expected in these areas within the next several years. The ultimate goal—full calculations of chemical reactions—will probably remain beyond reach in the near future.

COMPUTATION OF THE REQUIRED NUMBER OF SAMPLES FOR A CHEMOMETRIC ANALYSIS WHEN USING VIBRATIONAL SPECTROSCOPY

Although near-infrared analysis may often require larger numbers of samples than midrange infrared spectroscopy, resorting to chemometrics implies that Beer's law is not applicable. If few (or no) peaks are usable for a simple regression equation or the matrix is unstable (note: by matrix, I also mean to include temperature, viscosity, color, and refractive index along with chemical moieties), chemometric approaches almost certainly will require more samples than a system conforming to Beer's law.

A rule of thumb (5) is that you must have ($n \times 10$) + 10 samples to perform a reasonable feasibility study or an interim calibration. (In this case n is the number of constituents measured.) So, for a system wherein three constituents are predicted, (3×10) + 10 = 40 calibration samples are needed. Of course, this assumes some control of the reaction process. This low-end figure might work best with a very tightly controlled process, for example, a pharmaceutical granulation process that is heavily regulated and validated for a large number of lots.

If the chemical process contains raw materials that are more varied than in a pharmaceutical process where the temperatures and/or the pH are uncontrolled, then more samples are definitely required.

Several years ago, I consulted with a polymer company that used the effluent from one process as the feed-stock for a second. Monomer (from the initial reaction), sulfuric acid, water, alcohol, and acetone were uncontrolled and unmeasured. In this case, I had to recommend 1200 calibration samples! If you are unable or unwilling to monitor several factors such as temperature, pH, etc., then you may use the power of partial least squares (PLS), principal components analysis (PCA), or the "dreaded" neural network (NN).

As with any potential client, I caution you to first try simple math (yes, K.I.S.S. applies to all branches of science). If the system bends to the will of a simpler math approach, by all means listen to the numbers. Matrix approaches in the hands of amateurs can be dangerous: PLS can begin to model noise, fingerprints, and phases of the moon, and it was a neural net that determined that the Iranian passenger plane shot down several years ago by the U.S. Navy was a fighter jet! The more tightly you model any system, the greater chance you take on being able to predict only one of your calibration samples.

While third-party software is getting better and stronger, our understanding seems to be static. Most manufacturers of vibrational spectroscopy equipment offer software courses ... take them, unless you are really sure that you know what you are doing!

Sources of error can be lurking in the instrument, samples, or software. Caution is recommended! Ask someone who knows!

REFERENCES

1. E. Betzig and R. J. Chichester, *Science* **262**, 1422–25 (1993).
2. X. S. Xie and R. C. Dunn, *Science* **265**, 361–64 (1994).
3. D. A. Higgins and P. F. Barbara, *J. Phys. Chem.* **99**, 3 (1995).
4. D. A. Higgins, P. J. Reid, and P. F. Barbara, *J. Phys. Chem.*, **100**, 1174–1180 (1996).
5. J. R. Workman Jr., in *Handbook of Near-Infrared Analysis*, D. A. Burns and E. W. Ciurczak, eds. (Marcel Dekker, New York, 1992).

CHAPTER III.1

PURGAMENTA INIUNT, PURGAMENTA EXIUNT

As you may recall, I recently moved, and I'm getting used to living in sunny Maryland instead of New Jersey. Funny, I always thought that the left lane was for the faster cars and that you couldn't make a left turn from the right lane... Oh well, when in Rome, look Italian, I guess. Before I get into the first of my troubleshooting tips, I would like to briefly address the title of this month's column: "Garbage In, Garbage Out."

Recently, I overheard a conversation in which a marketing specialist from a spectro-scopic instrument company was discussing software. He commented that the displays for all spectroscopic instruments should be snazzier because that's what FT-IR users are used to and like. Not that they should work better, mind you, only look better. This got me to thinking about chemical education vis-a-vis instrument training.

Over the years, educators have been faced with the challenge of introducing an increasing number of topics to students in the same four-year time span. In the short run, it is easier to purchase automated instruments because they are faster to set up and run in the allotted laboratory period. The ubiquitous computer remembers the flow rates, scan rates, temperatures, and so forth, so that students can carry on with the "real" lab experiment. By that, the professor usually means injecting a sample into a liquid or gas chromatograph, or watching the teaching assistant place it into an FT-IR system. The resultant chromatogram or spectrum is then given to the students to interpret. This leads to an incredible lack of understanding about what exactly happens inside the instrument.

As illustrations of this type of education in practice, let me share with you a couple of stories. I once worked with a young Ph.D. who complained about high back pressure in his high-performance liquid chromatograph. When I asked what he was running, he informed me that he was cleaning his column with methanol. Further questioning disclosed that the previous mobile phase con-tained a phosphate buffer. When I asked the young doctor about the solubility of sodium phosphate in methanol, the truth dawned upon the lad ... precipitation. He knew the instrument but ignored the chemistry. Recently, a would-be customer asked me if a spectrum that I had messed up (I removed the probe before the scan was complete) could be improved upon using the software. I hesitated for about eight milliseconds, then answered, "Software cannot make bad data better."

This leads me to the title of this column. Instrument manufacturers have a responsibility not to confuse pretty colors on the screen with good, robust data. No amount of computer wizardry can make up for a noisy monochromator, high stray light, or poor sample preparation. I must implore my fellow analysts not to "ignore the man behind the curtain." We must always keep in mind that the first imperative is to choose the best instrument for the job. Then, and only then, add a software package that allows for the best use of that instrument.

FT-IR SAMPLE PREP

I offer as a start in our troubleshooting series a humble document on KBr pellets. Yes, even in the midst of our "space age" sophistication in FT-IR instrumentation, the sample introduction can be a problem. Still the mainstay of educators, the salt pellet is the most widely used and least understood sample preparation method for infrared spectroscopy. I was lucky enough to have taken a Coblentz Society short course in the year 19.. , well, suffice it to say, several years ago. In any case, the whole course was on pellets (Ps) and mulls (Ms) and was a real eye-opener.

The biggest problem I have encountered in both P& M preparation is that the samples are seldom ground finely enough. If a Wiggle Bug grinder is used, too often a plastic container is also used for speed and cost considerations. Here, the grinding time is critical: Too short and you have large sample particles, too long and you grind the plastic into the sample mix. The former shows up as a "measles" effect on the pellet: little spots throughout the pellet and a spectrum with a sloping baseline. The latter gives a ghost polymer spectrum.

Often students are shown that mull-making involves mixing some powder and a few drops of Nujol, and then grinding the two plates together. (Honest! Would I make up something like that?) Instead, the sample material should be smeared on the agate mortar prior to any addition of oil, thus giving a fine dispersion, which will yield a smooth spectral baseline.

The vacuum that is applied to the pellet maker (if that sounds foreign to you, I already know your problem) will, upon pressing, desiccate the system. If (latent) water is present, the resultant pellet will resemble a chocolate chip cookie. The blotches indicate the presence of water, either from the KBr or the sample. And, of course, every spectroscopist knows what water will do to a spectrum.

A half-moon effect (or "happy-face smile") will result if the powder is poured into the die unevenly. The thickness of the wedge-shaped pellet will vary, accompanied by varying degrees of transparency. Too much material in the die will result in an opaque pellet, while too little powder will result in a brittle pellet.

The point of this lesson is that as much effort should be put into the sample-handling aspect of training analysts and technicians as in worrying about the correct instrument settings. In conclusion, reread the chapter title for the punch line.

CHAPTER III.2

A LOOK AT VALIDATION FOR SPECTROSCOPIC INSTRUMENTATION: THE FDA HAS TAKEN THE LEAD

Although I hate to be a pothole in the road to success, the subject of this chapter has risen as a dragon to battle. For those in the pharmaceutical industry, validation of manufacturing processes and analytical methods is nothing new. What has been growing over the past decade, however, is instrument and computer validation.

As those of us in the industry are aware, even a small nonvalidated portion of a process may be enough to initiate a Food and Drug Administration (FDA) response ranging from a Form 483 observation (misdemeanor warning letter) to a full product recall. I have teamed up with Ron Torlini, manager of regulatory compliance at DuPont–Merck Pharmaceuticals (Wilmington, DE), to focus on the validation issue, which may become a major problem in the near future. Portions of this chapter were presented by Torlini at a PharmAnalysis Conference (1).

Although the FDA was quiet in the 1980s because of budget cutbacks and the overall non-enforcement attitude of the executive branch, several scandals and an aggressive FDA commissioner have changed all that. Although some may say that the FDA has gone too far in the other direction, we must be prepared to live with the new FDA. The agency is now heavily into documentation of every level of pharmaceutical life. Every nuance must be recorded and verified, or it is deemed nonexistent.

Even in a well-regulated laboratory environment, analysts may be prone to skipping the documentation step of writing in their books. Lab work is interesting; writing in a notebook is not. Guidelines exist (along with gentle prodding by the FDA) for what should be written in a notebook during both method development and routine analyses: solvents (and lot numbers), column types, pipette and glassware volumes, weights, all observations and readings (such as spectra and chromatograms), even sample calculations, and, of course, final results.

Unfortunately, clear-cut regulations for instrument validation are not as readily available, differing widely from company to company. Spectroscopic methods traditionally include wavelength-specific instructions, comparison with a standard, some type of linearity check, solvent specificity, and concentration instructions. Usually, chemical procedures have more safeguards than an instrumental method. Depending on which department you query, the extent of calibration of the instrument varies greatly.

Quality control (QC) departments often have the most stringent rules for calibration of equipment. The theory used to be that technicians were less trained or experienced than methods development scientists and so needed more guidance.

In reality, QC analysts use their equipment for a far greater number of assays in a given week than the R&D scientists. It is the research types who are usually guilty of developing a method on noncalibrated instruments, liquid chromatography columns with pasts more secret than Spiderman, and outdated solutions. Log books? Obviously only for QC types ... beneath the dignity of real scientists.

The sad truth is that few instruments are scrupulously monitored, calibrated, and subject to routine maintenance. The setup in the QA department at a former employer of this columnist was so rigorous that the personnel entrusted with the work were vigorously recruited by instrument companies. However, this is usually the exception. The theory often seems to be "as long as it passes system suitability, don't bother it." Each part of any instrument, be it hardware or software, must involve, at a minimum, general challenges for success. The challenges must address the following:

- Performance verification;
- Content;
- Documentation;
- Archiving;
- Training.

Reference standards should be evaluated prior to use. Various reagents and secondary equipment, such as temperature probes, must be included in the validation program. Trends in maintenance and repair records must be documented and historical performance checked. Cleaning instructions must be detailed to ensure that no residues are left on injection systems or windows.

Eventually, all spectrophotometers will be checked for linearity with National Institute of Standards and Technology (NIST) -traceable interference filters or solutions, the wavelength accuracy will be evaluated using NIST-supplied rare-earth oxide standards, and the instruments will be mechanically evaluated at prescribed times. Eventually, almost all analytical instruments will be linked to computers for both control and data manipulation. Unfortunately, computer operators and programmers do not have such clear-cut guidelines as analysts do.

Initially, the FDA told us that all computer systems used to calculate values must be validated. When we asked for guidelines, we were told, in essence, "do something and we'll tell you if it is enough." I would guess that the agency didn't have any clearer a picture of what was needed for a "necessary and sufficient" validation of a computer system than the programmers and end users.

One problem was that while all programmers have flow sheets for the programs, no two programmers approach the problem in the same manner. To further confuse the issue, "debugging" a system is often not documented. That is, a programmer will change one or several lines of the program to make it work without proving that the changes will not cause the answers to vary from the previous version. Software is the "brains" of the analytical instrument, and validation must incorporate the attributes of software (2) (see Table I). Companies should not assume that supplied software has been validated by the manufacturer, just as they should never assume that incoming raw materials are within company specifications.

Table I. Attributes of hardware and software

Hardware
CPU Floppy disks/hard drive Modems Wiring
Software
Source codes Calculations Data input/output functions Reports

The design documentation of any analytical system should incorporate the following:

- Hardware and software specifications;
- System/data flow;
- Software interface documentation;
- Hardware interface documentation;
- Input/output design documentation;
- Security specifications;
- Data structure;
- Control and update consistency.

This is but a bare outline of some points—the FDA will be looking for documentation on every aspect of design and debugging of any system.

It is not enough for an algorithm to add two and two and arrive at four. The source code must be available to prove that the program didn't use squares, square roots, and cosines in the process of adding 2 + 2. In other words, how we arrive at the answer is as important as the accuracy of the answer itself. Although at present we have no standard procedures for calculation data, they will surely follow in the future. As with gas chromatography packings and fittings in the pioneer days, codification allowed better work to be done. Of course, many programmers will whine about losing their freedom, but they will eventually come into the fold ... if they want to sell instruments to pharmaceutical manufacturers.

Some manufacturers of equipment have already anticipated the need for validated software and are working in that direction. Hewlett Packard (Palo Alto, CA), for example, offers a printed package for validation as well as being willing to allow the FDA access to source codes. NIRSystems (Silver Spring, MD) is rewriting its proprietary software in a Windows format for easier revalidation in case of updates. As stated previously, not only must the program be reproducible, but each function must be documented. This level of validation is of a assured that our electronic standards are "traceable" as well. Most references will be mere magnetic pulses on a tape or disk, so an analyst (and the FDA) must be sure that button "A" will produce spectrum "1" every time with no variations.

In the realm of process control, assay times will be compressed from minutes and hours to seconds and fractions of seconds. We are depending more heavily on our friends, the computers, to be our hands and eyes in real-time analyses of the manufacturing processes. We must have systems for which the software is as reliable as the instruments themselves. This could possibly be the first time that the FDA is totally correct ... and timely, to boot.

The big payoff to companies and industries not even remotely involved in pharmaceutical production is better instrumentation. Whenever the FDA or EPA changes detection limits or acceptable limits of various trace materials, instrument companies rush to improve their products. Thus the next generation of instruments is even better than expected. Painful or not, regulation is basically a good thing.

REFERENCES

1. R. P. Torlini, *Proc. PharmAnalysis*, June 1993, East Brunswick, NJ.
2. R. F. Tetzlaff, *Pharm. Tech.* **16**(5), 70 (1992).

CHAPTER III.3

CHEMICAL INTUITION IN ANALYTICAL TESTING: A CASE STUDY

Elisa Lyth and Lily M. Ng

Occasionally, a kindred spirit produces a nifty paper that sounds like something I would like to say (but didn't). One such paper by Elisa Lyth and Lily M. Ng (Department of Chemistry, Cleveland State University, Cleveland, OH 44115) states that (1) simply running an analysis isn't enough, and (2) there are coincidences or the world truly is holistic. I like the detective work involved in this chapter, which proves (again) that good analytical chemistry is forensic by nature.

Lyth is a doctoral candidate in chemistry and works as an independent consultant on household lead contamination. Now an assistant professor, Ng earned her Ph.D. from the University of Pittsburgh. Her research interests include FT-IR spectroscopy of the gas–solid and liquid–solid interfaces. Without further ado, Gentle Readers, I present the chapter (please, no talking during the show).

Fourier transform infrared (FT-IR) spectroscopy is used by analytical testing laboratories as a "fingerprinting" technique for identifying industrial chemicals. When the chemical is a contaminant found on a commercial product, the contaminant is usually extracted by a solvent such as chloroform before it can be analyzed spectroscopically. The infrared spectrum obtained is then matched with published spectra of known or suspected compounds. This process is considered routine. However, exclusive concentration on fingerprinting without careful consideration of sample preparation techniques can be detrimental to the analysis and lead to erroneous results.

Chemical knowledge and intuition can help in designing the appropriate analytical experiments to solve the problem. The case discussed here shows that chemical intuition and careful experimentation can also prevent the need for potentially expensive solutions.

THE PROBLEM

Polyethylene lids destined for ice cream containers were found to be contaminated by a film covering their top side. The film prevented printing inks from adhering to the surface, and by the time the ice cream container reached the checkout counter, the printing had come off the lid. This problem was not uniformly present in all of the lids manufactured at the same time. Nor was this problem noted in the tub portion of the container.

The film was visible on the top surface of the lids only, whereas the bottom surface appeared to be free of contamination. The lids, which were pressed in a mold treated with a silicone lubricant, did not appear to be contaminated at the time of manufacture. The problem was noted only when the printing began coming off at the supermarket.

The manufactured lids were routinely packed in plastic bags, which were then placed in cardboard boxes for storage. Lids were stored in a warehouse under unknown conditions for weeks to years before printing.

Infrared spectra of a chloroform solution of the film obtained by a consulting analytical lab (Lab A) indicated that the surface contaminant was probably a result of the degradation of the coloring vehicle, supplied by Manufacturer B. Manufacturer B requested our help in identifying which of the coloring vehicle's four components caused the contamination. The manufacturer was prepared to change the formulation and manufacturing processes, switch to new suppliers of the coloring vehicle components, and discard the stock of the "culprit" components and contaminated lids. All in all, a costly solution.

EXPERIMENTS

It was assumed that the previous analysis was fundamentally correct. If the coloring vehicle was present in the contaminant, then exposure to heat might be a causative factor. Because only part of the lot was affected and the contamination was not apparent at the time of manufacture, it was assumed that heat exposure during storage caused some part of the coloring vehicle to degrade and form a surface deposit.

On that assumption, we performed a sequence of analytical developments to identify the contaminant. Four components of the coloring vehicle, (a), (b), (c), and (d), were supplied by Manufacturer B. The infrared spectra of these components and the samples following were recorded using an ATI/Mattson (Madison, WI) Galaxy 2020 FT-IR spectrometer. The samples of the liquid components were prepared by placing a drop of liquid between two KBr disks. The disks were polished between each procedure because of the difficulty of removing the entire sample from the surfaces. Data analyses were done using the First software (ATI/Mattson). Several methods were used to extract the contaminant for analysis.

Experiment 1. The first method involved extracting the contaminant found on the top surface of polyethylene Lid 1 by applying ~ 10 ml chloroform to the surface. The solution was immediately removed from the lid, pipetted onto a KBr window, and a spectrum of the residue obtained. The procedure was repeated on the underside of Lid 1.

Experiment 2. Because chloroform is a powerful solvent, the possibility existed that the vehicle in the bulk of the lid could be extracted out with the surface contaminant. We decided to try a "gentler" method of extraction. A tissue, which had been soaked in chloroform to remove extractable chemicals, was used to obtain a background spectrum. The surface of Lid 2 was wiped with the "cleaned" tissue. A spectrum of the contaminant on the tissue was also recorded.

Experiment 3. We speculated that if the contaminant consisted of one or more components of the vehicle, exposure to heat may bring more contaminant to the surface. Lid 3 was wiped with chloroform to remove the contaminant from the surface. The lid was placed in an oven at 40°C, and after 3 h the temperature was increased to 60°C. After an additional 3 h, it was determined that the lid should remain in the oven overnight and be examined the next morning.

Experiment 4. The contaminant on Lid 4 was removed by wiping the surface with a piece of polyethylene. The polyethylene was rinsed with chloroform, and the residue was concentrated by evaporating the solvent. The residue was transferred to a second piece of polyethylene and mounted in a sample holder. A spectrum of the polyethylene background was subtracted from the spectrum of the residue on polyethylene, yielding a spectrum of the residue only. ATI's ICON library of reference spectra was searched for possible matches.

RESULTS AND DISCUSSION

The spectrum obtained in the first experiment did not resemble the data provided by Lab A. The testing lab must have soaked the lid in chloroform over a long period of time, thus dissolving the coloring vehicle in the solvent. The resulting spectrum would lead to the erroneous conclusion that the contaminant film resulted from degradation of the coloring vehicle. We were obliged to design three additional experiments to identify the contaminant.

The spectrum of the chloroform extraction of the bottom of Lid 1 showed no vibrational feature, indicating the lack of contamination. If the coloring vehicle was present because of degradation, we would expect to find it on both surfaces.

The top surface of the injection mold was treated with a silicone lubricant. A spectrum of a silicone spray was obtained to determine whether such a lubricant might account for the film. It was determined that the contaminant was not a silicone material.

Infrared spectra obtained on the heated lids have no distinctive absorption peaks. Heating did not result in the formation of a visible film. Testing indicated that the contaminant was not a substance used in the manufacturing process.

The spectrum of the residue obtained in Experiment 4 is shown in Fig. 1a. The background spectrum of polyethylene was subtracted from it to give Fig. 1c. Distinct vibrational features appeared at ~3432, 2130, 1648, and 1277 cm^{-1}, and small vibrational features were evident at 1386 and ~1100 cm^{-1}.

Figure 2 compares the spectrum of the residue to components of the coloring vehicle. It can be easily seen that the residue is not chemically equivalent to any of the four components. In fact, a search of the ICON library of reference spectra showed that the spectrum of the residue shown in Fig. 2 is almost identical to the infrared spectrum of mixed pyroligneous acids (Fig. 3). These acids result from the burning of plant materials and are not expected to be present in any phase of food preparation or food container manufacturing.

We suspected that the contaminating film resulted from the lids being

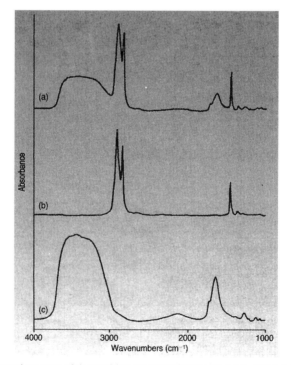

Figure 1. Infrared spectra of the residue on polyethylene film (a), clean polyethylene film (b), and the residue (c). Spectra (c) is the result of subtracting (b) from (a).

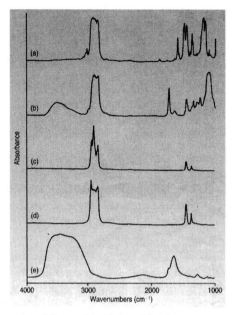

Figure 2. Infrared spectra of the residue (e) and the four components (a–d) of the color vehicle.

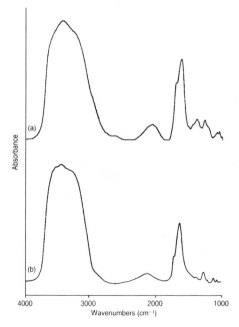

Figure 3. Infrared spectrum of the residue compared to infrared spectrum of pyroligneous acids.

exposed to smoke of some kind. We asked Manufacturer B about the possibility that some of the workers may have been smoking cigarettes when the lids were being packed, which would explain why only some lids were contaminated. To our surprise, the manufacturer informed us that a fire had occurred in the section of the warehouse where the polyethylene lids were stored.

CONCLUSIONS

The presence of pyroligneous acids in the contaminant film recovered from the nonprintable lids indicated that the lids had been affected by exposure to smoke during packing, shipping, or storage. This conclusion also explained why only some of the lids were affected whereas the containers were unaffected. Because the surface contaminant was water soluble, simply washing the lids with water would solve the contamination problem.

Routine testing using infrared spectroscopy can overemphasize "fingerprint-ing" compounds as the means to identify industrial contaminants. In fact, the method used to extract the contaminants from the manufactured product is very important, and if not done properly, can lead to erroneous results. Using chemi-cal intuition to design the appropriate analytical experiments and to interpret the data obtained can help manufacturers identify problems expediently. And, as in this case, costly modifications of manufacturing processes can be avoided and discarding products deemed "unusable" by previous testing can be prevented.

REFERENCES

For general information about FT-IR spectroscopy, see:

1. D. C. Harris, *Quantitative Chemical Analysis*, 3rd Ed. (W. H. Freeman, New York, 1991).
2. J. R. Ferraro and K. Krishnan, eds., *Practical Fourier Transform Infrared Spec-troscopy* (Academic Press, New York, 1989).

CHAPTER III.4

TWO VALUABLE TOOLS FOR INSTRUMENT EVALUATION: JCAMP.DX AND PRICE TAGS

As I was thinking of a topic I wanted to cover in a chapter, I realized that my approach might be a little bit like Andy Rooney's. I have wanted to share my views on a couple of subjects with my fellow scientists for a while now, and my thoughts were galvanized while watching, of all things, TV ads for discount appliance stores—a thought process worthy of Rooney, I think. Of course, he gets paid a great deal of money to wonder out loud; I just hope to stimulate discussion in the spectroscopic community.

APPLES AND ORANGES

In commercials, each discount appliance store "guarantees" the lowest prices and offers customers substantial gifts if they can find a lower price somewhere else. How can they do this? Well, for one thing, each store is the exclusive distributor of certain brands of appliances and can offer the lowest price for those brands. For the widely available brands, there are so many options and models that comparison shopping is almost impossible.

Fortunately, scientists don't have to contend with this type of hype. Or do we? Over the years, I have been bombarded with instrument manufacturers' claims of the lowest noise levels anywhere. (Although I still haven't seen an offer of double my money back if the claim proves untrue.) The problem is that no two vendors measure noise in the same manner. For instance, in trying to compare HPLC detectors, I've found noise measured using all sorts of wavelengths, with air or mobile phase (or water) filling the cell, and with static or moving solutions. Because the detector in HPLC is really a small, specialized spectrophotometer, I began to wonder about various manufacturers' noise claims for such things as IR, near-IR, and UV/vis spectrometers. To help me decide which claims are closest to the truth, I looked up some textbook definitions and examples of noise and signal-to-noise ratios.

One of the best discussions of noise I found was in a new book by Ingle and Crouch (1), which covers the subject quite nicely. The authors break down noise into two categories. Additive noise (often seen at 0% transmittance) is independent of the signal and is what I most often think of as "instrument noise." This noise can come from the detector, amplifier, and background (these can be lumped under "dark current," for those who remember that term). The noise carried by the signal is called *multiplicative noise*; this term usually refers only to analytical signal flicker, which is directly proportional to the signal. Modulation, time-domain filtering, signal averaging, and ensemble averaging of the signal are some of the suggested means of reducing noise.

An older article from Chaykowsky and Moore (2) gives practical ways to eliminate stray light, reduce bandwidth, and make other corrections for the instrument. The only trouble with suggestions such as these is that, in most cases, we can't design or modify the instruments we purchase. Instruments generally are sold with factory-set configurations. And we are discouraged from modifying our current instruments because our employers tend to frown upon our tinkering with $70,000 instruments for seemingly esoteric reasons. So we are back again to trying to sort out information supplied by vendors that is not easily comparable. What are we poor practicing spectroscopists to do?

JCAMP.DX

I'm glad I asked. Recently, I have been playing with something called JCAMP.DX (from the Joint Committee on Atomic and Molecular Physical Data, a subcommittee of the ASTM), which is a standard file structure that can be used to convert spectra (UV, IR, near-IR, NMR, and so forth) into a standardized format. The data are stored as labeled fields of variable length using printable ASCII characters (3). Labels, axis nomenclature, and constituent values all may be stored in a JCAMP.DX file.

Why is this so good? Now instrument purveyors can back up their claims. All that is required is that the potential customer supply vendors with the same material for scanning on the respective instruments under consideration. The vendors will then return the spectra in JCAMP.DX format, and the chemist may compare them at his or her leisure. For a given standard, one may compare wavelength accuracy, repeatability after numerous scans, and, of course, instrument noise.

For near-IR reflectance instruments, a porcelain disk, Teflon powder, or some other standard may be used. The supplier may tune, adjust, or otherwise "tweak" an instrument so it performs at its best and then run the sample according to agreed-upon guidelines. The scans, in JCAMP.DX format, can then be evaluated by either a neutral party or the customer using any or all of the suppliers' noise programs. The customer will then have the facts (no truth, only facts!) to work with.

Of course, prepurchase comparison is just one of JCAMP.DX's many potential applications (4–6) in the exchange of spectroscopic data. Although the opportunity for this type of clear communication about an instrument's capabilities vis-à-vis other instruments is available now, I wonder how many dealers will manifest the cooperation necessary to meet this challenge. I have been impressed that all the near-IR instrument manufacturers accepted the format so easily and quickly. Now I anxiously await the response of the rest of the spectroscopic world.

A ROSE BY ANY OTHER PRICE ...

Speaking of inexpensive instruments (as I will in Chapter IV.B.2), I am pleased to see the number of sophisticated, inexpensive instruments that have come on the market in recent years. I will mention a few of which I am aware, but mean no offense to those I miss. And no endorsement is intended because I have not even had a chance to touch most of these "toys." I don't intend to assert that these low-cost instruments are the technological equals of the super-charged research-grade models, but they do offer an alternative that is important to me as a college teacher at a small school. Even a university as prestigious as Princeton (just a few exits down the New Jersey Turnpike) has limited funds for new equipment. In addition to schools, a fair number of small- to medium-sized labs have recently come into existence in response to the growing need for environmental testing. These "mom and pop" type operations cannot easily afford a large number of $100,000 purchases.

One of the leaders in checkbook-friendly equipment has been Buck Scientific (East Norwalk, Connecticut). After starting out as a repair firm and then refurbishing older instruments, Buck now has its own product line. Imagine! Dispersive IR, UV/vis, and atomic absorption spectrometers for less than $10,000. However, the first low-cost FT-IR spectrometer I remember seeing was introduced a few years ago by Perkin–Elmer (Norwalk, Connecticut), a huge company. That introduction started an industry-wide trend toward "smallish" FT-IR instruments. Once the technology has been established by leaders in a given area, they or others can use laser-scored gratings, mass-produced computer chips, and readily available software to produce low-cost instruments with specifications that were undreamed of 10 years ago. Another example of this trend is the diode-array UV/vis spectrometer produced by Hewlett–Packard (Palo Alto, California). By taking the control out of the instrument and allowing a minicomputer to control it externally, HP was able to reduce the cost to the 7-kilobuck range. These are only a few examples of the trend to small but powerful instruments.

Again, I am not saying that the low-cost instruments are always as good as their more sophisticated counterparts. I am merely saluting the manufacturers—both those cited and others leading this trend—who are providing these bargains for us smaller customers. I have always taught in my analytical chemistry courses that it is wasteful to be more accurate than necessary. If you only want to know whether the water in your swimming pool is acidic or basic, you do not need a $3000 pH meter capable of reading to five significant figures. Litmus paper will do. Many industrial customers make vanity purchases—a sophisticated instrument is purchased for a mundane measurement, much to the chagrin of the operator who has to calibrate, clean, and maintain the monster. Of course, the sales commission is bigger for big-ticket items; so it is tempting for sales representatives to oversell to a customer.

As we Americans have taken to cheap, efficient automobiles such as the Hyundai, Honda, and Yugo for ordinary, everyday driving (who wants to haul bricks in a Jaguar?), so will we take to the new generation of smaller, well-built instruments for everyday analyses. Hats off to the many manufacturers who are at the forefront of small instrument design.

REFERENCES

1. James D. Ingle, Jr., and Stanley R. Crouch, *Spectrochemical Analysis* (Prentice Hall, Englewood Cliffs, New Jersey, 1988).
2. O. C. Chaykowsky and R. D. Moore, *Research & Development* (April 1968), pp. 32–36.
3. Robert S. McDonald and Paul A. Wilks, Jr., *Appl. Spectrosc.* **42,** 151 (1988).
4. Paul A. Wilks, Jr., *Spectroscopy* **1**(9), 49 (1986).
5. John Coates, *Spectroscopy* **1**(11), 14 (1986).
6. John Coates, *Spectroscopy* **2**(4), 14 (1987).

CHAPTER III.5

NOBODY KNOWS THE TROUBLES I'VE SEEN ...

The format of this chapter will be an eclectic mix of problems that have either been suggested or that I have faced myself including the solutions that corrected them (1). Yes, gentle readers, it's troubleshooting time!

I will begin with what may seem trivial, unless of course it happens to you. At one former place of employment, a service technician was called from a supplier to repair an instrument that "would not turn on." The expense of his travel time and time at our company came to several hundred dollars. Upon replacing the fuse, the instrument worked just fine! Such incidents are not so rare. On several occasions, when I ran an instrument group in industry, the problem was merely an unplugged instrument.

Although today's instruments are far more sophisticated than those of one or two decades ago, they are, to some extent, far simpler. Like the old televisions, instruments contained vacuum tubes, dials, wires, cogs, and gears. Although these components may have been simple to replace, there was often more cause to replace them. The one good thing resulting from this state of affairs was that the average chemist was also a pretty good repair person. Then came transistors!

Though I would not wish to return to the "days of yesteryear," new instruments cannot be repaired in real time. What happens is either the analyst or a field repair person presses a diagnostic button, the readout indicates which board to replace, and voila, the instrument is up and running! One consequence of this is that the average chemist feels helpless in the face of a breakdown. They have little or nothing to do with the installation and shakedown; so minor problems can stop them dead in the water.

TROUBLESHOOTING

I would like to suggest several "little" things we still have some control over, but could indeed make our lives easier despite their simplicity:

1. Always check the plug. It sounds obvious, but somebody may have needed that outlet more than you.

2. Check the line for current. A voltage meter or simple current tester will tell you if you have current to your baby.

3. While you are checking the line current, check the voltage. If you are in any way attached to a production facility, there is a chance that the voltage is as high as 140 VAC! Why? When the voltage is kept high, the machines draw less amperage, allowing for efficient working conditions. An active line conditioner is always a good idea with any instrument where a computer or printed circuits are involved.

4. The main power fuse of most instruments is located close to where the power cord enters the chassis. This is to give maximum protection to the "innards" in case of a power surge. Having a spare fuse or two isn't a bad idea. However, if a second fuse blows in rapid succession, a more serious problem is likely. (Now is the time to call in the cavalry.)

When I taught courses on instrument troubleshooting, the only time I deducted full credit on a test question was when the student left it blank. No answer meant (to me) that, in effect, the person sat on the floor and cried rather than trying anything. This is quite unbecoming to a sophisticated generation of scientists.

A word on KBr pellets. If you must, for financial reasons, use a porcelain mortar and pestle, never, never. never, wash it out between uses. Any time you use the darn thing, grind with some dry, clean salt to clean it. If you must wash it out, bake it in a vacuum oven overnight to dry off the moisture or solvents. You will find that many of your extraneous peaks will suddenly disappear.

Chloroform evaporation. John Coates (now with Perkin–Elmer and a consulting editor to *Spectroscopy*) commented on the chapter by Lyth and Ng (2). He observes that the evaporation of chloroform can cause two problems: condensation and alcohol peaks. Almost every solvent uses latent heat to evaporate. That is, it cools the surface it is on, much like perspiration on a person. This cooling may, in turn, condense water on the sample. Water, as you know, is the highest absorber in the infrared (IR) spectrum. Thus extraneous peaks will appear (not to mention fogging of the salt plates).

The second problem comes from the need to stabilize halocarbons from free-radical reactions. Either light alcohols or alkynes may be used for this, with alcohols being the first choice of most solvent manufacturers. (Chromatogra-

NOBODY KNOWS THE TROUBLES I'VE SEEN ...   **337**

phers take note: This same problem can make silicon-based, normal-phase separations unpredictable, depending not only on whether but also on how much
alcohol is present in your halocarbon. I recommend the brands employing an
alkyne wherein you can quantitatively and reproducibly add your own polar
modifier.) Because the chloroform (or methylene chloride) is more volatile than
the alcohol, it will leave an enriched surface film of alcohol behind. The OH
groups of alcohols are second only to water in extinction coefficients in the IR
and near-IR spectrum.

John also points out that using a solvent to extract surface materials is alto-
gether unnecessary. An attenuated total reflectance (ATR) measurement could
have been performed. I am also a fan of taking measurements on materials in
their native states. The act of dissolving or grinding any sample destroys the
matrix and, thus, some information.

The measurement of particle size using diffuse reflection is obviated by
milling or dissolving the sample. The analyst must decide before performing
the analysis whether matrix or "natural-state" information is worth saving or
whether the speed gained by destroying the sample is more propitious.

REFERENCES

1. E. W. Ciurczak, *Instrument Troubleshooting: A Basic Guide* (Stevens Institute of
   Technology Press, Hoboken, NJ, 1981).
2. Chapter III.3, this volume.

CHAPTER III.6

SOME MORE UH-OHS FOR THE VIBRATIONAL SPECTROSCOPIST

I have learned to live by the tenet that "nothing is so obvious that it can't be over-looked." Both mid-range infrared (mid-IR) and near-IR instruments essentially measure heat. The heat is radiated by the source and detected by the detector. Simple, no? Then why do so many laboratories place their instruments in front of windows or beneath air-conditioning vents? Even if the detector per se is insulated, the sample may heat up or cool off. If the instrument itself is heated or cooled much beyond "normal" operating temperatures, changes could occur in the precision optical bench tolerances. In plain English, metal expands and contracts. The baseplate of an instrument is made of metal. The mirrors, gratings, and windows are attached to the baseplate. The positioning of these components is critical. Heat changes these distances. Any questions?

CURRENT EVENTS

One small detail I omitted from the previous chapter: Power cords and computer cables make great antennae. One case in point involved the detector on an HPLC (still a spectrometer, you know!). We were sure we had bubbles or a dying lamp: The baseline was all over the place ... but only sometimes. As it turned out, the power line to the heavy-duty ultrasonic bath in the next room was parallel to the *LC* power line! The spiking that occurred was induced through the wall.

A similar event occurred with an unshielded rf generator for an acousto-optic tunable filter (AOTF). I noticed the source when I picked up "clicking" on my radio, which was set at 93.3 MHz. The working range for the AOTF in question was 60–110 MHz. Even *I* could see the connection here. (In case you don't know, this was a violation of FCC rules despite the low output rf power involved.)

MECHANICALS VERSUS INTERFEROMETERS

For those of you who haven't seen the heat, er, uh, light and are still having grating or prism-type instruments, a word of caution: scan speed (OK, two words). Because IR instruments are so frequently used for qualitative identification of species (one strong point of vibrational spectroscopy), scan speed must be specified in the procedure. Often the resolution achieved by slow scans is overlooked in the rush to meet a deadline. A 2-min scan may appear to be the same as a 22-min scan, but it isn't. A grating instrument, equipped with a computer library, is still only as good as the scan itself.

STRAY LIGHT

Virtually all modern instruments are stray-light free. Having said that, I will still issue a warning about stray light. Karl Norris has been mentioning the problems of dusty windows in reflectance spectroscopy for some time now (2). Manufacturers use windows to shield reflectance detectors from dusty samples; however, the window itself becomes a source of stray light as it gathers a layer of dust. The problem manifests itself with highly absorbing compounds.

The instrument is first "referenced" against a nearly perfect reflector. Dust means nothing because ~ 100% of the light is reflected back to the detector. Normal samples, that is, with absorbences < 1.5 AU, present little problem as well. The difficulty arises when a highly absorbing substance (water, carbon-black-containing materials, and so on) is scanned. Virtually all the light is absorbed by the sample, allowing the small amount of reflected light from the window to dominate the spectrum.

In practical terms, an instrument designed to be linear above 3–4 AU is suddenly limited to > 1 AU. Worse yet, the characteristic of an overload ~ a flat-top peak ~ is missing, often leaving the spectroscopist stymied as to the cause of the poor peak shape.

FIBER-OPTIC WOES

Fiber optics are one of the best additions to spectroscopy since the advent of the grating. But (there's always a "but") they can be misused as any other tool. The obvious problems, such as bending and heating the fiber causing distorted spectra, are well known. Less attention is paid to the effects of room light on the fibers. Room light, if derived from artificial light, can cause 60 Hz noise, stray light in certain bands, and overall less sensitivity. Only totally opaque sheathing is acceptable for normal usage.

Dirty connections need not be dripping with mud or chemicals to severely affect fiber-optic performance. Fingerprints and room dust are sufficient. Another problem found in fiber-optic-based instruments is "filament droop."

As the light source ages, the filament tends to sag or droop. This causes the light to strike the fiber optic at angles beyond the acceptance angle of the fiber. This eventual loss of power is especially critical for single-beam instruments. Further, a dual-beam setup fails to compensate for this subtle loss of sensitivity.

Dipping probes is one of the results of fiber optics. It is good practice to remember that the source and receiver of the light in these probes are quite small. The beam is often aligned in air or water for a reference intensity. The refractive index of virtually anything else is different and will cause a baseline shift (3). It was seen that both refractive index and temperature strongly affected the baseline of the spectra and could even be used as a quantitative measure. The point is that these physical parameters must be accounted for in any work with fiber optics.

MISCELLANEOUS TIDBITS

I neglected to mention in a previous chapter one other diagnostic tool for pellets (4). If the initial movement of the baseline is in an upward direction, this usually means that the powder wasn't ground finely enough before pressing. This seems to occur more often with users of the plastic grinders as analysts despair at finding polymer peaks in their spectra and a shortening of their "grind time."

Jerry Workman reminded me (5) that the world is becoming covered with a monolayer of silicone microspheres. This is from the overuse of silicone lubricant sprays. The SiOH and SiH bands that used to be rare are showing up more often. This is due to more sensitive instruments and the increasing love affair with microsampling. As the sample becomes smaller, the silicone bands become more important.

REFERENCES

1. E. W. Ciurczak, *Instrument Troubleshooting: A Basic Guide* (Stevens Institute of Technology Press, Hoboken, NJ, 1981.)
2. Paper presented at *Seventh Diffuse Reflectance Conference*, Chambersburg, PA, August 1994.
3. E. W. Ciurczak and D. E. Honigs, *The Measurement of Refractive Index of Liquids via NIRS Utilizing a Single Fiber Optic Probe*," paper presented at Pittsburgh Conference, New York, March, 1990.
4. Chapter II.A.2, this volume.
5. J. Workman, personal communication. September, 1994.

CHAPTER III.7

LEARNING TOOLS: CHEMISTRY EDUCATION AND MOLECULAR MODELING

While I was contemplating this chapter, I tried to remember how I felt when I finally got to use my first calculator. After I learned the basics and spelled out (upside down) words, the novelty wore off. It was, after all, a tool, not a toy. The same is true of computers: They are mainly tools, not toys. This started me thinking about education in general, and chemical education specifically.

I taught at the college level for more than a dozen years, and during that time I attempted to emulate the teachers who, I believed, did the best job of explaining the hairy points and helping me understand the subject. So as I thought over my years of teaching, I was somewhat surprised to remember that I often chose to include practical rather than esoteric examples to illustrate the subjects at hand—unlike most of the professors I had as a student.

A letter by F. A. Cotton (1), the eminent inorganic chemist, decried the government's imposition at graduate schools and elsewhere of a more "applied-research" approach. He pointed out that basic research can lead to practical applications, but that it may be decades before the initial research yields an application.

I agree that academic freedom is necessary to keep the steady stream of Nobel prizes coming to the United States (the only trade imbalance in our favor). However, the fundamental work performed by the professors and their students tends to spill over into the classroom. The arcane points covered in most classes are often of little practical use and seldom serve to prepare the students for applying their knowledge while plying their trade. The sad fact is that most colleges use the undergraduate curriculum to prepare students for graduate school and graduate schools prepare students to move into academia.

The fact that more than 90% of graduates with BS degrees will immediately or eventually end up in industry or government labs seems to be little noted. Schools seldom acknowledge that more than 50% of the chemists in the United States are either outright analytical chemists or *de facto* analytical chemists. Despite that, many colleges do not even offer analytical chemistry as a major. To my eye, the classroom work is totally theoretical and seldom has any connection with the lab, which is usually taught by an overworked and underprepared grad student. Despite the fact that most of us will be working in the lab, it is considered the duty—nay, the punishment—of a true teacher to be assigned to lab duty.

The professors either will not or (dare I say the obvious?) cannot teach from a practical point of view. The lecturers either deem topics of applied (more than four letters, but still will get your mouth washed with soap—excuse me, sodium dodecyl sulfate) chemistry beneath them or they simply have no experience with the topics. Aside from a summer internship or two, few college professors have ever ventured outside the ivy-covered walls. Maybe sportscasters needn't have played the game (like Howard Cosell), but it would help these molders of minds to have some knowledge of "the real world" before they set the curriculum. The sad fact is that I learned almost all of my applied science outside the classroom—at home or at work.

I also cannot remember how many times I've asked a biology major what she (I taught mainly at a women's college) would do if she weren't accepted to medical school—and I almost always got the same look as early explorers did who suggested the world might be round. This brings up another sore point about our colleges: Like generals who are still planning for the last war, chemistry departments are still training for the last generation. Seldom, if ever, do departments and colleges take a hard look at what students can actually do with their degrees once they're granted. Although the industry recognizes that it is easier to teach a chemist to program than to teach chemistry to a programmer, schools are still starting new computer programming majors without an academic major or minor. It is now commonplace for education students, for example, to major first in a subject such as English or math, then take education courses.

It's ironic that, in a country that prides itself on free trade, the colleges are still in the Middle Ages. In a free-trade scenario, when customers fall off, you improve the product. In our schools, unfortunately, you relax the requirements for a degree and pretend that all those courses weren't necessary anyway. It seems to be a race between the NCAA and our chemistry departments as to which one can lower the entrance requirements faster.

Not only do we continue cranking out more chemists, but we can't seem to find jobs for the ones we have. Have any of you ever wondered why the government cries about a shortage of Ph.D.s while at the same time the unemployment rate is so high (2)? (Perhaps cars aren't the only thing built better overseas ...) If we're professionals, why do sanitation workers in New York start at a higher rate of pay than chemists with bachelor's degrees? Supply and demand? If there were truly a shortage of chemists, we might be driving the same types of cars as MDs and lawyers.

When there is a glut in the business world, manufacturers diversify or go bankrupt. In academia, however, to stay in operation, departments consider it a sacred task to continue turning out unemployable graduates. Perhaps it is a survival technique because, deep down inside, the professors know there aren't enough jobs even for them if they close up shop.

This does not mean that there aren't a large number of dedicated, underpaid educators out there. Most teachers at the college level know their trade. I'm only saying that the ivy on the walls seems to insulate some of them from more than heat and sound. There is a disdain of the practical that is thick enough to cut with a knife. My plea is to mix in a little "nuts and bolts" science with the quantum theory—after all, the real world is where most of us are forced to live.

A NEW TOOL

The (real) subject of this month's column is one example of the use of high-tech computing for both theory and applications. I have found something that will help industrial chemists and students alike: computerized molecular modeling.

There is a number of types of molecular modeling programs available, but I will share one manufacturer's goods with you. The company is Molecular Simulations (Sunnyvale, CA), and they have a series of programs available to help both applied and research chemists.

The program that caught my attention is called QUANTA. With interactive access to a subroutine called CHARM, it can be readily used for small molecule modeling. I asked the vendor to draw one enantiomer of menthol and one enantiomer of sec-butanol. This was done in a "stick-model" manner in two dimensions. The program then formed the molecules into the proper spatial model (yes, three dimensions). As the polar regions were brought into each other's vicinity, the overall energy reading registered the change in the upper right corner. With a click of the mouse, the vendor chose an iterative algorithm to find the least energetic (most stable) combination of the two molecules.

He then flipped the menthol to its optical isomer and repeated the experiment. The energy difference between the two systems was −4 kcal/mol. This agreed roughly with my experiments (3) in which the OH peak shift in the near-IR is used to calculate the energy differences. This got me to thinking (a feat under any circumstance) that here was a strong tool for spectroscopists. The prediction of structure from vibrational data has been used for years (I did some in 1970 in an undergraduate project). With the molecular modeling software, we can "guesstimate" parts of a structure or complex as we watch red or blue shifts in the spectra. Energy can be directly computed by the shift, measured in wavelengths, then converted to kcal/mol, and so on.

Using a program such as this, chemists and students can sketch structures with all the accepted angles and van der Waals forces to give a visual representation. They can run a docking study and estimate the energy decrease or increase and then run the experiment for confirmation. If they cannot perform a particular reaction on the computer, then perhaps the actual reaction can be put at the end of the experimental list. Working from the other direction, students and chemists can use experimental data to guide the molecules on the screen toward one another from the correct direction, observing the resultant total system energy as a confirmation of a hypothesis.

Following is a brief list of some of the attributes and abilities in the 2D molecular entry portion of this software:

- Sketches small molecules with keyboard and mouse;
- May use all elements (all physical properties in memory);
- Automatically assigns chemical and physical properties;
- Calculates partial atomic charges;
- Automatically converts from 2D to 3D;
- Specifies coordination complexes;
- Aligns functional groups axially or equatorially.

In the 3D mode it

- Creates and edits by fragments;
- Interactively modifies conformations and chiral centers.

Some of the features I found interesting were its ability to display the van der Waals dot surfaces, show electrostatic potential surfaces, and color the molecular surfaces by shape. The software shows intermolecular distances between the atoms of the two molecules where they are in "contact." My tests were, of course, trivial when compared with the many possible molecular biology applications.

Some of the other uses and programs for this software are:

- Molecular comparisons;
- Structure analysis and refinement (X-PLOR program) by electron-density maps;
- Brownian dynamics simulations (UHBD program);
- Comparing nuclear magnetic resonance (NMR) data and structural prediction (NEOSYSIM program) by generating an expectation nuclear Overhauser effect spectroscopy (NOESY) spectrum;
- Helix modeling for polymeric structures.

I could go on, but I think that you get the picture. Figure 1 should help

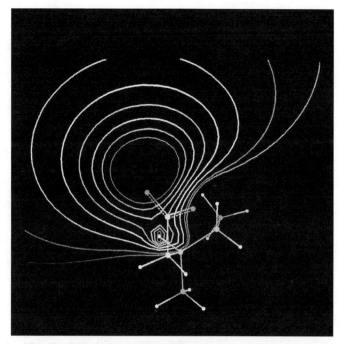

Figure 1. Visualization of the electrostatic potential in a plane around a molecule.

visualize how a simple model appears on screen. Considering how I started this doggerel, I want to close with the assurance that I don't believe all changes or theoretical models are bad ... just the improper use of them *in lieu* of real chemistry in the classroom. Too early an application of a calculator and the student will never learn to add or subtract.

REFERENCES

1. F. A. Cotton, *C& E News* **70**(42), 4–5 (19 October 1992).
2. J. J. Lagowski, *J. Chem. Ed.* **69**(2), 947 (1992).
3. E. W. Ciurczak, dissertation (in writing), Seton Hall University, first submitted 1993.

CHAPTER III.8

WHAT'S IN A NAME? AND REFLECTIONS ON A DUSTY SURFACE

I have attempted in my writings to simplify and debunk the officious tones used in our branch of science. Medical personnel are perhaps the worst acronym droppers in existence, but can somewhat be excused because they need quick responses in emergency situations. Chemists, and spectroscopists, in particular, tend to use smoke and mirrors instead of straightforward names. New graduates and members of other branches of science are intimidated from reading papers by the casual dropping of esoteric terms **without definitions!**

It is not only good science, but proper grammar to define an abbreviation before using it in any text. This is but one of my pet peeves, but, perhaps, one of the top five of all time. Of course, the broad stretches and liberties taken with the names of the techniques just to coin a clever name are pretty lame, even for a chemist.

I have spent time working at Goddard Space Flight Center (Greenbelt, MD), and I've discovered that NASA has more acronyms than a mongrel has fleas.

ANALYZING ACRONYMS

Acronyms: fact or fiction? (I always wanted to start that way.) But, before I point a finger at the government, I'd first like to take a hard look at spectroscopists. (*Especially, I might point out, in retrospect, the NMR people!*) In my campaign against the arcane and esoteric running rampant in our field, I now attack abbreviations and acronyms. Here goes ...

As I sat here nice and **COSY** (**co**rrelation **s**pectroscop**y**), I became a little **NOESY** (**n**uclear **O**verhauser **e**nhanced **s**pectroscop**y**). Instead of **SLAP** (**s**ign **l**abeled **p**olarization **t**ransfer)-ping a **CD** (**c**onvoluted **d**ifference or circular dichroism) into the stereo for a little **DISCO** (differences and sums of traces within **COSY**) music, I decided to write this **INEPT** (insensitive nuclear enhanced by polarization transfer) little **NOVEL** (nuclear spin orientation via electron locking). Now don't **PANIC** (polarization nuclaire [Dynamique] induite chemiquement), although I may be **INADEQUATE** (incredible natural abundance double quantum transfer experiment) for the task, my **FOCSY** (foldover-correlated spectroscopy) manner will make you **SMILE** (swift method for in vitro localized excitation in NMR). In fact, I may even get a **CHORTLE** (carbon–hydrogen correlations from one-dimensional polarization-transfer spectra by least-squares analysis), a **HAHA** (Harmann–Hann condition or match), or even a **HOHAHA** (homonuclear Hartmann–Hann spectroscopy) from you with my **SHARP** (sensitive, homogenous, and resolved peaks) observations.

Now, before you begin to **INCH** (indirectly bonded carbon–hydrogen correlation), **WALTZ** (wideband alternating-phase low-power technique for zero-residual Splitting), or **TANGO** (testing for adjacent nuclei with a gyration operator) away from this **DEFT** (driven equilibrium Fourier transform) **DESPOT** (driven-equilibrium single-pulse observation of T_1), allow me a **SWIFT** (stored waveform inverse Fourier transform) word of explanation. My **PET** (positron emission tomography) peeve has been the number of **GD** (grated decoupling), **SMART** (simulation of magnetic resonance pulse experiments in the time domain)-aleck abbreviations and cute acronyms that go undefined in text! However, before you get the **CRAMPS** (combined rotation and multiple-pulse spectroscopy) from my **SECSY** (spin echo correlated spectroscopy) usage of **DOUBTFUL** (double quantum transitions for finding unresolved lines) terms even **DANTE** (delays alternating with nutations for tailored excitation) wouldn't subject his minions to, I will **PROGRESS** (point-resolved rotating gradient surface-coil spectroscopy) to the subject of today's discourse.

Just one more observation before you **SLAP** me: I, the vibrational spectroscopist, drive an **SHO** (simple harmonic oscillator), of course. At this point, you are no doubt shouting "no **MAS** (magic angle spinning)," so I'll quit.

SOME DIFFUSE REFLECTIONS

Both midrange infrared (MIR) and near-infrared (NIR) spectroscopists are engaged in a technique called "diffuse reflectance" (DR). In its simplest form, DR simply means impinging the sample beam on the (usually) powdered sample from the normal (perpendicular to the surface) and collecting the reflected light at some angle, often 45°C to the normal.

Integrating spheres of the proper material (gold or some Teflon^{TM} analog) are also used to collect scattered light. The light is then focused onto one or more detectors imbedded into the sphere. Manufacturers have used both methods successfully for years. There is little that an analyst can do about the (usually well-built) hardware in terms of troubleshooting. Though we would like to place blame on our tools, the fault (apologies to the Bard) lies with our sample handling, not our instruments.

Even though thoughtful essays have been written on the physics of diffuse reflection (1), and the effects of particle size on reflected light have been well documented (2–5), we persist in pouring the sample into some type of holder and placing it in the light beam. When we produce less than excellent results, the cry of "I told you it wouldn't work!" can be heard echoing through the halls. For some reason, when an HPLC method doesn't give results on the first try, analysts will try another column or mobile phase. When (particularly) a near-IR method doesn't work the first time, the technique is seen as flawed!

MAXIMIZING YOUR DR METHOD

Let us look at a few helpful hints that just might save your IR–near-IR DR method. (Similar hints may be found in Ref. 5.) One mistake that many analysts make is in packing a closed cell. Because density is a major factor in light penetration of a sample, a repeatable amount should be placed into each cell. Weighing is one method when powder is involved. With fabric or yarn, a set length or area could be chosen with a specific, well-written procedure to ensure reproducibility.

In cases where samples must be ground, a specific particle size range must be maintained. An appreciable range of particle sizes may call for some sort of spectral manipulation and normalization (7). Software packages that perform the (*adjustment for*) particle size are commercially available (Chapter III.6 covered some IR–near-IR software). The *caveat* here is that the particles must have a Gaussian distribution. A bimodal mix will give low values (the smaller particles fall between the larger to give the appearance of a "smooth" surface.)

Although an instrument may have a dynamic range of up to three absorbance units (AUs) or more, one anomaly of DR is that the a sample may act saturated at <1 AU. In the case of a pelletized sample, a significant portion of the light may be transmitted through the sample or simply diffused throughout the sample. In either case, higher wavelengths appear to be "saturated." In Ref. 5, the sample must be of infinite depth for DR to be effective. In many cases, this is only a few millimeters. For samples such as nylon pellets, it could be several centimeters! White powdered materials are the easiest to measure, whereas polymers tend to be translucent, making it difficult to obtain good spectra.

In general, the sample should be rotated in some manner to average the signal. This is the main purpose of the integrating sphere, but the same results may be seen with rotating sample holders. As a rule of thumb (*a sexist term, originating in colonial times, wherein a man could beat his wife with a stick "no thicker than his thumb." Anything larger could be construed as abuse and land the husband in the stocks*) as much of a sample should be scanned as is practical. This, of course, may be carried to extremes, which would obviate the advantages of DR measurements *vis-a-vis* time savings.

The one rule that should be obeyed is this: If you did it to your standards, do it to your sample! Physical differences may dominate chemical variations in any DR measurement. Do not take the sample prep for granted; it is the critical part of any analytical measurement.

REFERENCES

1. J. M. Olinger and P. R. Griffiths, *Handbook of Near-Infrared Analysis*, D. A. Burns and E. W. Ciurczak, eds. (Marcel Dekker, New York, 1992), pp. 13–36.
2. P. Kubelka, *J. Opt. Soc. Am.* **38**(8), 448 (1948).
3. G. G. Stokes, *Proc. Roy. Soc.* **11**, 545 (1860–1862).
4. K. H. Norris and P. C. Williams, *Cereal Chem.* **61**, 158 (1984).
5. P. Kubelka and F. Munk, *Zeits. f. Tech. Physiks* **12**, 593 (1931).
6. P. C. Williams, *Handbook of Near-Infrared Analysis*, D. A. Burns and E. W. Ciurczak, eds. (Marcel Dekker, New York, 1992), 281–316.
7. E. W. Ciurczak, R. P. Torlini, and M. P. Demkowitz, *Spectroscopy* **1**(7), 36 (1986).

CHAPTER III.9

QUESTIONS AND ANSWERS ON PROCESS ANALYSIS: TIPS FOR SELECTING AND CALIBRATING IR INSTRUMENTS

Jerry Workman and I presented an American Chemical Society (ACS) short course on practical near-IR at the national ACS meeting in Chicago (the first time a near-IR course has been offered at an ACS meeting!) I mention this for two reasons. First of all, it shows that the ACS is successfully broadening its semiannual national meeting from a pure academic conference to a symposium that includes topics of interest to industrial chemists as well. Second, our class posed some very good questions.

This chapter is based on a couple of questions that pertained to the use of near-IR analysis in the process environment. For clarity, I will take the liberty of paraphrasing the questions and combining Jerry's and my comments into a single composite response.

RATING THE GRATING FOR PROCESS ANALYSIS

Question: Which is better for process work, an interferometer-based instrument or an instrument with a grating?

Answer: This question is akin to asking, "Do you bring your lunch or take the bus to school?" The student posing this question didn't include all the possible choices of monochromators (*or situations wherein they will be used*), such as diodes, interference filters, and acousto-optic tunable filters (AOTFs). The technique should be chosen based on the sample type or the assay being performed. Do we use the mid-IR or the near-IR? Do you use transmission, reflectance, or some combination? You must first specify whether the sample is a solid, powder, slurry, semisolid, liquid, or gas.

If you are working in the mid-IR range, then the weak light sources and strong absorbencies endemic to the region are greatly obviated by an inerferometric-type instrument. An interferometer (such as the one used by Michelson in the nineteenth century to measure the speed of light) has two advantages over a dispersive monochromator: its high light throughput and multiplexing capabilities.

The radiant power passed by a monochromator (for a continuum source) is proportional to the square of the slit width. The slit allows only a narrow band of wavelengths to impinge upon the sample. An interferometer, however, has no slit. (Actually, it has an orifice or circular waveguide, but it is much larger than a conventional slit.) Thus a far larger proportion of the source light is allowed to pass through the monochromator and, thus, the sample. This capability is called *Jacquinot's advantage*. It may range from a factor of 10 to more than 200 (1).

The second advantage of the interferometer is the Fellget or multiplex advantage (1). Basically, all the light transmitted by the interferometer simultaneously reaches the detector and results in an enhanced signal-to-noise ratio (S/N). Any given detector has a noise level indigenous to its type. The S/N may be enhanced by increasing the level of the signal because the noise is defined by the physics of the detector. The light passed by the interferometer is far more intense than through any dispersive monochromator. This S/N enhancement is especially important in trace analyses or cases where precise levels must be maintained within narrow limits.

In some instances, an interference filter may work almost as well as an interferometer. The light from the source encounters fewer lenses and mirrors and the bandpass is larger than in a grating instrument. This results in a relatively large signal, but one with less resolution. Most single-filter instruments are used for simple measurements such as moisture.

In the near-IR range, the sources are quite a bit more intense than those operating in the mid-IR region; so the high light throughput could actually damage a light-sensitive substance. (*I have gotten grief for this statement, mostly from FT-IR manufacturers, but I stand by my first impression. My experiences with pharmaceutical materials make me shy of bright lights coming in contact with drug substances.*) An interferometric instrument also achieves greater wavelength accuracy with the longer mid-IR wavelengths than those of the near-IR region. Near-IR work does not require a device to enhance light intensity. This gives the analyst the flexibility of using diodes, AOTFs, or gratings in production, if the near-IR region is chosen.

Near-IR detectors are usually constructed of semiconductor materials such as lead sulfide (PbS), indium arsenide (InAs), or indium gallium arsenide (InGaAs), to name a few. These semiconductors are usually 2–3 orders of magnitude quieter than detectors used in the mid-IR range. The larger S/N, enhanced by greater signal and lower noise, is common to most near-IR instruments. Thus a multiplex advantage is usually unnecessary. (*Note: several mid-FT-IR instruments introduced in recent years have noise levels comparable with NIRs, but my comments about the signal strength still apply!*)

To briefly answer the initial question: If you need to penetrate an opaque sample, you should use near-IR spectroscopy. If you need to measure the surface, or analyze a clear liquid (of short pathlength), or measure a gas, then a mid-IR technique might be your best alternative. And if you're working in the mid-IR region, then why not explore a Fourier transform–IR (FT-IR) spectrometer, the most common type of interferometric device used in this range?

CALIBRATION STRATEGIES

Question: How can we assure proper calibration transfer between instruments, and how can we be sure our calibration is constant on a single instrument?

Answer: The second part of this question addresses the crucial point. Because most IR and NIR instruments, especially those used in process applications, are single-beam by nature, wavelength accuracy and light throughput are of vital importance. The National Institute of Standards and Technology (NIST, formerly NBS, Gaithersburg, MD) has no single standard available for wavelength accuracy in transmittance. A reflectance standard exists (*although denied by NIST, we used this standard at Bran + Leubbe and NIRSystems for years ... apparently it was being tested and, for some reason, found wanting. It DID, however, later letters from NIST notwithstanding, exist and was used regularly!*) (#1920), but it has yet to attain wide acceptance.

Based on polystyrene, certain wavelengths (for example, 1680 nm) are considered benchmarks. The problem with using a polymer as a standard is that no two batches of any polymer are identical. NIST *was* expected to release another version of the standard within a year. I have not yet seen it (*nor will I in the near future!*); therefore, I can neither praise nor disparage it.

I recommend placing the onus of establishing proper calibration on the manufacturer, who should be able to return any instrument to the condition in which it was delivered. Basically, when you develop methods for complex mixtures (in either the mid-IR or near-IR regions), the chemometrics require reproducible parameters. It is more important for the instruments to perform in a reproducible manner than for them to represent absolute values of some mythical standard. That is, it doesn't matter that the 1680 nm of your instrument may be 1682 nm against an absolute standard (if one existed), as long as it *always* reports that point as 1680 nm. Because production instruments are not used for fundamental research, it is wholly unlikely that discrete peaks are sought. Rather, the purpose of scanning instruments in the production environment is to identify patterns in product quality.

In short, if you produce the same lumps at the same spots with the same intensities, you have an instrument that should hold a calibration. Merely tell the manufacturer of the toy (to quote Captain Picard), "Make it so!" It then follows that, if all your instruments have been built to the same performance specifications, you should be able to transfer a calibration with minimal statistical magic.

LAST WORD

To put production control into perspective, there must be a partnership between the customer and instrument manufacturer. You can no longer purchase a piece of electronic wizardry and go about your business. The manufacturer of that equipment *should* know more about it than any user. It is folly not to use the instrument manufacturer's applications group to assist in setting up and calibrating an assay. It is also no shame to employ a good consultant if your ongoing work would benefit from extra expertise.

REFERENCES

1. J. D. Ingle, Jr., and S. R. Crouch, *Spectrochemical Analysis* (Prentice Hall, Englewood Cliffs, NJ, 1988), p. 84.

CHAPTER IV.A.1

A REVIEW OF THE THIRD INTERNATIONAL CONFERENCE ON NEAR-INFRARED SPECTROSCOPY

It is no secret that one of my foremost loves (aside from eating) is near-infrared spectroscopy (NIRS). In 1990, I had the chance to attend the Third International Conference on Near-Infrared Spectroscopy (ICNIRS) in lovely Brussels, Belgium. In addition to fine food and drink, there were some rather interesting papers presented at the conference, which was organized by the International Committee for Near-Infrared Spectroscopy and the Belgian Administration of Agricultural Research, under the sponsorship of the Belgian Ministry of Agriculture, the National Board for the Marketing of Agricultural and Horticultural Products, and the Community Bureau of Reference, a commission of the European Community.

In Europe as in the United States, NIRS is still thought of primarily as an agricultural tool. Because I believe there is a need to emphasize the great versatility of the method, this review will concentrate on the sections of the ICNIRS program that addressed applications of NIRS in more diverse situations such as chemical and medical analyses. Indeed, beyond a wide variety of interesting agricultural applications, the program included presentations on the history and future prospects of NIRS, instrumentation, data processing, and applications in such industries as food and beverages, paper, pesticides, petrochemicals, polymers, and so forth. I will cover some of the most interesting presentations, in addition to new NIRS instrumentation and a few insights from some of the well-known authors in attendance.

PROGRAM HIGHLIGHTS

An interesting suggestion was put forth by A. M. C. Davies (Norwich NIR Consultancy, Norwich, United Kingdom). Tony reminded us that the traditional (if 25 years makes a tradition) wavelength range for the near-IR spectral region is 1100–2500 nm. The recent "hot" region of 750–1200 nm (the range of silicon detectors) has many names: near-near-IR or short-wave near-IR, for instance. He suggested we simply call the region the Herschel-IR region in honor of F. W. Herschel, the discoverer of the near-IR segment of the spectrum.

Karl Norris, who is recognized as the "father of near-IR spectroscopy," has performed some interesting work with a team from Johns Hopkins University since retiring from the U.S. Department of Agriculture. He and his co-workers (M. Williams, M. Farrari, D. Wilson, R. J. Traystman, and D. F. Hanley) have used NIRS to study the brains of cats and dogs *in vivo*. In these experiments, near-IR radiation is impinged upon the skull of a (sleeping) cat or dog through a fiber-optic probe. The diffusely scattered light is collected through a second similar probe. Spectra in the range of 700–1400 nm were generated. Contributing to the spectra were bone, fat, protein, and water, as well as the measured hemoglobin absorbances. The researchers were able to follow the changing signal as the animals breathed air of varying percentages of oxygen. Although these analyses were performed on the bare skulls of the animals, the virtual absence of muscle on the human scalp should allow this technique to (eventually) become a routine hospital test.

Jeff Hall, a bright newcomer on the scene (CME Telemetrix Inc., Waterloo, Ontario, Canada; *currently NIRSystems, Silver Spring, MD*), presented some work that he and several others (J. Samsoondar, P. J. Brimmer, and A. Pollard) performed on human fecal matter. Even though Hall took some good-natured teasing about his choice of sample matrix, attendees recognized that this is an important source for diagnostic information that has been widely ignored because of its less-than-desirable form. In these experiments, the sample is mixed with isopropanol and water, then mixed to a slurry, which can be easily scanned in a commercial near-IR instrument. The fat content and its form can be used in the diagnosis of various digestive disorders. For example, steatorrhea, or excess fecal fat, may be caused by maldigestion or malabsorption. This is easily measured in the near-IR technique described, which provides an agreeable alternative to the current arduous and unpleasant lab procedure.

Jennifer Griffin and W. Kohn (Ciba-Geigy AG, Basel, Switzerland), Lloyd Fox (The Upjohn Company, Kalamazoo, Michigan), and Brian Davies (Glaxo Manufacturing Services Ltd., Bernard Castle, United Kingdom) all spoke of techniques for spectral identification of pharmaceutical raw materials and actives.

Griffin used a new Fourier transform instrument equipped with a bidirectional fiber-optic probe. The software utilizes principal component analysis (PCA) to rapidly identify the material in question. Fox attached a fiber-optic probe designed at his facility to a rapid-scanning grating instrument. Spectral matching, using either a vector dot product or PCA approach, is used to identify the material, while PCA is used to evaluate the quality or purity of the substance. Davies spoke of the updated PRISM software package developed in his laboratories. The discriminate analysis section is only one portion of a system wherein the identification, purity, and suitability for the process are evaluated.

Your humble reporter presented a paper in which the art of tablet dissolution was examined using NIRS. One of the more significant results of this work (I think) was the quantification of substances in the 10^{-4} M range. Significant spectral features may also be seen in this concentration range. As compared with the relatively bland UV spectrum of acetaminophen, at least six wavelengths exist where the material may be quantified. The multiactive dosage forms used showed reasonable results, as well.

Willem Plugge (Gist-Brocades, Delft, The Netherlands) described his use of NIRS in the process control of ampicillin trihydrate. In lieu of the colorimetric USP XXIII procedure, NIRS is used as a rapid method for determination of potency. Since a "standard potency" is assigned to any material falling within the accepted range, NIRS is at least as good as, and several times faster than, Karl Fisher titrations for determining moisture content of the ampicillin.

TRENDS IN NIRS INSTRUMENTATION

A few trends in NIRS instrumentation were apparent at the conference's exhibition of instrumentation and in papers presented by scientists from many of the major NIRS vendors. In general, NIRS manufacturers are continuing to emphasize the method's wide applicability to industrial analyses, particularly process control.

As far as on-line instrumentation is concerned, Bran + Luebbe, LT Industries, and NIRSystems all spoke about their rapid, easy-to-use systems. It was acknowledged that the instrument best suited for any industry will depend on the material(s) being analyzed, ambient conditions, and the ability of the manufacturer to design and configure instruments for specific applications.

In my opinion, the best part of all the NIRS instruments described is the data security now included on instruments from all NIRS vendors. This has long been a sore point with pharmaceutical manufacturers. The Food and Drug Administration calls for accountability beyond what is supplied by typical "research-grade" software. The user limitations and secured equations found on the current generation of NIRS equipment are a good beginning.

As an old industrial type, I am heartened by these displays of commitment to industry, particularly to the process side. I believe it is in that area where near-IR spectroscopy will make the greatest impact on industry.

CHAPTER IV.A.2

"WHAT'S NEW IN SPECTROSCOPY?" A REPORT ON PITTCON '92 IN NEW ORLEANS

The most significant thrust of ultraviolet (UV) spectroscopy at PittCon 1992 was in the direction of multivariate analyses. That is, advanced algorithms such as partial least squares (PLS) and principal components regression (PCR) were used to examine the entire spectrum instead of single wavelengths or combinations of wavelengths (multiple linear regression). Most major manufacturers and independent software manufacturers were offering packages designed to be used with UV instruments. The strength of smoothing UV spectra was seen in recent years when derivatives were used to enhance "hidden" information. In complex mixtures, factor analyses (for example, PLS or PCR) can be used to extract information from spectra in which Beer's law is not strictly followed. In general, the instruments have reached a point where the analysts are seeking new frontiers: complex mixtures on a routine basis.

Because the need to address the complexity of near-infrared (NIR) spectra has been evident for a long time, multivariate programs have been offered by major producers for years. Added to the list of potential calibration techniques this year were neural networks (which I don't fully comprehend, but recognize the power thereof). A number of disparate paper topics demonstrated the power of NIR combined with advanced algorithms, including measuring the octane number of gasoline, nutritional value of food, BTU of natural gas, actives in drug products, and various properties of soil.

Near-IR is being presented more often as a method of process control in food, chemicals, petrochemicals, polymers, pharmaceuticals, paper, and textiles. Moisture may be determined in dried pharmaceuticals through the bottoms of vials, blood substituents studied *in vivo*, and the condition of arteries determined in living patients.

The instruments have reached a point where the old rule that NIR was good only for analyzing amounts in the percent range has been shattered. Polymer additives in the 100-ppm range are routinely analyzed in pellets, and water in solutions may be determined to low ppm levels. It's an exciting time for spectroscopy, folks!

CHAPTER IV.A.3

A REPORT FROM THE SIXTH INTERNATIONAL DIFFUSE REFLECTANCE CONFERENCE

Emil W. Ciurczak

I sometimes enjoy meetings that I have to attend. In the case of the Sixth International Diffuse Reflectance conference (IDRC), held August 9–14, 1992, at Wilson College in Chambersburg, Pennsylvania, I actually wanted to attend. At least 10 countries were represented, from parts of the world as far flung as Japan, the Commonwealth of Independent States (CIS, formerly USSR), Australia, the United Kingdom, Israel, Hungary, and Canada, among others.

The informality of the meeting lent itself to open (that is, "heated") discussions. Conventional sessions of oral presentations were held each morning and evening, with afternoons set aside for special events, both scientific and recreational.

One major theme of the conference was introduced on Sunday evening by Isaac Landa (LT Industries, Rockville, MD) when he spoke of process applications of near-infrared spectroscopy (NIRS). Near-IR has matured to the point in which plant managers are willing to entrust their multimillion-dollar processes to commercial systems (produced by at least 12 companies, at last count). The reproducibility of assays from one instrument to another is one of the reasons for the rapid acceptance of the technique.

Jean Daly of MARS Confectionery (Slough, U.K.) discussed the continuing success in measuring chocolate and product "goodness." MARS began its program back in 1979, and it continues today. Michael Guandaev (Joint Venture Interagmtch, Moscow, CIS) told of his experiences with hundreds of instruments spread over a land mass as large as the United States! The instruments are kept in calibration in the face of the disintegrating infrastructure of the former USSR. In all cases discussed, spectra generated on site using a standardized sample are often sent over phone lines to a central laboratory for "fine tuning" of the instrument in the field.

Successful networks were also described by David Ryan (Central Soya Foods Corporation, Ft. Wayne, IN) and Pierre Dardenne (Ministere de L'Agriculture, Libramont-Chevigny, Belgium). The food and agriculture people aren't that much smarter, just more practical than "real" chemists. The food industry is rife with difficult and time-consuming analyses; so the promise of rapid, non-destructive methods is a breath of fresh air.

Continuing in the vein of process applications, new and "newish" instruments were described for the production environment, David Honigs (Katrina, Hagerstown, MD) described an instrument that uses light-emitting diodes in the 800–1100-nm region (more diffuse transmittance than reflectance, but who's counting?) to perform 100% testing on candies—first determining that each has the requisite cherry, then testing to make sure there isn't a pit in the cherry.

James Psotka (Perten Instruments North America, Reno, NV) described a rapid filter instrument. Because the "real world" does not wait for discrete samples to be taken and analyzed, but rather rushes past at many tons per hour, measurements must be taken in a few seconds or less. This upgrade of a classic approach appears to be both rapid and simple (that is, "inexpensive").

Isaac Landa (LT Industries) spoke about the company's multiplexed diode-array instrument. The 256-diode instrument operates in the 800–1760-nm range and can read seven channels sequentially, allowing it to monitor as many as six processes with one channel for reference.

Brimrose Corp. of America (Baltimore, MD) displayed its acousto-optic tunable filter (AOTF) instrument. [AOTFs were described in Spectroscopy, **5**(1), 14 (1990).] Although the instrument was only being used to rapidly discriminate between wine, scotch, and gin (did I say "only"?), the speed of the measurements was rapidly apparent. The configuration demonstrated is best suited for liquids, but can easily be used for reflectance measurements.

While near-IR spectroscopy is rapidly being adapted to process control, the technique is also being accepted as a partner to Fourier transform infrared (FT-IR) spectroscopy by some research teams. One entire session was devoted to the effects of water on the spectra generated. Mutsuo Iwamoto and Hideyuki Abe (National Food Research Institute, Tsukuba, Japan) presented two papers in which they attempted to interpret the many forms of water present in solid samples: free, H bonded to sample, H bonded to water, and so forth. The previously uninterpretable spectra are beginning to be deconvolved!

The other major thrust of the technique was in biotechnology. Some amazing work has been done and is certainly proceeding, as described below.

David Wilson (Johns Hopkins University, Baltimore, MD) introduced us to a large body of biological data (sorry, I didn't even have to think about that one!) on blood chemistry. Using a diffuse reflectance probe, his group has followed blood chemistry (mainly oxygenation) during and after strokes. It appears that the progress of a large number of chemical reactions may be followed in this manner.

Bob Rosenthal (Futrex, Inc., Gaithersburg, MD) described the massive amount of clinical work that has gone into his glucose-in-blood analyzer. The tiny instrument uses a through-the-fingernail near-IR beam to determine the amount of glucose in the patient's blood. The system is calibrated by results obtained using the current intrusive techniques, which require a blood sample. Although not yet approved by FDA, this instrument holds much promise.

Jeff Hall (NIRSystems/Mt. Sinai Hospital, Toronto, Canada) shared some of his work in clinical applications of NIRS, including analyses in whole blood, serum, urine, feces, cerebrospinal fluid, breast milk, bone, and tissue. It would appear that NIRS can and has been tried on nearly every common clinical test. The presentation described methods that were at least as rapid as current tests and often much less labor intensive.

Rob Lodder (University of Kentucky, Lexington) outlined the work he is planning to perform with his new near-IR camera. As is his wont, Rob is working on high- and low-density lipoprotein (HDL and LDL) buildup in arteries. He described his methods for recovering light from diffuse transmission measurements. It involves some clever optics, which I believe he would rather describe in his own paper(s).

With names like Peter Griffiths, A. M. C. Davies, Karl Norris, Phil Williams, Fred McClure, Woody Barton, and Ian Murray added to the presenters already cited, you can understand why I am always excited to attend this "humble" little meeting in the cornfields of Pennsylvania.

CHAPTER IV.A.4

"WHAT'S NEW IN SPECTROSCOPY?" PITTCON '93 IN ATLANTA

As promised earlier, I am naming names for this review. I guess workers in the field are less touchy to my humor than instrument manufacturers. Oh, well, it is short and pithy, at least.

As usual, there were numerous papers on near-IR spectroscopy at PittCon again this year, in addition to the toys on display at the exhibition. The two themes that appeared in near-IR spectroscopy at the exhibition were remote measurements and smaller, simpler, less expensive instruments.

In the technical sessions, I did notice more "basic" papers, such as those given by the students of Prof. Chris Brown (University of Rhode Island) on water and hydrogen bonding. One student, Zhenfeng Ge, compared normal and malignant cells (from PAP smears) by near-IR spectroscopy. It was no surprise that Bill Fateley's group at Kansas State University had some interesting data to report—they are using a looooong-path FT-near-IR instrument to monitor traces of volatile organic compounds at industrial sites.

I also heard several papers on pharmaceuticals. Andy Duff of Merck spent his time outlining the problems a development chemist will encounter with solid dosage forms. His words were echoed by John Pope of LT industries, who described how the presentation of the tablet accounts for almost all of the "noise" in any tablet analysis.

In these presentations, less was said about the results than about how to take measurements. This back-to-basics approach is healthy, because many analysts have been running down the "software-is-everything" path for too long. The majority of authors made the point that sample presentation is more important than a description of which of the many algorithms was used.

Dave Tilotta and his research group at the University of North Dakota have been doing interesting work with the emission spectra of relatively low-temperature gases. Simple molecules such as water, methane, and nitrous oxide were heated to 300–400°C and the emission spectra taken. Surprisingly good spectra were demonstrated at that (relatively) low temperature.

Dave Wetzel and Albert Eilert of Kansas State University showed the versatility of their acousto-optic tunable filter (AOTF) instrument in measuring rapidly changing systems. Wetzel used a cyclohexane-to-benzene gradient to demonstrate the rapid response of the AOTF equipped with an IgGaAs detector. Eilert used the system as a detector for a supercritical fluid chromatograph. The CO_2 mobile phase lent itself nicely to quantitative work.

Of course, petrochemicals still have a strong place in on-line near-IR analysis systems. Several vendors addressed octane measurements and control of blending systems. In all, I was impressed by the variety of applications. Near-IR isn't just for food anymore.

CHAPTER IV.A.5

NEAR-INFRARED TOPICS FROM THE 1994 PITTCON

This year a large number of near-infrared (near-IR) papers addressed health-related topics, both pharmaceutical and biomedical. Robb Lodder (University of Kentucky Medical Center) spoke about using near-IR through video imaging and a remote fiberoptic probe to assess the oxidation level of carotid artery plaque of patients undergoing endarterectomy. These levels of oxidation were correlated with age, but not with gender or level of smoking.

Timothy Mills (University of Kentucky) spoke on magnetohydrodynamic acousto-resonance near-IR (MARNIR), which is used to quantify proteins and peptides. Yi Zou, also at University of Kentucky, used a dye laser pumped by the second harmonic from a Nd:YAG laser (512 nm) to generate visible-region photons. This light is converted to near-IR by stimulated Raman scattering. In his work, the first and second Stokes lines were used for near-IR absorption measurements.

Jim Drennan (Duquesne University School of Pharmacy) presented depth-resolved measurements of synthetic polymers, work related to his studies of new drug-delivery systems. John Kirsh presented comparisons of software used in near-IR between the bootstrap error-adjusted single-sample technique (BEAST), PCA, and a neutral network program.

Gabor Patonay (Georgia State University) spoke of using near-IR dyes as tracer compounds and reporting labels for DNA sequencing, fluorescent immunoassays, gene probes, and fiberoptic bioprobes. Guillermo Casay spoke of using cyanine and naphthalocyanine derivatives as fluorescent probes for heavy metals such as cadmium and lead. This work, at 700–780 nm, qualifies as near, near, near-IR.

Mark Arnold (University of Iowa) shared his progress on a nonintrusive technique for determining glucose in blood. Using FT-near-IR, he used increasingly complex matrices from phosphate buffer to blood serum. Not yet ready for the market; nonetheless, the data were encouraging. Jeff Hall (NIRSystems, Mount Sinai Hospital, Toronto, Canada) provided more examples of near-IR used in clinical testing of serum, plasma, whole blood, feces, bone, urine, and breast milk. Christopher Frank (Ohio State) used near-IR Raman to assay components of tissue samples such as lipids and proteins. The neat part was the discussion of diode lasers and tunable Ti:sapphire lasers as sources (784 and 830 nm).

An industrial application of note was presented by W. F. Arendale (University of Alabama, Huntsville). Using fiberoptics, his group was able to perform analysis of rough surfaces. The oxidation of the surface of Al and a stainless steel alloy was followed as a function of time, relative humidity, temperature, and surface preparation.

As a general observation, the number of food and polymer papers was minimal, whereas the theoretical, mathematical, pharmaceutical, and biological papers were definitely on the upswing.

CHAPTER IV.A.6

NEAR-INFRARED TOPICS FROM THE 1995 PITTCON

Because I didn't present a paper at PittCon for the first time in about 15 years, I had more time to look at other *artistes* of the arcane art of near-IR. A quick observation is in order: most of the posters and papers were from academia or instrument manufacturers. Although this is not bad, I was concerned about the lack of industrial and agricultural users presenting new work.

One poster about cyclosporin identification (Sandoz) was good, but was only an update of work done in the 1980s. Posters on the excitation of dyes in the near-IR, which are then read at different wavelengths, may not be considered near-IR methodology. Two of Jim Drennen's students, John Kirsch and David Wargo (Duquesne University), spoke about monitoring film coating of pharmaceutical tablets and following the homogeneity of powder blends, respectively.

Graduate student Stacey MacDonald (Indiana University) had some interesting preliminary work with optically active materials, and the team of Dave Wetzel (Kansas State University) and Arnold Eilert (Brimrose, now Bran + Leubbe) gave two interesting papers on accousto-optic tunable filter (AOTF) instrumentation. Of course, Matsuo Iwamoto (Ministry of Agriculture, Forestry, and Fisheries, Japan) received the Hirschfeld award and shared his work on hydrogen bonding in water.

Papers were also given by representatives of Brimrose and NIRSystems on process work and new instrumentation. Although some very good work comes from applications groups of instrument companies, the number of trade shows covered each year often doesn't allow the very able scientists to come up with original work each time.

Again, the researchers from Georgia State (Gabor Patonay and Maryam Daneshvar) presented some fine work with biomolecules. And as before, Chris Brown (University of Rhode Island) discussed some excellent work, this time on sol-gels and hydrogels.

Either there has been a decline in interest (not apparent in the number of students at the near-IR courses given) or near-IR has become truly accepted. That is, it has reached a plateau where the legal departments of the (notably, pharmaceutical) companies have disallowed papers on the premise that they contain proprietary information. As with HPLC 10–15 years ago, when a method is seen to be useful, it isn't allowed to be "given to the enemy." Thus, HPLC papers from industry often represent methods passed over for betted (non-presented) methods. I fear that the state of near-IR is too fragile to survive the silence of being held as proprietary information.

It is possible, of course, that the temporary slowdown in the chemical industry caused fewer chemists to attend or present near-IR papers this year. After seeing four, five, or six sessions at past shows, the appearance of only two NIR sessions worried me. One other minor problem was the tiny rooms assigned to NIR sessions. It is possible that the low number of submitted papers caused the organizing committee to believe that these sessions would be lightly attended. Several NIR papers were placed in chemometrics sessions, and the ASTM section meetings on NIR and chemometrics were scheduled directly against these very topics!

In 20-plus years, I have seen the exposition outgrow all but the largest venues. This would be wonderful if it were more manageable. Getting older and less inclined to walk long distances, I have a suggestion: Divide the theater, putting spectroscopy and related exhibits on one side and chromatography and friends on the other, with the multi-product firms right down the middle. Perhaps this would help the average scientist see the show in less than a month. Less doable might be two specialty shows, but I can't see that happening.

The number of papers has also increased tremendously. (Are there really 1500 good papers ready to present each year?) The sheer number forces the committee to make the abstract deadline fall six to seven months before the conference. If a researcher has some really hot data, he or she isn't going to wait more than half a year to present it—it gets published. The papers presented at PittCon may often fail to resemble the hopeful abstracts submitted the year before. Many papers seemed to be: a) included to promote a new product, or b) submitted to justify a trip to PittCon to the bean counters in corporate finance.

My guess is that many "outside" specialists would gladly assist in the selection of papers for presentation—if we were allowed to participate. In the case of presented papers, less is more. I still love PittCon; I only hope it doesn't become so large that it becomes unmanageable. But, as one of the designers of the Titanic might have said, "Bigger is not always better!"

CHAPTER IV.A.7

THE GOOD OLD DAYS, OR WHAT?

When Editor Mike MacRae mentioned that the issue in which this chapter appeared represented the 10-year mark for *Spectroscopy*, I was sent down Memory Lane (Exit 14A). I remember meeting the first editor, Heather Lafferty, as she was hawking her new, glossy magazine at the Eastern Analytical Symposium in the fall of 1985. My first (less than) enthusiastic response was, "Oh good, another free journal." However, here was an editor willing to consider near-infrared (near-IR) as a legitimate technique; so (of course) I bought into the idea of the publication.

I was lucky enough to be published in Volume 1, Number 1 (1)! Honestly, I was in a hurry to be published because I wasn't sure the odds favored yet another magazine succeeding in a crowded field already dominated by such established scientific publications as *American Laboratory*, *Research & Development*, and *Laboratory Equipment*. However, the fact that the articles were refereed gave a shot of respectability to the magazine. Three successive editors ran the publication for the first few years, a course usually fatal for a neophyte journal. However, each contributed something worthwhile, building on the strong points of the previous editors. In other words, "change for change's sake" wasn't the order of the day.

Where was the "art" of near-IR spectroscopy at that time? I was working at Sandoz Pharmaceuticals in East Hanover, NJ, playing with a 500C scanning near-IR spectrometer from Technicon (now Bran + Luebbe, Buffalo Grove, IL). Theirs and Pacific Scientific's (now NIRSystems, Silver Spring, MD) model 6250 were the cream of the crop in 1985. Of course, now there are more scanning models than you can shake a fiber-optic probe at.

When the salesperson from Technicon (John Sylvestri) approached me with a near-IR instrument, it sounded too good to be true. So I tossed him out. Only the "suggestion" of my associate director brought him back with some technical

373

support for a second try. I have never been sorry that I bought into the concept of near-IR, primitive as it was at that time.

Lest you think I cast aspersions on the hardware, let me correct the image. Both aforementioned instruments were quite advanced for that time. It was the limitations of the computers of the era that limited the utility of near-IR spectroscopy. The HP 1000 (Technicon) and Northstar (Pacific Scientific) computers were later used as boat anchors. The problem lay in the nature of the spectra, not the instruments that generated the spectra.

Near-IR spectra (as I have mentioned *ad nauseam* over the years) consist mainly of lumpy, overlapping peaks with few, if any, absorbances directly attributable to individual vibrational modes. In the midrange IR, many peaks are "standalone" and can be interpreted without the aid of computers. A typical undergraduate can find a carbonyl, amine, or hydroxy peak in an IR spectrum without recourse to a computer. Thus the midrange IR had a good head start on the near-IR by 1985.

The commonly used software at the time was provided by the instrument manufacturers. There weren't enough instruments in use to tempt private software writers to get involved. The programs themselves were merely variations of the multivariate, linear regression equations developed by Karl Norris at the U.S. Department of Agriculture in Beltsville, MD. There were more philosophic than scientific differences at that time on how to generate a usable equation for analyses.

Technicon didn't have a good derivative algorithm; so the staff insisted that derivatives weren't necessary. Pacific Scientific did derivatives well; so its staff insisted that only derivatives should be used. Which one was correct? Yes.

At that time, long before principal component analysis (PCA) or partial least squares (PLS), both companies' approaches were mostly good. Derivatives could level some, but not all, of the baseline sloping due to particle-size variations or surface texture. Absorbance maxima were also enhanced by derivation. Although Pacific touted a "single-wavelength" equation, in truth, performing a second derivative utilized three wavelengths for each data point.

The best combinations program from Technicon often picked combinations that truly simulated derivatives by picking adjacent wavelengths and assigning alternating + or − signs to the coefficients—the hallmark of a second derivative. The software correctly assumed less skill on the part of a routine user and chose the wavelengths for the analyst.

The main reason I chose the Technicon equipment was their principal chemist, Howard Mark—yes, the very one who writes for this magazine. I wanted to perform discriminant analysis, and he assured me that his program would be ready "any time now." In truth, I became a beta test site for his new software in 1982. My HP 1000 (dual floppy, no hard drive) was "souped up" with a 9-Mb (now the capacity of most computers' RAM) Winchester hard drive, roughly the size of a large VCR. I was in hog heaven . . . and the software worked! Mark introduced the world to the Mahalanobis approach in near-IR (2, 3).

In fact, as the Technicon "lab rat," I was allowed to present my work using the completed program at one of Technicon's yearly (well-run and informative) meetings (4). This and another paper on using the software for pure liquids (5) led to the *Spectroscopy* article in 1986 (1).

I mention this work as a prelude to my first-ever comments on software. I was recently given a copy of the good Dr. Mark's latest software package to evaluate (6). Guess what? He hasn't lost it! Where the original Discrim (clever marketing name) program took three to four days to give an equation for just five or six materials, the new package takes minutes—a combination of well-written software running on a 486 machine. I can only imagine how fast it would scream on a Pentium™ chip!

The software also includes the venerable "step-up" approach. This, simply put, starts with the single best wavelength, calculates a linear regression equation, and prints. It then moves on to the best two, three, etc., wavelengths and calculates the multiple linear regression (MLR) equations for each group. The statistics are superb, giving more information than you will ever need to justify which equation you choose to use.

Lest you think I am on the take from Mark Electronics, I will mention that the best thing to happen to near-IR spectroscopy over the past 10 years has been the proliferation of numerous, excellent third-party software programs. I can only cover several of the better-known packages with which I have personal experience.

I used Pirouette (Infometrix, Seattle, WA) a few times and like the graphically represented statistics. The SIMCA (soft independent modeling of class analog) analysis and MLR portions of the software were relatively easy to use and lent themselves to novices who are opposed to reading the manual. In fact, one of my high school students read a chapter and commented, "Oh, yeah, that's clear enough." There is hope for me, then.

Grams 386 (Galactic Industries, Salem, NH) has been around for a few years and is constantly being upgraded. I'm a little more familiar with Grams because I used an earlier version for atomic absorption work. Galactic has just issued a PLS/plus package (which I haven't had a chance to try yet) to go with its current MLR package. I like how Grams allows you to specify which company's software was used to generate the spectrum. The Grams program automatically translates the data to its own code before calculations begin.

Of course, Unscrambler (Camo, Trondheim, Norway) was introduced by Harald Martens several years ago and is still one of the best programs around. It initially depended almost entirely on PLS, but current issues feature MLR and some very strong statistics programs. Again, the data are graphically represented in any number of ways, making the life of the analyst easier than ever.

Unscrambler, as well as the other packages mentioned, is offered by a growing number of instrument companies to complement their own home-grown software. From the point of view of an applications person, the software updates paralleling the hardware improvements have been far more dramatic than the hardware improvements themselves. That is saying quite a bit, considering the introduction of the Fourier transform, the acousto-optical tunable filter, and diode-array instruments. The instruments have become smaller, faster, and more rugged. Even prices seem to be slowly dropping. There are at least a dozen new manufacturers of near-IR instruments out there now, and more spring up each year. Although I have all the respect in the world for the hardware, I am forced to admit that the software made near-IR a respectable technique.

I can't call the 1980s the "good old days" because every year was good. The instrument companies made the best instruments possible with current available technologies at every step. I have been allowed to work with numerous companies as a consultant and enjoyed all the toys.

Each year of the past decade was the best. Then, to stretch the point, which year/editor/volume of *Spectroscopy* was best? All of them! The magazine's growth has paralleled the changes in the spectroscopic industry quite nicely. The publication has managed to be commercially viable without selling out its editorial principles. Congratulations. I'm proud to have been even a small part of *Spectroscopy's* first decade.

REFERENCES

1. E. W. Ciurczak and T. A. Maldacker, *Spectroscopy* **1**(1), 36 (1986).
2. P. C. Mahalanobis, *Proc. Nat. Inst. Sci. India* **2**, 49 (1936).
3. H. L. Mark and D. Tunnell, *Anal. Chem.* **57**(7), 1449 (1985).
4. E. W. Ciurczak, "Qualitative Analysis of Pharmaceutical Raw Materials via NIRA Spectroscopy," paper presented at the 7th Annual Symposium on NIRA, Technicon, Tarrytown, New York, 1984.
5. E. W. Ciurczak, "Discriminant Analysis of Pure Liquids via NIRA," paper presented at the 9th Annual Symposium on NIRA, Technicon, Tarrytown, New York, 1986.
6. H. L. Mark, "Qualitative Analysis with Wavelength Search" and "Multiple Regression with Wavelength Search," Mark Electronics, Suffern, New York.

CHAPTER IV.A.8

REVIEW OF THE 8TH INTERNATIONAL DIFFUSE REFLECTANCE CONFERENCE (1996): OR "ROUND UP THE USUAL SUSPECTS!"

I just finished a refreshing week (August 12–16, 1996) in beautiful, bucolic Chambersburg, Pennsylvania. I attended the 8th IDRC, a biennial event of some note. The majority of the world's "name" spectroscopists (dealing in diffuse reflectance and, largely, NIR spectroscopy) were on hand: Phil Williams, Peter Griffiths, Karl Norris, Woody Barton, Fred McClure, and a host of other of the same ilk. The conference has never attracted as many as 200 attendees, but that is the idea of a Gordon-style conference: intimacy. A scientist attending the meeting gets the chance to meet, greet, and eat with the best of the best. The location, Wilson College (a 300-student woman's college in Chambersburg, PA), is ideal for the face-to-face interaction among the people attending. Along with morning and evening sessions, there is MORE than enough togetherness to overcome shyness of the newer spectroscopists.

One finds that Peter Griffiths has an infectious laugh and will guffaw at jokes made at his expense ... and will discuss virtually any topic with clarity. Karl Norris has a worthy suggestion for any topic and shares with anyone who asks. Fred McClure's folksy drawl can't hide the decades of experience. No one is "a star" in this peaceful setting: Equals are here to exchange ideas. Graduate students inform the "masters" of breakthroughs and, in turn, learn "ancient" spectroscopic wisdom from them. Aside from being co-chair (with Lois Weyer, who really is running the show … I'm just another pretty face!) in 1998, I can't wait to sleep in an old dorm room, eat in a cafeteria, and attend talks at 8:00 am! Without television or radios, a person can think and *must* confront other human beings at an intellectual level (it was great to be able to use polysyllabic words again!).

WHEAT, WHEAT, WHEAT!

While there were nowhere as many papers as usual on agricultural applications, *per se*, the founders of the conference were, after all, food scientists. Thus, one may always expect a goodly representation of grain applications. In fact, pure "applications" papers, overall, were less numerous this year. The emphasis was on theory and applicability of diffuse reflectance in spectroscopy.

Sandra Kays (of Woody Barton's group, Russell Research Group, USDA, Athens, GA) presented a classic paper of hard work and a well-designed experiment. She worked for two years on analyzing total dietary fiber in all kinds of breakfast cereals. Not exactly my field, but the careful planning was a model of how to choose parameters and reference methods carefully.

Sample presentation of wheat was the topic for A. Kettry (Perten Instruments North America, Reno, NV). It would seem that after all this time, the manufacturers of equipment still need to help the scientist with presenting his/her sample to the instrument. Since no one knows the tools better than the manufacturers, maybe we should begin to listen to them!

NEW MATH APPLICATIONS, OR "GO FIGURE"

Since diffuse reflectance (especially in the NIR) spectra can be difficult to interpret in a meaningful manner, math algorithms used are critical. In the past few years, the number of software introductions by both instrument manufacturers and "third-party" vendors has increased exponentially (nice timing on this pun, eh?). The newest, sexiest approach, known as neural networks (NN), was mentioned a number of times.

Paul Gemperline (East Carolina U., Greenville, NC) again did his best to explain the applicability of neural networks … and most of the people understood. I cannot claim to be a math guru, but I am beginning to think that there is a future in NN. I must remind the adventurous that it was an NN that told the sailors on an American ship a few years ago that an Iranian Airliner was a fighter plane! The one drawback I notice is that NNs tend to tell the user what he wants to hear. (I'm sure some management types in QC settings will love this.) Paul gave us a break by comparing NNs with partial least squares (PLS), thus giving us some point of reference for sanity.

Riccardo Leardi (Ins. di Anal. e Tech. Farm. ed Alimen, Genoa, Italy) introduced (most of) us to the world of generic algorithms. In a manner of NN, the generic algorithms allow more freedom to the software in selecting portions of the data as important. I suggest contacting him for more information on the subject ... or someone like Howard Mark through this journal. (I always call Howie for my math problems ... he is also a consultant for detailed work! Howie ... a free plug!)

Two other topics that cannot be covered in a meaningful manner in a short review are wavelet transforms, as presented by Allen Moser (E.I. DuPont de Nemours & Co. Inc., Wilmington, DE) and CART by Stephen Sum (U. of Delaware, Newark, DE). Again, for reasonable explanations, contact either. I will also try to sweet talk either or both into writing for this column.

Annie, Get Your Software!

One of the least planned and, often, most interesting evenings is the "software shootout." The format is simple: Anyone may participate by accepting several sets of reflectance (NIR) spectra. The participants then may use any software they choose to squeeze information from the spectra. The audience votes on the best presentation, and the winner receives a nice prize. This year, Susan Foulk (Guided Wave) won the honors (hence the "Annie" in the heading, get it?)

It was interesting to watch different scientists approach the problem in various ways. It helps reinforce the idea that there is seldom *one* way to do any analysis! That is an idea that professors and analytical group leaders should take to heart!

... Anything but the Near-IR!

There sometimes arises the idea that only NIR is used in reflectance, often because that's all we hear about. This whole session was NOT about NIR. Charles Walthall (RMSL, USDA, Beltsville, MD; as Robbie Alomar would say, "within spitting distance from my house!") talked about the tribulations and rewards of remote sensing. It is not common knowledge, but it takes quite a bit of skill and imagination to glean meaningful data from airplane flyovers using sunlight for a source! The angle of leaves alone causes major color shifts in a spectrum. Angles of light, time of day, speed of the craft all must be accounted for. I was impressed.

Art Springsteen (Labsphere, Inc., North Sutton, NH) gave us all we could handle about integrating spheres and the "single beam substitution" error involved therein. Art spoke about some of the standards for diffuse reflectance, such as sulfur (which I learned as sulphur) and Spectralon™.

Linda Kidder (NIH, Bethesda, MD) shared some of the research from her group on Raman and infrared imaging. The microscopic work at extremely small areas was really interesting. And Peter Griffiths added his voice to the chorus of Raman users with some tidy work with fiber optics his group performed. The difference here was that it was done on bulk materials instead of small samples.

Observation (How Can You Have Conclusions about a Meeting?)

Small focused meetings will always have a place in the world of megameetings like PittCon. Meeting and sitting down with an author or authority at a large meeting is logistically difficult. At a small get-together, an "A moment of your time, good Sir or Madame!" and a beer will usually suffice. Given my druthers, I'll go for the little meeting anytime ... even if the eating is a little more restrained.

CHAPTER IV.B.1

A NEW BEGINNING AND A WRAPUP OF WHAT'S NEW IN INSTRUMENTATION

Greetings from New Jersey (I can just hear the traditional question: "Which exit?"). In all seriousness, though, I am lucky, enough to be in the center of the pharmaceutical, polymer, and chemical industries and within hailing distance of many major instrument companies. Does all this activity make a molecular spectroscopist a happy camper? You bet! Given the task following John Coates, I will have to use the contacts formed during my 20 years in industry and my 10 (now, over 17) years of teaching.

Although I am now and always will be interested in hardware, I have a special place in my heart for accessories and techniques. Software is also becoming increasingly important in spectroscopic analyses, but sample preparation and presentation remain important and should not be ignored. In this chapter I share some of the introductions of new products for molecular spectroscopy made since last year's Pittsburgh Conference.

INFRARED SPECTROSCOPY PRODUCTS

Bomem (Wood Dale, Illinois) recently introduced the Michelson MB FT-IR, a low-cost high-performance instrument. Using its current technology, the instrument has a new optical bench and is purported to be more flexible than previous models. Using a more powerful Sprouse Scientific Systems (Paoli, Pennsylvania) data package, it can also use Sadtler Research Labs (Philadelphia, Pennsylvania) spectral libraries.

Nicolet Instrument Corporation (Madison, Wisconsin) introduced several products for FT-IR analysis. The 510P FT-IR spectrometer is designed around the IBM PS/2 microchannel bus and the OS2 operating system. The instrument's software is menu-driven and mouse-controlled for simplified operation, and the manufacturer's chemometrics algorithms are included. The System 800, a benchtop research-grade FT-lR system, reportedly provides high resolution, fast data collection and processing capabilities, high signal-to-noise performance, and stable operation. The system accommodates accessories for FT-Raman, microspectroscopy, and hyphenated techniques combining FT-IR with supercritical fluid and gas chromatographies. On the subject of hyphenated techniques, Nicolet is also offering new interfaces for linking its FT-IR systems with the separation capabilities afforded by SFC and the quantitative weight-loss information provided by thermogravimetric analysis. Other infrared products from the company include a line of four low-cost IR spectrometers manufactured by Philips Scientific and a methods development system for the 8200-series chemical analyzers. Nicolet also reports having introduced a series of IR spectral reference libraries for IBM PC-compatible computers and the PC/IR spectral analysis software, which is described later in this chapter.

Perkin–Elmer Corporation (Norwalk, Connecticut) introduced several additions to its 1700X series of FT-IR instruments. A far-infrared version of the instrument is available to cover the 720–30 cm^{-1} spectral range using a single beam splitter, and a near-infrared version provides coverage from 15,800 to 2700 cm^{-1} with its standard optical configuration. Both instruments are compatible with the 1700X-series standard accessories. The model 1725X benchtop system reportedly rivals research-grade systems in its range, resolution, and sensitivity. Fully upgradeable, the system is PC-controlled and uses mouse-driven window software. The company also introduced the model 1760X research system for operation from 10,000 to 370 cm^{-1} with a single beam splitter. This system is said to achieve a signal-to-noise specification of 3000 to 1 for a 4 s measurement and a maximum resolution of 0.5 cm^{-1}. The company's new 1720X system is designed for ease of operation by occasional users, but reportedly provides performance and expandability comparable to research-level systems. Like the other instruments in the series, it features a sealed and desiccated optical system that does not require purging. A TGA interface was also introduced.

A new GC/FT-IR accessory from Bio-Rad, Digilab Division (Cambridge, Massachusetts), was also introduced this year. The Tracer accessory is designed to offer a smaller and less expensive alternative to matrix isolation equipment. The company reports that the accessory is based in part on work performed by Peter Griffiths of the University of California, Riverside, and involves the condensation of GC eluates at liquid nitrogen temperatures on a moving plate. After the eluates form a narrow track, their spectra are obtained with an FT-IR microscope. According to Bio-Rad, the detection and identification limits are comparable to those achieved with GC-MS techniques. Also from Bio-Rad is the Micro/IR7, a system for routine IR microscopy applications. The system consists of an FTS 7 FT-1R spectrometer, a microscope interface department, and the UMA 300A transmittance and reflectance microscope accessory. The system provides objective magnification of 36×, a binocular viewer with a 10× reticule eyepiece, and circular and rectangular variable apertures. According to the company, the system is compact and rugged, requires no water cooling, and is available with a sealed and desiccated optical bench. The system is menu-driven and available with a range of application software.

UV/VIS-NEAR-IR PRODUCTS

Milton Roy (Analytical Products Division, Rochester, New York) has introduced a new diode-array spectrophotometer, the Spectronic 3000. This instrument has a spectral range from 190 to 900 nm with an absorbance range of -3.3 to 3.3 AU. It has an onboard IBM 286- and 287-compatible computer replete with numerous software packages, including a kinetics program. Accessories include heated and ambient flowcell/sippers, a six-position cuvette holder, and a gel scanner.

Bran + Luebbe (Elmsford, New York) has announced the addition of two fiber-optics attachments for its InfraAlyzer 500 systems. The EDAPT-1 uses a low-OH fiber-optic bundle, which ends with a self-contained probe incorporating a gold-coated integrating sphere and PbS detectors. The probe may be handheld, mounted on a stand, or guided by a robot hand. Used for large or bulky objects, the probe boasts a noise level comparable with the internal integrating sphere (<30 μAU). The EDAPT-2 is a transmission bench, which also can be retrofitted to the InfraA!yzer 500. A true dual-beam transmittance device, it allows various attachments to be used in its Cary-type compartment. The wavelength range of the bench is 650–2500 nm. This range allows the 500 system to be used in either the reflectance or transmittance mode.

Dickey-john (Auburn, Illinois), one of the first manufacturers of near-IR instruments, announced the introduction of its Instalab regression and data-collection program. This low-cost program runs on most PC-DOS and MS-DOS computers. The software is rapid and stores the best 25 calibrations, arranged in descending rank. The program is designed to work with the company's Instalab 600 line of near-IR (filter-type) reflectance spectrophotometers.

Guided Wave (El Dorado Hills, California) has introduced its model 260 UV/vis/near-IR spectrophotometer. The company reports that the instrument is faster than its current line and has better signal-to-noise and wavelength reproducibility. It was designed to be used with fiber-optic probes for remote and online work. Guided Wave also has introduced a new ATR probe, named GEM, for use with highly absorbing or colored materials. It requires no dilution liar measurements. The company also has unveiled a spectral matching program called Verifier, which may be used to identify incoming raw materials in the liquid or gas phase as well as solids.

Perkin–Elmer's contribution to the UV/vis world was the Lambda 2 UV/vis spectrophotometer, a double-beam system designed for applications in the environmental, pharmaceutical, and biochemical disciplines. Available with a range of accessories, the system can be operated as a standalone unit or as a PC-controlled system using Perkin–Elmer Computerized Spectroscopy Software. A cassette method-storage system permits transfer of data between instruments or laboratories. The company also has introduced two luminescence spectrometers for the measurement of fluorescence, phosphorescence, and chemi- and bioluminescence. The LS-30 spectrometer is a rapid-sampling system with a dual monochromator and a self-aligning 7-μL illuminated volume flow cell. The LS-50 spectrometer is designed for the selective analysis of many different sample types using a wide range of sampling accessories. Both instruments feature a pulsed xenon source.

LT Industries (Rockville, Maryland) announced the introduction of its Fiber-Line fiber-optic cables, probes, and multiplexer. These are designed to be used with the company's vis–near-IR Quantum 1200 instrument. The multipoint measurements allow control over numerous processes or multiple positions of a single process. The probes are usable to 500°C and 1000 psi. The multiplexer can handle eight channels and scan at a rate of five channels/s. The instrument *per se* can scan at the rate of five full spectrum scans/s.

Pacific Scientific (Silver Spring, Maryland) is offering a modular near-IR spectroscopy system for harsh production conditions. The options include single fibers for long-distance nondiffuse measurements and fiber bundles for closer, high-energy throughput applications. The beam may be standardized without moving the sample position, and the system is capable of 1–5 scans/s. The setup may be used on line or off line, as needed.

Because we are not limited to absorbance spectroscopy, we may consider the new model F-2000 intracellular calcium system from Hitachi Instruments (Danbury, Connecticut) in this group. This fluorescence system is designed to use Quin, Fura, or Indo dyes. A slew rate of 200 nm/s allows maximum data from short analysis times. As little as 0.6 ml can be measured in a standard 1-cm cell with the new horizontal beam geometry.

At this point, the chapter takes a twist (forgive me) with the model 62DS circular dichroism spectrometer from AVIV Associates (Lakewood, New Jersey). With a range of 175–600 nm and a step size of 0.01 nm, the instrument is exceptionally accurate. Data manipulation capabilities included can calculate the sum, difference, product, or ratio of two data sets. The algorithm can take the reciprocal, log, or antilog of data set x or y axis values. The system can also perform general fourth-order polynomial transforms and smoothing by least-squares polynomial fit up to the tenth order. Derivatives are, of course, included.

Spex Industries (Edison, New Jersey) also introduced a circular dichroism system, the Mark VI, for automated analysis of such products as pharmaceuticals, insecticides, and herbicides. The system features a spectropolarimeter covering the 175–700 nm range and a PC/AT-compatible computer for automation. Spex also came out with the 1877E Triplemate, a system that makes it possible to scan—independently or synchronously—the filter and spectrograph stages of

Raman instruments equipped with diode arrays, CCDs, or other multichannel detectors. The instrument reportedly increases the flexibility and precision of automated Raman experiments.

The Protein Secondary Structure Estimation (SSE) software from Jasco, Inc., (Easton, Maryland) is designed to be used with its models J-500 or J-600 automatic recording spectropolarimeters. These instruments and software cover the UV/vis–near-IR region of circular dichroism. Estimations may be performed on known or unknown concentrations using constrained or free fitting methods. Data may be tabular or graphic (cathode ray tube or x–y plotter).

Instruments SA (Edison, New Jersey) has introduced the CP 200 flat-field imaging spectrograph, which is reportedly the first spectrograph specifically designed for use with area and linear-array detectors. "The spectrograph is astigmatism-free, permitting point-to-point imaging, and it uses an aberration-corrected holographic diffraction grating for high signal-to-noise ratio.

ET CETERA

In some cases, the accessories or software may be more important than the instruments on which they are used. Most manufacturers produce excellent instruments *in toto*. The instrument designers, then, can concentrate on individual components of and additions to existing models. In this section, individual components are highlighted.

For starters, Nicolet has introduced an optimized accessory for FT-Raman spectroscopy. The self-contained accessory includes an internal Nd:YAG laser for near-IR excitation, a sample stage, optics, optical filters, and software. Other features include a large sample compartment with removable, prealigned baseplates; 180° on-axis backscattering standard excitation geometry, with optional 90° geometry; and a filtering system that provides for the rejection of the Rayleigh line and the simultaneous collection of Stokes and anti-Stokes scattered energy. During the year, Nicolet also introduced version 1.2 of the PC/IR workstation software for IBM PC/ AT, PS/2, and PC/AT-compatible computers. As mentioned previously, the program can search user-created or commercial IR spectral libraries in addition to several other functions. Complementing this product is a variety of FT-IR spectral analysis packages for Digital's VAX computers and VMS operating systems.

Balzers (Hudson, New Hampshire) has introduced new application-oriented, menu-driven analytical software, named Quadstar Plus, for use with its quadrupole mass spectrometers. The software allows for independent regulation of high-powered ion source voltages, field axis, resolution, mass scale, SEM voltage, and amplifier gain.

A new GaAs microchannel plate photomultiplier tube with a quantum efficiency exceeding 20% at 800 nm has been introduced by ITT Electro-Optics (Fort Wayne, Indiana). The company reports outstanding sensitivity in the 600–900 nm range.

Several interesting detectors have been developed by EG&G (Princeton, New Jersey, and Sunnyvale, California). From the Princeton Applied Research division comes the OMA detectors, the M1455A with 512 or 1024 elements and the M1456A with 1024 elements only. Compatible with standard equipment and newly developed intensifiers, these detectors are reported to enjoy greater signal-to-noise ratios and longer lifetimes. Also available is the Cool Cath cooling option for its 1420/1421 diode-array detectors. This retrofittable option lowers the dark counts and allows shorter sampling times. Its greatest value will be in the fields of Raman, luminescence, and fluorescence spectroscopy. EG&G's new OMA III CCD detector was developed for very small light intensities over very long time exposures. Additionally, the company has announced Data Acquisition Design (DAD) software for the OMA III M1460 detector console and M1461 detector interface controller. According to the manufacturer, the DAD package provides the user with spectral data acquisition flexibility and the ability to control all aspects of an experiment. From EG&G's Sunnyvale division comes the SR series of random-access photodiode-array detectors, which can scan the entire array or any portion of interest. Also announced was a new series of crystalline colloidal filters for use in Raman, pump-probe, fluorescence, and laser spectroscopy.

SLM Instruments (Urbana, Illinois) has developed automated, programmable cell changers for its 8000 C, 4800 C, and 48,000 C spectrofluorometers. The two- and four-cell cuvette holders have optional magnetic stirring and temperature control. The company reports that the products make unattended analysis possible through features such as automatic switching from one position to another in either clockwise or counterclockwise rotation and a user-selectable dwell period for a cuvette in the measuring position.

Perkin–Elmer introduced the Infrared Data Manager (IRDM), an infrared data manipulation software package designed for use with P-E's 1600-, 1700-, and 1700X-series FT-IR systems and the 800-series DIR instruments. The package incorporates a graphics–user interface designed to simplify the automation of routine tasks. A windowing environment and mouse control further simplify operation of such computers as IBM's PS/2 series and the Epson series. The IRDM package also can control specialized programs such as the PC Search software package, which was also recently introduced. PC Search is an expert spectral interpretation software package that uses both spectral matching and interpretation methods for qualitative analysis.

Amoco Laser Company (Naperville, Illinois) has developed a new IR microlaser for FT-IR and IR-Raman spectroscopy. The diode-pumped YAG laser produces in excess of 350 mW of linearly polarized light at 1064 nm. The unit is reported to have high stability and low noise and has won several awards for performance and innovation.

Axiom Analytical (Laguna Beach, California) has produced its first product, a deep immersion probe for chemical batch process monitoring. It employs a circular cross-section ATR element in conjunction with a dual conical optical coupling system. Collimated IR radiation is channeled to and from the head via a pair of optically coated metallic light pipes. It is designed for use with FT-IR units.

Shimadzu Scientific Instruments (Columbia, Maryland) has introduced several software packages for spectroscopic data processing. Programs for the company's IR-460 infrared spectrophotometer and RF5000U spectrofluorophotometer are designed for use on IBM PC–compatible computers. Based on Galactic Industries' Spectra Calc package, the programs' capabilities include parameter setup from the computer, data acquisition, and WYS/WYG display and plotting. Data processing functions include curve fitting, deconvolution, real-time interactive subtraction, and Array Basic programming. Also introduced was an IBM PC–compatible program for wavelength scanning, quantitative analysis, multiwavelength photometric analysis, and kinetics with the company's UV-160, UV-265, and UV-3000 UV/vis spectrophotometers. A six-color x–y plotter for use with the UV-160 also was announced.

Hamamatsu Corporation (Bridgewater, New Jersey) has announced an updated version of its HC210-series phototransducer. The device, model HC220, consists of an integrated photodiode high-gain–low-noise amplifier packaged in a cylindrical mount with a type-D connector for supplying power and transferring the output of the detector–amplifier assembly. The phototransducer is designed so the user can position it at any desired location in the instrument via a standard-size or rod mount. A variety of photodiode types covering the UV–mid-IR spectral range are available.

Laser Interfaces (El Macero, California) has introduced several new products. The LI1500A transient digitizer/A–D converter is an expansion board for IBM PC/AT and compatible computers, which contains a 12-bit 106 samples/s A–D converter with onboard buffer RAM. The LI-1600B 16-bit CAMAC interface is an expansion board for an IBM PC/AT or clone, which permits control and data transfer at a rate of up to 2 Mbytes/s. And finally, the LI-1700A photodiode-array system is a turnkey system for beam diagnostics in laser spectroscopy, allowing real-time pulsed laser diagnosis at rates as fast as 100 Hz.

New Methods Research (Syracuse, New York) is offering its SpecStation computer systems for reduction and analysis of NMR and IR spectral data. The systems range in size from desktop units to large high-performance workstations. Interfaces are available for most instruments.

Spectra-Tech (Stamford, Connecticut) displayed a new advanced analytical IR microscope at FACSS. The latest addition to the IR-Plan microscope line is designed to fit most FT-IR instruments and has some very nice design touches, such as laser lockout and variable-intensity transmitted and incident illumination. The company also introduced the QCIRCLE, an automated liquid sipper for FT-IR systems. It is a temperature-controlled device for turbid, viscous, or strongly absorbing liquids.

Crystal Technology (Palo Alto, California) produces custom acousto-optic tunable filters for various applications. An example is a TeO_2 filter that deflects an **E** vector parallel to the base to an output beam perpendicular to the base. And Linseis (Princeton Junction, New Jersey) is offering a line of recorders. The

L4100 series may be coupled with Uni-soft software to allow interfacing with any laboratory PC, while retaining an analog trace for documentation purposes.

CONCLUSIONS

There were two main sources of information for this chapter. We attempted to canvass the exhibition floors at 1988s major fall analytical meetings (the Eastern Analytical Symposium and FACSS), and, to make sure we reached manufacturers that might not have exhibited at these meetings, we sent a letter to our entire mailing list of manufacturers. Coordinating these product reviews is traditionally a demanding task and we (the editorial staff and I) have made every effort to be complete, but it is inevitable that we have omitted a few products introduced since the 1988 Pittsburgh Conference.

CHAPTER IV.B.2

PITTSBURGH CONFERENCE REVIEW: 1989 IN ATLANTA

Emil W. Ciurczak

I now believe that I have found something that makes filing my income tax return seem almost like fun. After sorting through all the new product releases for PittCon, the IRS seems like a genial taskmaster. The walk through the conference was estimated to be on the order of four miles, and the volume of circulars I perused did nothing to refute that claim. If the vendors weren't passing out all those freebies, I wouldn't have made it. Since I have been attending PittCon faithfully since 1971, I feel I can evaluate and comment upon the meeting. I am getting older while the conference is getting bigger. It is now too big!

Seriously, I wish that the disciplines could be segregated into spectroscopy, wet chemistry, chromatography, and etc. (including lab furniture, ACS insurance, encyclopedias, surveys, and book sellers). I would still cover the whole floor, but I would be able to see one topic per day and be able to keep my thoughts straight on one subject at a time. (And, yes, keep the apples.)

Oh well, in any case, here is a slew of new molecular spectroscopy instrumentation introduced at the conference. I will address the subjects in order of more or less increasing wavelength (as alphabetical order seems too fatuous even for me). As is our habit in the preparation of these product reviews, we have made every effort to be comprehensive, but it is probably inevitable with a show the size of PittCon that some worthy products will be overlooked. If I omit anything, please don't let me know ... the vendors surely will! In all seriousness, we will attempt to make up for any omissions in a future issue.

ULTRAVIOLET/VISIBLE

UV/vis has been a standby in the analytical laboratory for so long that it might be taken for granted. Some of the new introductions in this area give lie to that idea. For instance, Hitachi Instruments (Danbury, Connecticut) introduced its new model U-3410 double monochromator UV/vis–near-IR spectrophotometer. It can be used with various accessories including specular and diffuse reflectance cells. Optics measurements such as polarization characteristics of beam splitters and fiber-optic transmission may be made. Its wavelength range is from 187 to 2600 nm. Chelsea Instruments (London, United Kingdom) is offering a Fourier transform vacuum UV spectrophotometer. The instrument can read down to 185 nm on a routine basis. The new processor and memory capability allow a two million point Fourier transform to be performed in minutes. Pacific Scientific (Silver Spring, Maryland) introduced an interesting tool in the visible range. The Color Sphere spectrophotometer makes use of a diode array, a $d/8$ integrating sphere, and fiber optics to make both transmission and reflectance measurements of color. The instrument uses dual-beam technology and can measure color from the source or over a 5-ft fiber-optic cable. Another color measuring spectrophotometer was introduced by Minolta (Ramsey, New Jersey). The highly portable instrument has a self-contained keyboard, an LCD panel, an integrated printer, and an inboard 3.5-in. floppy disk drive. The measurements are made from a hand-held measuring head on a 51-in. fiber-optic bundle. The range is from 400 to 700 nm. Relatively sophisticated software is standard. Another color measuring device from MTS Colorimetrie (Osny, France) was disclosed by the French Technology Press Office (Chicago, Illinois). The instrument has the software to manually synthesize the correct shade from spectral memory, thus avoiding much physical trial and error on the part of the formulator. Although it is a relatively new company, MTS seems to have some good ideas.

The old Cary instruments are long-standing favorites of many spectroscopists. OLIS (On-Line Instrument Systems, Jefferson, Georgia) has kept the best part of these instruments—the optics—and added modern light sources, electronics, and detectors to produce an up-to-date spectrophotometer. Coupled with a PC/AT-compatible computer, the applications software offered by the company makes this reworked instrument an attractive alternative to a whole new system. Speaking of old familiar names, as a teacher I still use a bunch of old Spec 20s. Well, Milton Roy (Rochester, New York) has come up with an instrument at the other end of the spectrum (if you will pardon a cliché). The new Spectronic 3000 diode-array instrument promises to give other top-of-the-line instruments a run for their money. The spectra can be generated every 20 ms, and the instrument is IBM compatible. The instrument was intended to be an "upscale" model, but I find the price to be moderate (speaking as a school teacher, of course).

Did someone say "low cost?" If so, Pharmacia (Piscataway, New Jersey) is offering a low-cost visible spectrophotometer, the Novaspec II. Aimed at clinical and teaching labs, this compact instrument operates from 325 to 900 nm with a bandwidth of <6 nm. Another low-cost instrument is being offered by Secomam (Fords, New Jersey). The model S.1000 UV/vis instrument is capable of scanning up to 2400 nm/min. Numerous software packages are available, and it comes with a four-color printer.

Groton Technology (Waltham, Massachusetts) has introduced a new photodiode-array instrument, the model PF/3. The spectral range is from 190 to 900 nm, and the rapid scan rate of 48 to 300 scans/s is being advertised for kinetics work. The software for this instrument features Microsoft Windows compatibility. Does Perkin–Elmer (Norwalk, Connecticut) have anything new? Are fish waterproof? One item I thought particularly clever is the DPA 7 Photocalorimetric Accessory for the company's thermal analyzers. It is used to follow photocuring of polymers. Of course PE's recently introduced Lambda 2 UV/vis spectrophotometer was in evidence at PittCon. This unit can stand alone or can be PC controlled. It is expandable and comes with a plethora of software options.

Beckman (Fullerton, California) introduced its DU 64 equipped with its Data Logger. This version collects scans, performs various data reductions, and expands the 64's capacity up to 114 samples. It operates with IBM PC/XT, PC/AT, or PS/2 computers. The Data Logger may also be used with other Beckman DU 50- and 70-series instruments and the DU 6 and 7 models. Various options include a batch sampler, a gel scanner, a 50 μL cell, and various other software options (MPG Software for data manipulation, DU Data Comm for downloading information to the DU series, and Dissolution 6/12 Soft-Pac for online dissolution testing).

Software and Accessories

As has been the trend at recent PittCons, there were exciting introductions in the area of software and other accessories. For UV/vis products, a variety of new tools were unveiled. Hitachi introduced a UV/vis enzyme kinetic data system for the model U-2000. It has an automated six-cell transport along with advanced software. Graphic data can be overlaid and windows used for statistical work. Hitachi also offers Spectra Calc (Galactic Industries, Salem, New Hampshire) software for use with its instruments. Sadtler (division of Bio-Rad, Philadelphia, Pennsylvania) announced its new UV/vis PC Search software. Some features are full-spectrum search, peak search, name search, automated batch searching, molecular structure display, first and second derivative plots, and user-generated libraries. Sadtler sells a library with 2000 compounds for this system.

Varian Associates (Palo Alto, California) developed several programs for UV/vis instrumentation. The DMZ Concentration Program, based on the IBM PS/2, can store and retrieve up to 500 calibration curves. It can also store 500 sample data files, each containing 200 samples. It also controls the company's DMZ 200/300 systems. Varian's PC Control Scan Program is similar, but more rapid, and is designed for direct instrument control.

Spectra-Physics (Mountain View, California) introduced several new lasers. The 2016 line offers 6-W visible, 100-mW UV power in a miniature package. Its high-power diode-pumped lasers feature the 7200-L3 and Y3 with a series of interchangeable YAG and YLF heads with wavelengths from 532 to 1321 nm. Perhaps the highlight of the Spectra-Physics display was the new tunable Ti:sapphire laser, which is quite nice. It can deliver a peak output of 2.5 W and is tunable between 700 and 1000 nm. This range is about 3x broader than conventional dye lasers, and the output surpasses them as well.

Hamamatsu (Bridgewater, New Jersey) introduced a 16×16-element diode array consisting of 256 diodes. The response is from 400 to 1100 nm. Jelight (Laguna Hills, California) developed a Capillary Lamp. It produces $3\times$ the intensity of a standard double-bore lamp using mercury as a source. A programmed cuvette holder was shown by SLM Instruments (Urbana, Illinois). Available in two- and four-cell models, the changers have stirrers and temperature control.

Specac (Orpington, Kent, United Kingdom) displayed reflectance accessories that measure diffuse reflection and specular reflection and fit most research-grade UV/vis spectrophotometers. The company also displayed a variable-temperature analyzer that allows temperatures to be maintained or varied in the -190 to $250°C$ range with a reported stability of $\pm0.1°C$.

Most of the introductions from Hewlett–Packard (Palo Alto, California) were in the area of software. Pharmaceutical scientists will be interested in a dissolution testing system designed to help users comply with the drug industry's Good Laboratory Practices (GLP) guidelines by automating the rate determination at which active ingredients in solid-dosage pharmaceutical products are released. Operating with an HP Vectra PC, the UV/vis software uses Microsoft Windows and has provisions to prevent unauthorized modifications of these test procedures. HP also introduced MS-DOS software for its HP 8452A UV/vis spectrophotometer and Vectra PC. Designed for labs that need a reliable and straightforward system for routine analyses or method development, the program has options for general scanning, quantitation, and kinetics. It can display, overlay, and mathematically manipulate spectra.

FLUORESCENCE

There have been numerous "glowing" reports about this technique of late and, all bad puns aside, several interesting instruments for fluorescence spectrometry were displayed for the first time in Atlanta. Hitachi introduced its low-cost, scanning fluorescence spectrophotometer, the model F-2000. It needs only 0.6 ml of sample for a 10-mm cuvette. It comes with an inboard microprocessor and 9-in. CRT display. It can slew at 200 nm/s and has several programs included for spectral analyses. One available attachment is the new Intercellular Cation Measurement System. The system operates with a stirred microcell and can monitor two probes, such as Fura and BCECF, simultaneously.

Perkin–Elmer's new entry is the model LS50, a PC-controlled instrument that can be used with numerous accessories, including the FURA-2 for the analysis of calcium. Included is PE's Data Management Software, which not only controls the instrument, but contains data handling routines, application programs, and high-resolution graphics. It is compatible with JCAMP.DX, allowing other PC programs to be used on the spectra. It is also applicable to phosphorescence, chemiluminescence, and bioluminescence.

SLM/Aminco has produced the DMX-1000 Multiparameter Cation Measuring spectrofluorometer. It combines a high-speed scanning instrument with a light-chopped excitation system in a T-Optics configuration. This versatile instrument is designed to work with a microscope and digitizing camera, and it performs normal operation and rapid kinetic studies. Two other models, the SLM 4800C and SLM 48000S, are scanning lifetime and multiple-frequency lifetime spectrofluorometers, respectively. These may be used with automated programmable cell changers equipped with temperature control and stirrers.

Spex (Edison, New Jersey) introduced a filter cation measurement system. The model AR-CM is a compact instrument used to measure calcium and pH in single living cells. A dual-excitation pH probe (BCECF) and a dual-emission Ca probe (INDO-1) are available for the instrument. All experiments can be automated with current Spex accessories. The Spex Fluorolog-2 instrument is quite interesting. It detects fluorescence in the IR region of the spectrum. This is useful for analyzing materials such as Nd:YAG for lasers, measuring singlet oxygen in biological processes, and following bacterial photosynthesis. The instrument can also be used for stop-flow kinetics work. Enzyme rate constants are easily measured with data acquisition in the 1-ms range.

CIRCULAR DICHROISM

For lack of a better place, I will talk about circular dichroism (CD) here. OLIS introduced a double CD system that contains a true dual-beam absorbance collection system and uses source rationing to enhance the signal-to-noise ratio. Of course, rather sophisticated software is needed and is supplied with the hardware. The CD system from Spex is the Spex/JY CD6. The instrument routinely works from 175 to 700 nm and can be extended to from 800 to 1000 nm. Also offered are HPLC micro cells, magnetic CD cells, and heated and cooled cells.

Aviv Associates (Lakewood, New Jersey) has introduced the model 62DS. This microprocessor-controlled instrument can collect data versus wavelength, time, or temperature. It includes such software as the Protein Secondary Structure Analysis program (by the method of Yang). The nominal upper limit may be extended to 800 nm by changing optional photomultipliers. Accessories include stopped-flow cells, thin-film holders, and fluorescence detection systems. It certainly looks like CD is finally coming out of the closet!

NEAR INFRARED

Yes, dear readers, there is something between the visible and infrared regions. Since it was only discovered in 1801, it didn't get much press coverage; nonetheless, you might lend an eye to these manufacturers' new offerings for near-infrared reflectance spectroscopy.

Bran + Luebbe/Technicon Industrial Systems (Elmsford, New York) has committed to fiber optics in a big way. Its new InfraAlyzer 600 T/L is an in-line instrument with up to 300 ft of fiber-optic capability. Six remote probes can be multiplexed into the main instrument, with possible ranges of 800–1600 or 1600–2200 nm. The 600D has a remote optical head for online diffuse reflectance measurements and is based on the (diffraction) filter wheel. Both are controlled by an IBM PS/2 model 50. The standard model 500C can be updated with fiber-optic attachments such as the recently introduced EDAPT-1 remote probe, based on a small integrating sphere at the site of the measurement. The newly released EDAPT-2 is a transmission bench, based on a "Cary-type" cell holder. This is capable of 650–2500 nm scans. Cascade, Bran + Luebbe's newest software, links several programs. It reportedly recognizes the sample and calls upon the correct calibration equation for quantification.

Guided Wave (El Dorado Hills, California) introduced the model 260, a rapid scanning, remotely placed spectrophotometer. As with most of this company's instruments, the 260 works with remote fiber optics and can be used with conventional multivariate software or partial least squares (PLS) software.

LT Industries (Rockville, Maryland) announced the introduction of the Quantum 1200 line of near-IR spectrophotometers. The ScaNIR 1200 VIS/NIR is meant to be a lower-cost grating-based spectrophotometer. It boasts the ability to run five full spectral scans/s in either the visible (400–800 nm) or near-IR (800–2400 nm) ranges, and can be run in transmittance, reflectance, or transflectance modes. The 12001 model is built to withstand industrial environments while providing the same rigorous specifications as the laboratory model.

NIRSystems (Silver Spring, Maryland) is the new name for the near-IR instrument division of Pacific Scientific, recently purchased by Perstorp. Its introductions include a new top-of-the-line grating spectrophotometer, the model 6500. The instrument covers the spectral range of 400–2500 nm and is substantially smaller than the current model 6250. It can be used with various modules such as reflectance, transmission, spinning sample, sample transport, or fiber optics. Controlled by an IBM PS/2, it is supplied with a software package with capabilities for partial least squares, fast Fourier transform, and multiple linear regression. Several of NIRSystems' proprietary programs are also included. Several less sophisticated but still grating-based models were also unveiled. The model 4500 for food and agricultural applications covers the range from 1300 to 2400 nm and is designed for analysis of various powder and liquid samples. The model 5000 can make use of the various accessories offered by the company and operates in the "traditional" range of 1100–2500 nm, while the model 5500 is referred to as a "far-vis/near-IR" system, covering from 400 to 1100 nm, and is designed to be used with a remote reflectance probe. All are controlled by IBM PS/2 series computers. In addition to all the new software, NIRSystems has also introduced SCANT, a communications system designed to link and network its remote systems with a central computer over telephone lines. Among other functions, calibration transfer is said to be simplified.

A company named Geophysical Environmental Research Corporation (New York, New York) produces two portable "spectro-radiometers" working from 300 to 3000 nm. Their main purpose is geological studies, measuring components in marble, alunite, pyrophyllite, and so forth.

Guided Wave introduced the model 300 NIR–UV/vis mainframe spectrophotometer. Based on a single 500-μm core fiber, the instrument is said to match the specifications of most research-grade lab instruments. The instrument's strength is its ability to measure reactions and materials under factory conditions while protecting the instrument per se. A lower-grade instrument is the model 260, designed for both lab and pilot plant work. It uses a fiber bundle in lieu of a single fiber. The company has introduced the GEM ATR probe for use with highly absorbing dyes or colorants in solution (in the UV/vis range). For software, Guided Wave developed the Mini-Unscrambler PLS/PCR package for multivariate calibrations.

Accessories and Software

EG&G Judson (Montgomeryville, Pennsylvania) has announced the J 16 series of germanium arrays. Produced for NIRS, the arrays cover a range of 800–1800 nm. Excellent specs are listed for this product. EG&G Reticon (Sunnyvale, California) has announced a new line of diode arrays. These include the K series for HPLC and conventional spectroscopy, the R series for dual-image sensing, and the SR series of random-access linear arrays.

Specac introduced a line of near-IR fiber-optic systems for remote sampling analyses, with a choice of sampling heads for solid, liquid (static or flow), and gas phase. Configurations are also available for Fourier transform and dispersive spectrometers. The company also introduced solid and liquid sample handling accessories for near-IR FT-Raman systems. The solids holder is mounted on a 3 × 2-in. backplate to minimize alignment necessary to achieve maximum signal intensity. A bulb holder and cuvette holder are available for liquid samples.

Two InGaAs photodetectors were introduced by Epitaxx (Princeton, New Jersey). One is a large type, detecting from 800 to 1800 nm, while the other, a PIN type, is listed as "fast" and covers the same range. The company also produces an array detector containing 512 elements, covering the range of 1.0–2.6 μm, and it has announced that the 256-element, 1.0–1.7 μm diode array is currently available. Burle Tube Products (Lancaster, Pennsylvania) introduced its C83062E photomultiplier tube for positron emission tomography.

MIDRANGE INFRARED (AND RAMAN)

This is a region of the spectrum for which hardware and software is most abundant. This year, a number of rather clever innovations were introduced. In fairness, I randomized the order of presentation. Nicolet Instrument Corporation (Madison, Wisconsin) has a number of offerings in the field of FT-IR. The company's low-cost model 205 was designed for routine analyses in quality control, environmental, or teaching (yes!) labs. It is a self-contained unit built to be compact. The model 510P integrates Nicolet's 510 bench with the IBM PS/2 50Z. The advantage is clearly in its computing and deconvoluting abilities, as the bench is a standard. The model 8220 gas analyzer is a self-contained unit specifically designed for routine, repetitive gas analyses. While compact and simple, it does contain a 50-Mbyte Winchester hard disk for data and program storage. Nicolet's new research-grade system is the System 800. The instrument scans from 30,000 to 10 cm^{-1} with numerous options. The system accommodates FT-Raman, SFC-IR, GC-IR, and microspectroscopy attachments. Buried in the preceding was mention of SFC, or supercritical fluid chromatography. This is a nice plus and an area where much work will be done in the future. Nicolet also offers several packages for VAX computers. Virtually all the company's sophisticated software packages are now VAX system compatible.

Mattson (Madison, Wisconsin) introduced two new FT systems. The model G14 (6020 series) has 0.25-cm^{-1} resolution and uses an IBM 16-MHz 386-class computer with a 80387 64-bit math co-processor, 4 Mbytes of RAM, VGA color graphics, and a Microsoft Bus Mouse. Scanning from 6000 to 400 cm^{-1}, the instrument is said to possess speed that is ideal for kinetics work. All other Mattson software is also applicable. Mattson also added the model M34 to its 8020 series. This is a high-resolution optical null dual-beam system integrated with a Macintosh IIx. This instrument is reported to have 0.09-cm^{-1} resolution over the entire range from 6000 to 400 cm^{-1}. The speed is even higher with the addition of the Mac's 16-MHz 68030 32-bit processor with an MC68882 64-bit floating-point coprocessor. More my speed is the Galaxy series of benchtop models. These instruments range over a span of prices and specifications and are designed to be used with the user's own computer.

Laser Precision Analytical (Irvine, California) has introduced a number of new models of FT-IR instruments. The RFX-30 series is the smallest, with 2-cm^{-1} resolution. With a 16-bit 80286 CPU and 80-bit 80287 coprocessor, rapid work is possible. It comes with 1-Mbyte RAM and a 50-Mbyte hard disk. The RFX-40 series claims 0.65-cm^{-1} resolution. This is a medium-resolution instrument, "industrially hardened." It has essentially the same hardware as the 30 series. The high-performance models are the RFX-65 series. The 0.2-cm^{-1} resolution instrument comes with a 32-bit 80386 CPU and 80-bit 80387 co-processor. It has 2 Mbyte of RAM, a 1.2-Mbyte flexible disk, and an 85-Mbyte hard disk. It can be adapted to far-IR or near-IR ranges as well. Laser Precision's highest performance models are the RFX-75 series. These can be run at a resolution of 0.1 cm^{-1}. The data station is essentially the same as that for the 65 series. The company also offers an online system, the PCM-4000 series. These are rugged, real-time instruments for industrial work. The series EVM-6400 monitoring IRs are meant to be environmental monitoring stations, which can monitor up to 24 sampling locations for hazardous gases.

Bio-Rad, Digilab Division (Cambridge, Massachusetts) has introduced two new FT-IR systems. The FFS 7R is a low-cost model with full expansion capabilities. It is of modest size and price, designed for QC or teaching conditions. The FTS 60A, on the other hand, was designed to be a research-level instrument. The company reports that its air-bearing interferometer has been improved, the source enhanced, and the number of mirrors minimized. Normal resolution is 0.25 cm^{-1} (with upgrade to 0.1) and is seen over the full range of 15,000–10 cm^{-1}. All its accessories are compatible, as well as all its software.

Bomem (Wood Dale, Illinois) introduced a variety of products for infrared spectroscopy. The MB-series model MB122 spectrometer features Spectra-Tech's IR-Plan analytical microscope. An external sideport is used to couple the microscope to the spectrometer, leaving the main sample compartment free for normal FT-IR work and a microbeam compartment available for micropellet, micro-ATR, and diamond anvil cell work. Spectra-Tech's Q-Circle cell is also available on the sideport of the MB-series FT-IR systems for liquid samples, which are provided with a separate detector for the QCircle. Bomem also introduced a fast Fourier transform (FFT) booster board that provides a factor of five improvement in the time between acquisition of final data and display of results. The company displayed its combination FT-Raman/FT-IR spectrometer system, which uses one beam splitter to cover both the near-IR region (for FT-Raman) and the mid-IR region (for FT-IR). Alternating between the two methods is achieved by switch, and the laser system is enclosed to prevent operator exposure. An auxiliary sample compartment was introduced for cryogenic sampling with the DA3 research-grade FT-IR systems. Finally, Bomem unveiled a double-emission FT-IR system with interferometric spectral radiometry. This instrument can interface with telescopes and a blackbody source for radiometric calibrations, and radiometric software is available.

One nice addition to the field, not necessarily for superior performance but for price, is the new Buck Scientific (East Norwalk, Connecticut) dispersive IR. For only $7000, one receives a simple, microprocessor-controlled IR system. I am pleased that a number of companies have chosen to appeal to the low end of the market. While I drool for some of the outstanding instruments described in this chapter, I have a very modest budget at my school.

Automatik (Charlotte, North Carolina) has introduced an online, process control FT-IR for polymer production. The molten polymer is transferred to the IR cell and scanned. The instrument used is produced by Laser Precision Analytical. Standard hardware links and software are offered.

Geophysical Environmental Research also produces a portable, remote-sensing FT-IR spectrometer. With a 2-s scan time and a range of between 2 and 20 μm, this seems to be an interesting tool.

Assorted Accessories and Software Packages

Perkin–Elmer is offering an external beam facility for its model 1600 FT-IR. This device allows the analyst to work with large, unusually shaped samples. The company has also released a Raman accessory to expand the utility of the IR instrumentation. This is designed to work with the 1700X instrument. A movable mirror switches between functions. Perkin–Elmer also introduced the DRTA-9A diffuse reflectance/transmittance integrating sphere accessory for its Lambda 9 instrument.

Specac unveiled the Specaflow attenuated total reflection (ATR) system for a variety of FT-IR sampling requirements. The system can be configured to sample pastes, gels, syrups. At least two companies released new products in the field of GC-IR. The Digilab Division of Bio-Rad announced its Tracer system. The device freezes the sample for IR analysis. The system may be evacuated for trace analyses. It is physically placed between the FT-IR spectrometer and gas chromatograph.

A cell based on the more conventional light-pipe design was shown by Aabspec (Wantagh, New York), the model G5 gas cell. It may be used as a detached gas cell or part of a flowing system. A second cell, the EGA 500, can be heated to 500°C so as not to condense samples, which are then read in the gas phase. Heating is programmable with the Aabspec Programmer, used with viscous and mobile liquids, solids. The cell can be used with virtually any instruments. Other options allow liquid microsampling. Aabspec also announced its new UMA sampling, variable-depth profiling, and auto-150 FT-IR microscope accessory, built specifically for flow analysis. Specac also displayed a monolayer/grazing angle accessory for specialized FT-IR applications. Optimization is attained by making two mirror adjustments, and the transition between the two techniques is made by changing sample holders and optimizing the incident angle. Monolayer-mode applications include the study of surfactants, proteins, steroids, polymers, and phospholipids in biological membrane. Grazing-angle-mode applications include the study of thin-film semiconductor devices and coatings on reflective metal surfaces.

A range of IR photodetectors were introduced by Belov Technology Company (New Brunswick, New Jersey). The company, which can design detectors to customer specification, provides HgCdTe detectors and arrays, room-temperature detectors for CO_2 lasers, low-noise preamplifiers, and DTGS and TGS pyroelectric detectors. Sunstone (Dayton, New Jersey) displayed IR converters designed from beam imaging of light-emitting diodes, diode lasers, Nd:YAG lasers, CO_2 lasers, and other sources of infrared radiation. The company also produces UV-to-visible converters and a video system for remote beam diagnostics.

Axiom Analytical (Laguna Beach, California) has introduced a "deep immersion" probe for monitoring chemical reactions. It contains an internal reflectance element with a dual conical optical coupling system. It comes with light guides up to 8 ft long.

Laser Precision Analytical has introduced a thin-layer chromatography/FT-IR instrument. The apparatus accepts standard TLC plates, scans them in the diffuse reflectance mode, and uses proprietary software to calculate concentrations. Since I am working on this concept in the near-IR, I am excited to see it becoming reality. Laser Precision also produces an attachment for microscopy. This apparatus has dual eyepieces, and the specs on it are as good as I've seen to date. Another very good IR microscope is produced by Spectra-Tech (Stamford, Connecticut). Named the IR-Plan, this device was designed to be versatile and sensitive. Several proprietary techniques such as Targeting and Redundant Aperturing are used to enhance resolution. It would seem that this technique is also introduced in SpectIR, an advanced data reduction and spectral search software package. It may be run on all VAX computers (VMS or ULTRIX), on Sun workstations, and several UNIX-based minisupercomputers. It is suitable for networking and provides transfer and decoding from a wide variety of FT-IR spectrophotometers.

Sprouse Scientific Systems (Paoli, Pennsylvania) introduced a number of homegrown software packages for FT-IR systems: MicroTrace II, an off-instrument spectral search and analysis package; CQC, a QC package for pass/fail testing of formulations and raw materials; and IPCC, a quant/qual process monitor and control system. Cooperative development efforts resulted in three more packages: a molecular structure IR spectra display option for Chem Base (in cooperation with Molecular Design, San Leandro, California), a digitally searchable Raman Library (in cooperation with Spex Industries), and an industrially rated IR gas analysis module (in cooperation with Airgas Technologies, Radnor, Pennsylvania).

Lexel Laser (Fremont, California) developed a 500-mW, 752-nm krypton laser for Raman spectroscopy. Geophysical Environmental Research showed its airborne scanners, 64- and 24-channel imagers in the ranges of 2.0–2.5 μm, 300–3000 nm, and, in development, 300–1100 nm. The fact that these detectors can be used from the visible to the far-IR while in flight makes them very interesting.

Spectra-Tech (Stamford, Connecticut) introduced REACTIR Reaction Analyzer System model 1000 for following reaction kinetics in the IR. Buck Scientific introduced a new liquid vessel cell with variable angle. It may be used as a 4- or 8-reflection unit with any commercial IR unit. Buck's IR PLan is an analytical microscope, that offers excellent and competitive specs compared with other models on the market. Its Progressive Sampling line of holders was also highlighted. These can be used for measuring the reflectance of various forms of solid and liquid samples.

Vestec (Houston, Texas) introduced a liquid chromatograph–to–gas-phase detector interface. It is used for FT-IR, as well as for other detection devices.

NUCLEAR MAGNETIC RESONANCE

Bruker (Billerica, Massachusetts) introduced a new NMR system dedicated to solid samples, which is available with options such as phase shifting and an expandable number of rf channels. It comes equipped with 100-, 200-, and 300-MHz field strengths. Bruker also upgraded its low-field instruments; the model AC-E has been renamed AC-F and includes a process controller. Wide-bore magnets may also be added to the 100- and 200-MHz systems. The company's model AMX features its ASPECT 3000 with a 32-bit ASPECT X-32 UNIX computer. These instruments can be equipped with 500-MHz standard-bore or 600-MHz wide-bore magnets. The new software can run on a PC/AT compatible computer equipped with VGA graphics and a minimum of 1-Mbyte RAM and can be networked. One interesting attachment is the ACR-100 series of cell cytometers for analysis of bacteria and other microorganisms.

JEOL (Peabody, Massachusetts) introduced its model 270 CPF FT-NMR, which has several interesting features. Its 32-bit data acquisition processor and 1.25-Mbyte memory, coupled with a 40-Mbyte Winchester disk, allows for much data crunching. Most of its features have been spruced up a bit; so the system seems rather good.

A robot-controlled pulsed NMR for solid content analysis was displayed by Praxis Corporation (San Antonio, Texas). Uses such as analysis of solid fat content are the forte of the model SFC-900. It can perform multiple sampling with a robot arm attachment. Appropriate software is included. The Praxis II desktop NMR is designed for moisture content, oil content, degree of polymerization, ratios of solid fat to liquid fat, and so forth. The system is designed to be used with an IBM PS/2 model 30.

MASS SPECTROMETRY

Kratos Analytical (Ramsey, New Jersey) has introduced the new PROFILE GC-MS and LC-MS mass spectrometer. It is digitally controlled and operates from a mouse interface. It has a high mass range, MS/MS capabilities, a cesium gun, and continuous flow capabilities. Bruker Instruments developed a portable MS for environmental studies. The model MM-1 can continuously monitor 20 compounds simultaneously. Builtin samplers allow monitoring of dust, air, water, and solid surfaces. Although portable, it is claimed to rival instruments based in a laboratory.

VG Tritech (Danvers, Massachusetts) has built an MS-MS system based on its model TS-250. The Tribid system is a standard TS-250 followed by a collision cell and electrostatic analyzer, which are then followed by lenses that introduce the ions into an ix-only collision quadrupole (variable energies from 0 to 250 eV). The system comes equipped with advanced software that runs on a VAX station computer. The new AutoSpec MS features advanced software, VG's patented trisector (EBE) ion optics, and SIOS interface and VAX host processor. The OPUS data system is highlighted by advanced graphics and ease of use. Its resolution of >60,000 at up to 3000 Daltons is not to be ignored either. Also offered by VG Tritech are its Trio-1 and Trio-2 systems for GC-MS (benchtop) and GC/LC-MS, respectively. The Trio-1 uses an HP 5890A

GC for the front end. Its range is 1000 amu. The Trio-2 is a larger version with more options. The maximum mass of 200 amu is coupled with the processing power of the DEC Micro PDP-11/73 computer with optional 4-Mbyte memory. Higher-volume pumping and more software options are available. VG's Microtrace division (Cheshire, United Kingdom) introduced a glow discharge quadrupole MS system, the VG Glo Quad. This instrument is designed for direct analysis of inorganic solids.

New from Finnigan MAT (San Jose, California) is the model ITS40 ion trap system GC-MS. Attached to a Varian 3400 GC, the system uses quadrupole MS to trap and store ions. It can run at 10 spectra/s and uses a Compaq Deskpro 386/20e computer. Also introduced was the MAT 900 high-performance sector MS. Finnigan's conventional SEM detector is paired with the PATRIC (Position and Time Resolving Ion Counting) detector. The company's high-field magnet gives a mass range of 10,000 Daltons. At reduced voltage, the mass range may be extended to 50,000 Daltons.

Hewlett–Packard introduced the HP CLP PLUS software for the HP A-series GC-MS systems and the HP 3350A and 3359A laboratory automation systems. The program is designed to simplify and speed up the processing of data for total organics required by the U.S. Environmental Protection Agency's Contract Laboratory Program. The system performs MS analysis of volatile and semivolatile organics, as well as GC analysis of pesticides and Aroclors. HP reports that this is the first system to automate the entire process.

Several new instruments were shown by Extrel (Pittsburgh, Pennsylvania). The first was the ThermaBeam FAB LC-MS system. This instrument permits flow rates up to 1.0 ml/ min and can identify temperature-sensitive compounds in the 2000–4000 amu range. The second is the model 4250 AMU compact research quadrupole MS system with an upper mass limit of (you guessed it) 4250 amu. It can also be used with the ThermaBeam LC attachment. Its new EXT 1000 glow discharge quadrupole mass spectrometer is specially designed for the elemental analysis of metals and alloys to ppb levels. Its range is 1300 amu. The Questor ammonia analyzer is a process control design used for rapid real-time control of ammonia production. Faster than GC, it can be interfaced with 31 ports for measurements at numerous spots along the process stream. Extrel also introduced the EL 2000 high-performance data system and an environmental software package for its ELQ 400 GC-MS system. The data package allows maximum flexibility of its system with increased RAM and storage capability (170-Mbyte disk or 380-Mbyte tape, on line). Coupled with the new Colorfast mouse driven workstation, this system is extremely powerful.

A static SIMS system from Hiden Analytical (Canton, Massachusetts), the model 3S, was designed for surface studies of routine industrial samples. The quadrupole mass spectrometer is fitted with secondary extraction ion optics and energy filter. It is microprocessor controlled with its own keyboard. Numerous software options are included. Sciex (Ontario, Canada) introduced the API III LC/MS-MS system, based on atmospheric pressure ionization technology. Using a technique called HyperMass, peptides, proteins, and oligonucleotides with molecular weights up to 100,000 Daltons at femto- or picomole levels. The computer system is based on the Macintosh IIx microcomputer.

The Nermag division of Delsi Instruments (Houston, Texas) introduced a new benchtop mass spectrometer that is said to adapt the qualities of a research quadrupole to an instrument capable of routine analysis. The Automass spectrometer is available in four models with features such as electron impact, direct capillary column coupling, wide-bore coupling, packed column coupling, chemical ionization, negative ions, and differential pumping with turbomolecular pumps.

Accessories and Software Packages

HDS (Stone Mountain, Georgia) developed PSIDOM for chemical structure work. It has been incorporated into VG's trio-1 mass spectrometer. It contains more than 43,000 compounds based on the NIST library. The operating system for the Trio-1 is called LAB-BASE, and HDS is converting SMILES, a linear string of characters describing a chemical structure, into HDS format. The company claims the ability to do similar work with nearly any database.

Kratos Analytical also presented the new Mach 3 format conversion programs for converting data acquired for VG or Finnigan MS systems to a format that may be run on a Sun workstation. This allows multiwindowing and high-resolution graphics on the 32-bit workstation. Kratos has also introduced an FD/FAB (Field Desorption/Fast Atom Bombardment) ion source for its CONCEPT series of MS instruments. Switching between the FD and FAB is performed by the software. An additional source, FI (Field Ionization) may be added for such samples as high-boiling petrochemicals and rubber additives.

Palisade Corporation (Newfield, New York) has produced a PBM Mass Spec Library Search for IBM PCs. The new benchtop/PBM and PC Windows programs were introduced at the show. The PBM or probability based matching is based on the Wiley/NIST library. It provides structural displays for 100,667 of the reference spectra in the library (which contains 123,000 spectra). Various search modes are available. By the Plasmarray ICP allows tailoring of the selection of wavelengths for each sample preparation technique. Different emission lines for the same analyte in the lead bead and the nickel sulfide matrices are required to avoid spectral interferences. The Plasmarray optical and detection system incorporates the flexibility to change emission line selection easily in this case. This editor eagerly awaits more applications to be published using the photodiode-array optical design.

Leeman Labs (Lowell, Massachusetts) introduced its PS series of ICP/Echelle spectrometers. The "a" series includes the PS1000—a sequential system; the PS2000—a simultaneous system capable of determining up to 45 elements at once; and the PS3000—a combination simultaneous/sequential system. The most important advance in the PS series is the controller, which uses an IBM PC/AT-compatible computer plus an onboard computer-dedicated data collection, instrument operation, and source control. The controller automates the entire process from igniting the plasma, through to autosampler logic, to shutting down the system. The PS series now includes nitrogen purging for analyte lines below 190 nm. The random access autosampler can hold 6 standards, 7 check standards, and 120 sample cups. The power supply of the PS series is still based on the 40.68-MHz free-running generator.

Perkin–Elmer (Norwalk, Connecticut) introduced its model 320 Laser Sampler, a laser system for the direct elemental analysis of solids, as an accessory for the Perkin–Elmer SCIEX ICP-MS system. The laser sampler incorporates a high-power, high-pulsed laser that automatically vaporizes (ablates) a preselected portion of a solid sample. The vaporized sample is transported with a flow of argon directly to the plasma of the ICP-MS system. The system is reported to be safe in that no laser light exposure is possible. The entire system, including laser parameters, is computer controlled. The laser beam diameter is adjustable, allowing the operator to select optimal conditions for each sample type.

Perkin–Elmer also introduced an IBM PS/2 interfaced to the Plasma 40 ICP system. The computer allows for color graphics, incorporation of artificial intelligence, and up to nine spectra to be simultaneously overlaid for comparisons in nine different colors.

Shimadzu Scientific Instruments (Columbia, Maryland) introduced two new optical emission products, including a line of pulse distribution analysis (PDA) arc-spark emission spectrometers with data processing capabilities. The new line features completely new designs in polychromators, source, stands, electronics, and data processing. The HPSG-330 multiwaveform source eliminates the need for sample preburns, and burn time is reduced to 5 s. A two-burn analysis is completed in 22 s. Shimadzu claims that no regular factory maintenance or service contract is required because all maintenance can be performed by the operator.

Shimadzu's GDLS-5017 glow discharge lamp source spectrometer was introduced for coated weight, coating content, and coating thickness analyses of plated metal products. The unit features a compact, self-cleaning glow discharge source, a 0.5-m polychromator, and an 80386 data system with color graphics. Most elements, including H, N, and O gases in a metal coating, can be analyzed routinely. The speed of analysis is reported to be between 20 and 120 s for 1030 μm coatings, depending on coating composition and desired depth resolution.

Spectro Analytical Instruments (Fitchburg, Massachusetts) introduced a new sequential ICP spectrometer and two new portable emission spectrometers. The Spectroflame ICP model S features a direct wavelength drive that accomplishes wavelength selection with an accuracy of 0.0008 nm. The wavelength selection is said to be typically achieved in <1 s for any wavelength, analyte signal, or background. The model S does not use a peak search routine and does not make 5–7 measurements to compute peak maxima. Therefore, analytical throughput, expressed in elemental determinations per minute, is reported to be 3–5 times faster than peak-search sequential ICPs. The model S can be upgraded to a combination simultaneous/sequential spectrometer.

Spectro also introduced the Spectroport I and II portable emission spectrometers for the identification and/or verification of metals in the field. The hand-held spectrometers measure the alloying content of metals via a dc arc. The Spectroport II provides a higher level of software utility with interchangeable optical modes.

Thermo Jarrell Ash (Franklin, Massachusetts) introduced three new optical emission products. The PolyScan 61E spectrometer, a simultaneous/sequential spectrometer for elemental analysis of water, oil, and other materials. The spectrometer consists of a 0.75-m polychromator and a 0.75-m scanning monochromator. The monochromator and the polychromator share the same ICP source and sample introduction system and are controlled using the familiar Thermo-SPEC software. The entire system, including the rf power, gas flows, and sample introduction system are controlled via an IBM PS/2 model 50 computer. The plasma torch can operate in a low-power–low-argon-flow or a high-powered high-argon-flow mode.

The AtomComp 81 spectrometer is a direct reading arc-spark optical emission system based on the 0.75-m optical design with an improved sample stand and updated solid-state dc arc power supply. The AtomComp 81 is operated using an IBM PS/2 microcomputer and ThermoSPEC software. The ISC-150 intelligent sample changer is a peripheral device that is compatible with TJA's ICP and AA systems. It can accommodate up to 150 samples plus an auxiliary rack for blank controls, standard preparation, and matrix modification. The preparatory functions can be executed by interfacing the ISC150 with the PS-150 Prep Station.

CEM Corporation (Matthews, North Carolina) introduced the model MAS-300 microwave ashing system that is said to fully ash a wide variety of materials at temperatures up to 1200°C in 5–10 min. The system includes a quartz-fiber sample carrier that cools rapidly, allowing samples to be reweighed 30 s after being removed from the furnace. Crucibles made of ceramic, graphite, or platinum can also be used for ashing. The MAS-300 does not require special high-voltage or high-current electrical connections and reportedly does not consume as much energy as a conventional muffle furnace.

Although actually designed for use in gas chromatography, a new atomic emission detector from Hewlett–Packard (Palo Alto, California) is significant enough to spectroscopists to merit mention. Reportedly the first totally automated multielement atomic emission detector for GC work, the HP 5921A can selectively detect any element except helium. The detector can be used as a means of identification or as a screening tool for mass spectrometry and infrared detection. Hewlett–Packard reports that the instrument can detect oxygen in a variety of matrices, and is useful in the analysis of petrochemicals, foods and flavors, gas products, and environmental samples.

ATOMIC ABSORPTION PRODUCTS

Fisons Instruments introduced its ARL/GBC 906 atomic absorption spectrophotometer, which is described as a fully automatic, multielement AA spectrophotometer. The entire system is controlled by an IBM PS/2 computer and can analyze up to 12 elements automatically. The 906 stores graphics traces for all samples in a multielement run. An optional System 3000 automated graphite furnace is available.

Thermo Jarell Ash introduced its dual-channel, single-beam Smith–Hieftje 21 atomic absorption spectrometer. The 21 is compatible with the CTF 188 graphite furnace and incorporates an automatic baseline compensation capability for drift reduction.

Shimadzu Scientific Instruments introduced its AA-680 atomic absorption spectrophotometer featuring an eight-lamp turret, color video display, and automatic selection of optional elemental analysis parameters. The AA-680 can be equipped with an optional graphite furnace accessory.

Analyte Corporation (Grants Pass, Oregon) introduced its model 16F atomic absorption spectrophotometer, which represents a very unique design. The flame AA system can analyze up to 24 elements in sequence in a single sample, taking approximately 4 s/element. The 16F is equipped with a 16-lamp turret, which can move from one lamp to another in about 1 s. The monochromator is able to move from one wavelength to another in <1 s. The flame conditions are controlled automatically by a computer. The 16F employs a unique movement of the burner system to give the effect of a double-beam design from its single-beam optics. The fast elemental analyses design allows for the incorporation of an internal standard to reduce transport interferences

Varian Associates (Palo Alto, California) introduced two accessories for its line of atomic absorption spectrophotometers. The Mark-VI burner heads employ a flared slot design, which reduces burner blockage during analyses of high-solids samples. The SpectrAA/PC Report Manager provides the ability to transfer AA data to the PC during or after analysis; to customize commands, report headers and footers, sample labels; and to perform weight/volume correction and Lab Note storage. The Report Manager incorporates a sample-label match facility that allows the analyst to choose individual samples, standards, and QC samples from an analytical run to generate separate reports. Lotus MACRO worksheets are included for processing data with Lotus 1-2-3 version 2 software.

X-RAY AND SURFACE ANALYSIS PRODUCTS

Rigaku (Danvers, Massachusetts) introduced the model 3270/3271 sequential wavelength spectrometers, which are based on IBM PS/2 computers and include a complete software package with advanced and versatile computational capabilities. An online automatic analysis program is designed for samples with complicated matrices or materials for which it is difficult to prepare standards. The program includes a semiquantitative analysis method that identifies the elements present and calculates their concentration without the use of a standard; a free quantitative analysis method that calculates the concentration of designated elements when standard samples cannot be prepared; and a group quantitative analysis method that calculates the concentration of designated elements when standards are available. A full range of matrix correction models are available for empirical calibrations. Reports can include calculation of mean values, standard deviations, plotting, XR (average-range) control charts for check samples, accuracy checks, and analyses of variance (t-test).

HNU X-Ray (Oak Ridge, Tennessee) displayed an energy-dispersive microanalysis system that incorporates an IBM PC/AT-compatible computer. The System 5000 microanalyzer includes an upgradable workstation with powerful analytical software and an optional image analysis capability. The system is also compatible with available PC software for report generation, word processing, data bases, and so forth.

Siemens Analytical X-Ray Instruments (Madison, Wisconsin) introduced the R3m/V Series II x-ray diffractometers, designed to provide flexibility and improved performance in small-molecule single-crystal structure determinations. The instrument uses a 3-kW medium-frequency x-ray generator and features a four-circle goniometer housed in a modular radiation enclosure. Operation is automated by a microprocessor-based instrument controller linked to a MicroVAX 2000 host computer equipped with SHELXTL PLUS software for system control, data reduction, and structure solution. Siemens also introduced Spectra AT software, which can combine theoretical and empirical alphas for maximum flexibility in the analysis of complex spectra. The fully integrated, menu-driven software requires only one standard. Users can have 64 elements in one program with up to 24 alphas for each element. Alphas can be selected by value or by major trace selection.

The XRF Division of Kevex Instruments (San Carlos, California) introduced the Sigma online system for process control and monitoring and for field applications. The system provides a modular energy-dispersive x-ray fluorescence (EDXRF) subsystem that combines speed with nondestructive and noncontact elemental analysis. The Sigma consists of a PSi Peltier-cooled solid-state detector and electronics, a portable x-ray source tube with integrated high-voltage power supply, a recirculating cooler for the source tube, and an IBM PC/AT-compatible computer, Toolbox II software, and an MCA board.

Philips Electronic Instruments Company (Mahwah, New Jersey) showed its new integrated scanning electron microscope (SEM) and energy-dispersive spectroscopy (EDS) system, the PHAX-SCAN. The system incorporates the company's series 500 SEMs and is said to be user friendly and to provide high operating speeds for such applications as full quantitative analysis, quantitative digital x-ray maps, digital linescans, particle analysis, unattended analysis, 3D metrology, library sorting, and random analysis. The instrument typically can complete an analysis of more than 2000 particles in <20 min, and its process capabilities can reduce data up to 300 times faster than other systems, according to Philips.

Vacumetrics' Vacuum Division (Ventura, California) introduced the M.A.E.S. data system for x-ray photoelectron spectroscopy (a.k.a. ESCA), Auger electron spectroscopy, and other electron spectroscopies for elemental analysis of surfaces. The system, which performs peak extractions and area measurements, is fully automatic and can process a 50-peak spectrum in 2 min. Functions and features include automated background fitting, peak synthesis, interactive operation, and automatic element identification. The system operates on IBM-compatible PC/XTs, PC/ATs, or PS/2s, and complete Auger and ESCA databases are provided.

CHAPTER IV.B.3

INSTRUMENT SERVICE SURVEY

As promised, dear readers, this is the long-awaited reader survey on instrument repair service. First, I must say that I do not intend to disparage or promote any instrument company through this tally. Indeed, I am collecting these data in the hope that they may lead to better sales and servicing of instruments.

As the Pittsburgh Conference approaches, many of us have visions of sleek, sparkling instruments dancing in our heads (I'm writing this a week before Christmas, after all). In the excitement of updating our laboratories, we rush off clutching purchase orders with nary a glance at instrument vendors' records for repairing and servicing the expensive apparatus they sell us. By a company's record, I mean the frequency its instruments needed repair and the length of time they stayed repaired. The many facets involved in this seemingly simple query will become apparent when you get to the survey questions.

I chose the 25 questions of this survey to develop a feel for the quality of followup service provided by instrument suppliers; however, I am also interested in determining the average customer's usage habits. I encourage readers to photocopy the survey page from the magazine, complete the questionnaire, and return it to my attention either at the address at end of this column or at this publication's editorial office (Spectroscopy, P.O. Box 10460, Eugene, Oregon 97440). The survey is intended for instruments used on a regular basis, not just a few times per week or month, and two things must be kept in mind while filling out the questionnaire. First, determine if the breakdown was the instrument company's fault or the user's. That is, was the problem a design flaw, the lack of a properly written training or instruction manual, or poor construction? If not, and you acted against written warnings or commonsense precautions, keep in mind that a lack of maintenance on the part of the user does not constitute a flaw in instrument design.

Second, try to recall the response of the seller in terms of speed, competence, and price. The former point is especially important. Proper use of an instrument's instruction manual is an often overlooked, but vital step in learning to use an instrument to its maximum efficiency and safety. I feel comfortable saying that a lot of maintenance problems that crop up in many laboratories are the result of a well-meaning operator taking matters into his or her own hands before consulting the book. Although it is sometimes cumbersome to do so, reading the owner's manual is important. And from my perspective, having written more than one instrument operator's handbook in my time, I assure you it can take months of painstaking effort to produce a useful one. For no other reason, read your manual out of pity for the poor person who had to write it!

The more interesting results of this survey will appear in a forthcoming column. More detailed information will be made available to interested readers by hard copy or floppy disk after the results have been computed (there will be a nominal fee to cover processing and postage). Although I will provide general data for the public, responses naming specific companies or their employees will not be published or otherwise made available to anyone other than appropriate representatives from the company in question. More information about obtaining a copy of the results will appear in a future column.

I encourage your participation in the hope that we can all learn a little more about one another's unique experiences in this vital area. Please return the completed survey by 16 April 1990.

Questionnaire

1. What types of spectrometers do you use frequently? (Check all that apply.)

__ Dispersive IR   __ FT-IR   __ Filter Near-IR   __ Grating Near-IR
__ Raman   __ NMR   __ UV/Vis   __ Filter fluorometer
__ Scanning fluorometer   __ Mass spectrometer

2. What models/suppliers of each do you have (and use)?

3. On average, how often must you replace expendables (lamps, etc.)?

4. Do the replacements last as long as the originals?   __ Yes   __ No

5. Are service calls usually necessary?   __ Yes   __ No

6. The last time you ordered a part, how long did it take to arrive? (After purchasing got around to ordering it, of course.)

7. Was the operator's manual any help the last time you attempted to fix an instrument, or did you ask someone else for help?

__ Manual   __ Other person   __ Fixed it by luck/skill

8. What kind of training have you received from the supplier?

 __ Good   __ Fair   __ Poor   __ None

9. During your last purchase, was your sales representative helpful in choosing the proper model and/or options?   __ Yes   __ No

10. Were you ever sold an instrument that did not perform the task for which it was purchased?   __ Yes   __ No

11. If the answer to question 10 was "Yes" (be honest now), was it because (check one)

 __ The sales rep misrepresented it.
 __ The sales rep withheld information.
 __ The sales rep didn't know how to answer my questions.
 __ I wasn't sure what questions to ask.
 __ I assumed that I knew the capabilities.

12. When an instrument breaks down, what is the average response time

 __ before the service department returns your call?
 __ you receive the part to do the work yourself?
 __ a service rep calls on your facility?

13. When a service technician calls on your facility, does he/she

 __ have all the needed parts?
 __ know the model in question?
 __ have the knowledge to make repairs?
 __ explain how to avoid a similar breakdown?

14. When the repair is made, is the job done well? (In other words, does the instrument work as well as before the problem occurred?)   __ Yes   __ No

15. How much importance do you place on service when making a purchase?

 __ A great deal   __ Some __ Not much   __ None

16. Do you believe that your instrument reflects the general reliability of that company's other instruments? (That is, will one lemon spoil it for company XYZ?)   __ Yes   __ No

17. In general, is durability or versatility the main characteristic that you seek in a new instrument?

 __ Versatility   __ Durability

18. From your experience, is your most used instrument

 __ Versatile?   __ Durable?   __ Both?

19. Overall, how would you rate the instruments in terms of dependability and serviceability?

___ Excellent   ___ Good   ___ Fair   ___ Poor

20. Overall, how would you rate the repair service from the instrument manufacturer or supplier?

___ Excellent   ___ Good   ___ Fair   ___ Poor

21. On what do you base your judgment for question 20?

___ Personal experience with same/similar model
___ Experiences of colleagues
___ Surveys such as this
___ Manufacturers' claims/reputation

22. In what state is your facility located? _____

23. At what type of facility do you work? _____

24. How large is your facility? (people) _____

25. How large is your facility's annual instrument budget? $ _____

CHAPTER IV.B.4

1990 PITTCON PRODUCT WRAPUP, PART 1: INSTRUMENTATION FOR MOLECULAR SPECTROSCOPY

Emil W. Ciurczak

Another huge Pittsburgh Conference has come and gone. PittCon certainly has grown since I first started attending it back in the Cleveland years. From 5–9 March, 1990, the conference filled New York's sprawling Jacob Javits Center, one of the nation's few venues large enough to accommodate an event of this magnitude.

The Javits Center was alight with an exciting array of new "toys" for spectroscopists to see, touch, and covet. The objective of this chapter is to share the majority of these products with spectroscopists who either couldn't attend or, if they were there, couldn't see everything. As the title suggests, this column will cover instrumentation, accessories, and software for the molecular spectroscopy techniques: UV/vis, infrared (IR), Raman, circular dichroism (CD), fluorescence, nuclear magnetic resonance (NMR), and mass spectrometry (MS). I will start with instruments covering the shorter wavelengths, proceed to the radio waves, and conclude with NMR and MS products. Within each grouping, I will discuss instruments and complete systems first and then deal with accessories and software. Products within these sections will be listed alphabetically by company name.

Now for what has become an annual disclaimer for the PittCon wrapup article. We have taken every effort to make this review as complete and accurate its possible within our space limitations. Unfortunately, with an exhibition the size of PittCon some new products might have been omitted. Information for this column was obtained from literature supplied by the vendors, from personal inspection of selected products in New York, and from manufacturer press conferences. In many cases, the amount of material I was able to include about a particular product is proportional to the ease with which we were able to extract information from its supplier. With that out of the way, let's proceed with the wrapup. I hope you enjoy reading about the newest labors of the PittCon exhibitors as much as I enjoyed seeing and testing them.

UV/VIS–NEAR-IR PRODUCTS

Beckman Instruments (Fullerton, California) introduced several interesting programmable spectrophotometers based on the company's "step program" concept and the DU series 60 instruments. These spectrophotometers can create color characterizations. The program controls setup, the instrument, data acquisition, data reduction, and printout of results.

Buck Scientific (East Norwalk, Connecticut) presented the Cecil 500 series double-beam UV/vis spectrophotometer. It offers double-beam optics, method storage, curve storage, baseline memory, RS-232 serial port, derivatives, kinetics, and scan overlays. The instrument is also compatible with 25 accessories.

Hitachi (Danbury, Connecticut) displayed its new U-3501/U-4001 double monochromator, double-beam UV/vis–near-IR spectrophotometers. These compact instruments are based on a Littrow-type "premonochromator" and a Czerny Turner-mounted diffraction grating. Wavelength reproducibility is ±0.5 nm in the near-IR range. As with all of the latest "upscale" instruments, it runs on sophisticated software with window display, color graphics, and mouse operation. Numerous accessories are available from the manufacturer. Also shown was Hitachi's Fourier transform (FT) visible microscope photodiode-array spectrophotometer. The system, operating from 400 to 800 nm, reportedly measures particles down to 3 μm.

Another "colorful" entry came from Hunter Associates Lab (Reston, Virginia), which specializes in color-measurement instruments. The company introduced the handheld MiniScan, which allows color control of products as numbers or pass/fail readings. The company also produces the MatchMaker color formulation system and SpecWare, a software package.

Oriel (Stratford, Connecticut) introduced InstaSpec 111, a new diode-array detector for UV/vis spectrophotometry. It offers real-time analysis of pulsed and kinetic studies, high sensitivities, and low noise. It uses a fiber-optic data link and innovative software that contains an internal BASIC macrolanguage for custom programming.

Otsuka Electronics USA (Havertown, Pennsylvania) introduced a line of UV/vis/near-IR instruments tailored for applications in biospectroscopy.

Through fiber-optic technology, the systems can perform noninvasive *in vivo* experiments, collecting full spectra in 21 ms to 5 ns. The company also introduced the DLS-600 and ELS-800 systems, based on dynamic and spectrophotometric laser light scattering principles, respectively.

The Lambda 2 UV/vis spectrophotometer from Perkin–Elmer (Norwalk, Connecticut) has been upgraded and equipped with a routine kinetics package. The Lambda 6 UV/vis has been upgraded as well through the use of new software allowing instrument control from standard (MS-DOS) PCs. The new macrolanguage, called OBEY (based on Digital's GEM software), has Windows, a mouse-pointing device, and pulldown menus.

The S-750 spectrophotometer from Secnmain (Sarcelles, France) has a wavelength range of 200–1100 nm with a pathlength up to 50 mm. This standalone instrument was developed for colorimetry. The company's S1000G, an upscale, IBM PC compatible version with a CRT and numerous accessories, scans wavelengths from 200 to 1000 nm.

Shimadzu (Columbia, Maryland) showed its UV-2101PC UV/vis personal spectroscopy system that incorporates an IBM-compatible computer and Shimadzu software with the photometer unit. The instrument covers the range from 190 to 900 nm and −4.0 to 5.0 AU with up to 0.1-nm resolution. The company's UV-3101PC UV/vis near-IR system offers similar computer control in the 190–3200-nm range.

Varian Associates (Sunnyvale, California) announced the expansion of its Cary line of spectrophotometers with the development of the Cary 4 for UV/vis and Cary 5 for UV/vis–near-IR spectroscopy. The upper wavelengths for the instruments are 900 and 3300 nm, respectively. These instruments reportedly provide low noise, low stray light, and high resolution because of a new double-monochromator configuration and "out-of-plane" Littrow design. The Cary instruments incorporate a new operating language called applications development language (ADL) and, of course, are computer controlled.

Accessories

ARIES (Concord, Massachusetts) showed its model FF250 flat-held spectrograph/monochromator. This instrument has some impressive specifications, such as a slew rate of 400 nm/min at 0.01-nm increments, wavelength accuracy of ±0.03 nm, and precision of ±0.015 nm. Other features include a 25 × 50 mm flat focal field for compatibility with charge-coupled device (CCD) and diode-array detectors with no vignetting, keypad conversion from spectrograph to monochromator, and remote digital control of all functions by its handheld keypad or by PC.

Crystal Technology (Palo Alto, California) entered the field of acousto-optic tunable filter (AOTF) technology with its tellurium dioxide filters for the 490–980-nm range. The bandpass resolution ranges from 1.7 nm at 490 nm to 10.5 nm at 980 nm. A second AOTF is available for the 1.2–2.5-μm range with resolution ranging from 4.5 nm at 1200 nm to 18.0 nm at 2400 nm. The computer-controlled filters allow random bandpass access in <100 μs. The company has also fabricated what it calls the largest flaw-free KTP (potassium titanyl phosphate, $KTiOPO_4$) crystal in the world. Designed for a variety of optical applications, the KTP crystal can be used in optical waveguide devices such as phase modulators, amplitude modulators, and directional couplers.

Helma Cells (Forest Hills, New York) added new microcells to the company's impressive collection. The new cells have volumes down to 50 μL.

FLUORESCENCE SPECTROMETERS

Jasco (Easton, Maryland) displayed its FP777 fluorescence spectrometer that boasts a 90:1 signal-to-noise ratio.

Photon Technology International (South Brunswick, New Jersey) introduced a line of fluorometers. The totally integrated LS-100 (an "R&D 100" award winner) reportedly can record fluorescence and phosphorescence emission and excitation spectra in the femtogram range. The company's Alphascan and M-series fluorometers are modular and may be assembled for the application at hand. Software for the instruments is sophisticated and in living color.

SLM Instruments (Urbana, Illinois) is marketing its DMX-1000 multiparameter cation measuring spectrofluorometer designed for studies involving intracellular calcium probes, as well as for measuring fluorescence spectra, taking radiometric measurements, and conducting polarization studies. The system comprises a high-speed scanning spectrofluorometer with a light-chopped excitation system in a T-Plus optical configuration and incorporates an Image I/SL fluorescence imaging system from Universal Imaging Corporation. Accessories include an SLM-designed epifluorescence microscope interface for the Nikon diaphot microscope (equipped with camera/monitor) as a unit for fluorescence measurements of extremely small samples; an emission beam splitter; a stopped-flow accessory; an actinic front plate; and an automated programmable cell changer. This powerful package is capable of rather sophisticated imaging.

Spex Industries (Edison, New Jersey) brought its new FluoroMax spectrophotometer to New York. This moderately priced instrument is supplied with a PC and full software. It has a scan speed of 160 nm/s and reportedly can perform time-based measurements as fast as 1 ms.

NEAR-IR EQUIPMENT

Bran + Luebbe Analyzing Technologies (Elmsford, New York) introduced the first commercial near-IR instrument incorporating an AOTF. The InfraAlyzer AOTS (acousto-optic tunable scanner) is designed to perform on-line process control. The wavelength range is 900–1700 nm, running at 70 scans/s. Absolute wavelength accuracy is ±0.25 nm with a reproducibility of ±0.01 nm. Using an (optionally cooled) InGaAs detector, noise is reportedly 15 μAU at 1500 nm and 0 Abs. The company's new software includes Cascade, which identifies unknowns and automatically switches to the applicable product calibration.

D.O.M. Associates International (Manhattan, Kansas) proudly displayed its Hadamard transform spectrophotometers. In addition to near-IR, the instruments operate in the visible range as Raman instruments and stationary interferometers. The instruments have a slit array with 255 elements as a stationary Hadamard encoding mask containing no moving parts. Features such as Rayleigh line rejection and no phase corrections are two advantages of this device. I, for one, am happy to see this concept incorporated in a commercial instrument.

Guided Wave (El Dorado Hills, California) unveiled extended-range versions of the model 300 fiber-optic spectrophotometer. Now available in versions for UV and visible analysis, the analyzers feature a double-beam, single-strand fiber design and can multiplex up to eight sample points. Laboratory and process versions with custom, application-specific software are available. The company also introduced the Shuttle Probe, a single-strand fiber probe for inline process measurement. Qualities include consistent probe alignment and pathlength, as well as easy removal from the process stream when maintenance is necessary. Guided Wave also reported at PittCon that near-IR calibrations can now be successfully transferred between its analyzers and spectrometers made by its new sister company, NIRSystems (Silver Spring, Maryland). This development should simplify the calibration process for users of field-based systems, who can now develop calibrations in the laboratory on one instrument and transfer them to a different system in a remote location.

LT Industries (Rockville, Maryland) highlighted its new Quantum 12001 PLUS, the industrial version of the 1200 PLUS general purpose visible/near-IR analyzer. It operates at a speed of 5 scans/s in a real-time mode. The analyzer can be part of a closed-loop process, monitoring up to 10 components simultaneously from its optional NEMA 4X enclosure. Its InGaAs detector is optimized for the wavelength range from 800 to 1700 nm. LT also introduced several new software packages, including SpectraMetrix 1.70, which contains chemometrics methods and stepwise linear regression analysis: LighTcal Plus, a multivariate calibration program that offers partial-least-squares and principle-component regression methods: and Transparency (TRANS), which showcases the ability to transfer calibrations easily between all Quantum 1200 instruments. Also shown were the company's FiberLine fiber-optic probes and Multiplexer, which allows one instrument to monitor up to 16 lines. According to the company, the probes can operate in environments up to 500°C and 3000 psi.

Perstorp Analytical (Thornbury, Bristol, United Kingdom), which now includes NIRSystems and Guided Wave, hosted a multicompany booth at the conference. NIRSystems showed its standalone instrument equipped with a 40-Mbyte hard-disk drive for its operating software and data and operations files. This small, rugged unit is designed for plant use. The company's modular system for online processing may be equipped with various fiber-optic probes. An indoor–outdoor NEMA 4X container is offered with the unit. An accessory, the new DESIR drying unit, evenly dries a liquid on a GF/F matrix for introduction into a near-IR unit. Measurements of materials may be enhanced by 100 times or more. The company's new IQ^2 software allows for the identification, qualification, and quantification of routine raw and process materials. The software contains principle component and spectral-matching algorithms and can perform multicomponent quantitative analyses.

ACCESSORIES

EG&G Judson (Montgomeryville, Pennsylvania) introduced the J 16M series, a 128-element, 2.5 mm × 100 μm photodiode array covering the 800–1600-nm range. Available from EG&G Princeton Applied Research Corporation (Princeton, New Jersey) is the 1452 near-IR detector. This compact, multichannel-linear array, InGaAs detector was designed to work with the OMA III 1460 detector console or the OMA III 1461 detector interface. It covers the 800–1700-nm range and uses a two-stage Peltier cooler to maintain a stable operating temperature as low as 40°C.

Specac Limited (Orpington, Kent, United Kingdom) introduced a series of near-IR fiber-optic samplers for both dispersive and FT systems. The fibers cover the range between 30,000 and 2100 cm^{-1} using either zirconium fluoride or quartz WF material. The samplers may be customized for gas, liquid, or solid samples. Specac also introduced a variable-pathlength (17 m) multipass gas cell for trace-gas analysis. The long-path miniature cell is made of corrosion-resistant materials and has removable mirror systems for easier cleaning. The cell has a volume of 530 μL, permitting routine microgram sensitivity. Coated mirrors for UV through IR coverage are available.

MID-IR AND RAMAN PRODUCTS

Bio-Rad, Digilab Division (Cambridge, Massachusetts) released several new high-tech FT-IR products. The first is the FTS 7R spectrometer, a low-cost, expandable unit for various applications. It can be upgraded with external accessories, such as the UMA 150 FT-IR microscope or high-sensitivity fast detectors. It features the new EasyIR easy-to-use software package. If extended to the near-IR region, the instrument is designated FTS 7N. It may be used with a gas chromatograph or their new gravimetric analyzer, using the special properties of these techniques, or it may be equipped for use as a Raman spectrophotometer. Representing the research level of Bio-Rad's FT-IR line, the company introduced new products in the FTS 60A series, which feature dynamic interferometer alignment. The company reports that the FTS 65A model can achieve resolution down to 0.1 cm^{-1}. The spectrometers in the series cover the range from 15,000 to 10 cm^{-1} and come with sophisticated, research-grade software. Numerous sampling accessories are available from Bio-Rad, including diffuse reflectance, autosamplers for polymers, and fixed and variable specular reflectance attachments.

Bomem (Wood Dale, Illinois) introduced the DA8 series research-grade FT-IR system. The instrument can be configured to cover 55,000–4 cm^{-1} and can achieve 32–0.0026 cm^{-1} resolution. The systems include a VAX host computer and dynamic alignment for reduced alignment errors.

Bruker Instruments (Billerica, Massachusetts) introduced its new vacuum FT-IR spectrometer, the IFS 66v, which is flexible enough to range from the near to far IR. It is keyboard-driven and accommodates bolometers as well as room-temperature and liquid-nitrogen-cooled detectors. The new FT-Raman microscope accessory may be used to study extremely small samples. This FRA 106 module can magnify objects as small as 20 μm for viewing.

A low-priced, dedicated instrument from Buck Scientific is used for the detection of oil in water. Operating at 2930 cm^{-1}, the instrument was designed for EPA Methods 413.2 and 418.1. Also introduced at PittCon was a low-cost, dispersive instrument with a range of 4000–600 cm^{-1}. The Buck model 500 IR uses a digital readout, stripchart recorder, or a PC. This single-beam instrument stores the background scan and subtracts it from subsequent scans, appearing to give double-beam performance.

The model NR-1100 laser Raman spectrophotometer from Jasco is designed to resolve close peaks in liquids, gases, and solids. It contains a double monochromator and dual detector system for high and low scattering, and operates from 330 to 910 nm. Jasco's model J-720 CD spectropolarimeter was also unveiled.

Mattson Instruments (Madison, Wisconsin) introduced new and enhanced models in its Galaxy series of FT-IR instruments. The 4020 model G12 was designed for QA/QC work, but with upscale capabilities: 1-cm^{-1} resolution, an 80286 64-bit math coprocessor, 2 Mbytes of random-access memory (RAM), and a range of 6000–400 cm^{-1}. The model 2020, with 2-cm^{-1} resolution, was designed for more routine work, but the model 6060 research-grade instrument comes with all the bells and whistles one could ask for. It features 0.25-cm^{-1} resolution, a low-noise detector, and spectral software such as the company's kinetics package, which can track microsecond events. The model 8020 is a true dual-beam FT-IR spectrometer; its spectral range is from 25,000 to 40 cm^{-1}. This is accomplished by changing the beam-splitter and (prealigned) detector, which can be achieved without breaking the vacuum in the sample compartment. Also introduced was the model 6020 thermogravimetric analysis (TGA) FT-IR spectrometer. The heated 10-cm gas cell allows a large variety of samples to be analyzed. The company also entered the Raman arena with the addition of the Galaxy 4080 near-IR Raman system that covers the 10,000–4000 cm^{-1} range.

Midac Corporation (Costa Mesa, California) showed its "Collegian" system designed for academic applications. It is incredibly small, rugged (even my students haven't been able to destroy it yet), and accurate. Even when mistreated it can still perform up to its 0.5-cm^{-1} resolution potential. The system can be tailored to its intended application (for example, analyses of gas) using various cells and detectors. The company's High Tech software is provided with the system.

Nicolet Analytical Instruments (Madison, Wisconsin) has coupled its 500 series FT-IR spectrometer with a Macintosh 11 computer to produce the 510M spectrometer. The company's Nic/IR software uses pulldown menus and integration of data into spreadsheet or word-processing programs. Other software is available for IBM PS/2 systems as well as Nicolet's own model 620 workstation. This state-of-the-art workstation runs sophisticated mutlipurpose software. Nicolet's top-of-the-line System 800 FT-IR instrument is quite powerful, allowing for FT-Raman, emission, and microspectroscopy, as well as hyphenated techniques such as GC/RT-IR and supercritical fluid chromatography (SFC)/FT-1R. The 700 series is also a versatile system that can be adapted to GC, SFC, and TGA systems. The new model 8220 gas analyzer is designed for repetitive gas analysis. The unit is simple to use, and the system yields parts-per-billion sensitivity for up to 20 materials at each location. Also introduced was Nicolet's new specific infrared detection (SID) software for GC/FT-IR. It is extremely versatile, allowing for condition changes throughout a run. It can process up to 100 autoinjector samples and perform routine and nonroutine tasks.

Perkin–Elmer displayed its new analytical grade 16PC FT-IR spectrometer, which is powered by a DEC 80386-SX computer. It offers 2-cm^{-1} resolution throughout its range of 7800–35 cm^{-1}. As with many other new products at PittCon this year, it offers mouse-controlled pulldown menus and windows for user interaction. It has a color monitor, a 4Mbyte random-access memory (RAM), and a 122-Mbyte hard disk. PE also unveiled its low-cost FT-IR microscope. Fully compatible with its 1600, 16PC, and 1700x series, the unit can serve as a sample preparation, optical, or IR microscope.

Spectra-Tech (Stamford, Connecticut) stepped up its development of specialized FT-IR systems since last year's Pittsburgh Conference, and introduced the MonitIR 400, a system devoted to operation in pharmaceutical production labs. This is one of the latest in Spectra-Tech's series of dedicated instruments for industrial FT-IR. Others include the ReactIR 1000, which allows *in situ* monitoring of chemical reactions and has been given new optional low-temperature capabilities. Using a liquid recirculating cooler in conjunction with specialized reaction vessels, the option allows cooling of reaction mixtures down to $-70°C$. This provides increased ability to monitor and control a wider variety of reactions in organic chemistry and other applications. The ReactIR 3000, developed in cooperation with Autoclare Engineers (Erie, Pennsylvania), performs *in situ* analysis of catalytic reactions under high-pressure, high-temperature conditions. This new system combines two of Autoclave's EZESEAL high-temperature and high-pressure stirred reactors with application-specific FT-IR technology based on Spectra-Tech's internal reflectance IR probes. The ReactIR 500 was also introduced. This unit is a mobile version for process development laboratories.

Infrared Accessories

A new incandescent IR lamp was introduced by Analytical Products (Hayward, California). Suitable for the visible and near-IR regions, these incandescent lamps with sapphire envelopes have a color temperature of over 2400 K but operate at 39°C.

ARIES, which operates closely with Specac in supplying accessories for IR spectroscopy, also introduced the variable-pathlength trace-gas accessory described previously in this chapter. Other tools from this marketing collaboration include the new variable-temperature cell that ranges from −190 to 250°C using the Dewar principle for insulation. The constant-thickness filmmaker, which is used primarily for polymer preparation, can achieve reproducible film thicknesses of 15–500 μm. The company also introduced a diffuse-reflectance system for FT-IR spectroscopy that accepts micro (4 mm), standard (11 mm), and tilted sample cups. The fiber-optic remote sampling system is designed for batch and process monitoring. The company also produces a series of FT-Raman sampling accessories.

Axiom Analytical (Laguna Beach, California) added several new products to its line of FT-IR accessories. The SRX-330 noncontact FT-IR surface analysis system is based on isotropic specular reflectance and can be used with most FT-IR spectrometers on the market. The company's LFG series laminar-flow IR gas cells employ a tubular flow path, yielding approximate laminar-flow characteristics. They can operate at up to 150°C and 10 bar of pressure. The cells are equipped with the Axiom optical transfer system for remote operation. Also shown was the DPX-110 universal FT-IR immersion sampling and reportedly minimizes beam divergence. Axiom's TNX series tunnel cells are also used external to the main unit. Volumes range from 25 μL to 20 mL. Associated with these devices are the LSC series liquid sample controllers, which are used for precise temperature control and sequencing of sample, solvent, and dry air.

An FT-IR Brill cell is available from CDS Instruments (Oxford, Pennsylvania) for use with its Pyroprobe 2000 for in-beam thermal manipulation of samples for analysis by FT-IR. This gas cell allows direct sampling of pyrolyzed materials.

Several interesting products were unveiled by Connecticut Instruments (Norwalk, Connecticut), including a dual-mode heated transmission and attenuated total reflectance (ATR) accessory with a zinc selenide (parallelogram) prism. The cell may be heated to 250°C, and temperature is controlled by a self-tuning digital microprocessor. The company also showed its matrix-isolation integrating sphere for trace-gas sampling. The intergrating sphere is OFHC copper polished and then coated with gold. The company's low-temperature, variable-angle total-reflectance accessory has options for UV/vis, near-IR, and IR regions. Because the angle is variable from 5° to 85°, the sphere can measure specular, diffuse, and total reflectance.

Lexel Laser (Fremont, California) introduced Ramanlon, a new krypton, 500-mW Raman laser. Lasing at 752.5 nm, it is small, accurate, and inexpensive.

Minarad Scientific (Fairfield, Connecticut) introduced the Decentered infrared Sensor Circlet for Uniform Sampling (DISCUS) cell for long-path IR sampling. This 6-in.-diameter toroid employs refractive optical decentering and comprises a multipass folded beam with pathlengths up to 9.6 m. The cell is compatible with any type of IR unit.

Molectron Detector (Portland, Oregon) introduced the P-1 pyroelectric detector/amplifiers. The instruments are divided into the 700 series ($LiTaO_3$) and 800 series (doped TGs) and are used in FT-IR, gas analysis, noncontact temperature measurement, radiometric measurements, and IR spectrometry. Also offered is a line of IR thin-film grid polarizers on substrates such as KRS-5, Ge, CaF, BaF_2, and ZnSe.

Opsis (Old Greenwich, Connecticut) offers a series of remote-sensing air-quality instruments for environmental monitoring, namely, for control of emissions from municipal and private sources.

IR microscopy services offered by Spectra-Probe (Slingerlands, New York) are a sign of the times, indicating how far the FT-IR technique has progressed since its introduction just a relatively few years ago.

Spectra-Tech displayed additions to its Baseline series of low-cost FT-IR sampling accessories. The series includes various tools for sampling gels, pastes, liquids, powders, and films, as well as kits for ATR, diffuse reflectance, and 30° specular reflectance. The company introduced a version of its Sample Plan compression cell with diamond windows. The new cell permits crushing and flattening of hard materials such as minerals, rubber, plastics, polymers, and tablet-form drugs to optimum thickness for transmission analysis. The accessory works on Spectra-Tech's IR-Plan series of microscopes, the IRμs microspectroscopy systems, Analect (Laser Precision Analytical) FT-IR microscopes, or any other microscope with at least 14 mm of working distance with its objective. Diamond windows can be ordered for earlier μSample Plan cells. Speaking of the company's IRμs microspectroscopy system, Spectra-Tech was distributing an applications note at PittCon describing the instrument capabilities for applications such as molecular mapping of photo-oxidized polymers.

Sprouse Scientific Systems (Paoli, Pennsylvania) introduced a series of FT-IR spectral search libraries and databases offered in search library formats or in the original reference spectra as measured. Topics include minerals, fibers (microscope), epoxy resins (including curing agents and additives), gas-phase spectra of environmental chemicals, quantified high-resolution small gas molecules, and general organic compounds.

MISCELLANEOUS EQUIPMENT

C Technologies (Short Hills, New Jersey) introduced a fiber-optic probe for use in fluorescence or Raman spectroscopy. The six-around-one configuration is tapered to conserve energy by lessening the area of coverage and increasing the effective numerical aperture. The 400-μm fibers are available in lengths up to 5 m. EG&G Princeton Applied Research brought numerous electronic devices to New York for display, including the 5209 single phase and 5210 two-phase lockin amplifiers. The amplifiers are controlled by PC software supplied by the company. In addition, software is available for all the company's detectors; the 2000 OMA Spec general-purpose optical multichannel analyzer has been released to be run on all IBM compatibles. It includes features needed by spectroscopists, including multisensor capability.

General Electric Canada (Vaudreuil, Quebec, Canada) developed the SPCM-100-PQ (for general use) and the PQ-FC (fiber optic) single-photon counting modules. With a 400 1060-nm range, this silicon avalanche photodiode is quite interesting. Single-photon arrival times can be measured with an accuracy of 3 ns root mean square (rms).

Hamamatsu (Bridgewater, New Jersey) introduced its new MOS linear image sensor. The unit reportedly has good UV sensitivity and low dark current. It comes with 128, 256, 512, or 1024 elements.

Hi-Tech Scientific (Salisbury, United Kingdom) launched its Hi-Mix range of sample handling and mixing devices, which are used for rapid, precise and easy mixing of reagents and samples for a range of applications in IR, UV/vis, fluorescence, and general stop-flow work.

Instruments SA (Edison, New Jersey) introduced a variety of molecular spectroscopy products. A new monochromator, the HR250, features an interchangeable dual-grating turret in a 0.25-m focal-length Czerny Turner configuration. The monochromator features four-port capability, which permits simultaneous experimental setups. The company's Spectraview 2-D array detection system is a CCD-based detector that provides low dark current and high sensitivity in applications with extremely low light-level or high dynamic-range requirements. The system is said to provide high quantum efficiency, enhanced red response, and low readout noise. The company also introduced the Spectraview linear-array system.

Lambrecht Research Laboratories (Tucson, Arizona) presented the Stokesmeter for the measurement of light polarization. The instrument measures and displays polarization information in terms of four parameters of the Stokes vector.

PCP Inc. (West Palm Peach, Florida) introduced the Phemto-Chem 110 portable ion mobility spectrometer (IMS). With applications in chemical monitoring, security monitoring, quality assurance, chromatographic detection, and research, the IMS system can detect trace organic species at below parts-per-billion levels. The system provides virtually instantaneous response with chemical identification based on the mobility of the detected ions. It can be operated in the laboratory or in the field with a 12-V dc power source.

PL Separation Sciences (Church Stretton, Shropshire, United Kingdom) displayed the Otsuka ELS-800 electrophoretic light-scattering spectrophotometer. This system has been developed for the measurement of mobility, zeta-potential distribution, and size distribution of fine particles. According to the company, it is suitable for surface-characterization work. This system was described previously along with Otsuka's UV/vis–near-IR instruments.

GENERAL SOFTWARE

Dow Chemical (Midland, Michigan) updated its Camille data acquisition and control systems. Version 11 of the software is said to operate with most sensors and can run complex equations.

Galactic Industries (Salem, New Hampshire) announced its Discriminate program for use in routine identification work. Based on the concept of Mahalanobis distances, the program is useful for routine work. Also introduced was the PLSplus version 2.0 software for quantitative analysis based on partial least squares, which is a popular technique for analyses.

The science software division of Heyden & Son (London, United Kingdom) showed Spectraltie, a data station software package that is available for PC control of IR and UV instruments. Heyden also showed Chromafile software for diode-array instruments.

National Instruments (Austin, Texas) announced its LabWindows program development tools. This interactive software uses the windows concept to control the unit, collect data, and perform calculations. The company also introduced new data acquisition capabilities and increased GPIB functionality for Labtech Notebook PC-DOS software. This menu-driven package is used for data acquisition, control, and manipulation. There is also a version for PS/2 models. Several interfaces are offered to complement the software: an IEEE-488 interface, a multifunction analog and digital I/O interface, and 32-bit and 24-bit parallel digital I/O interfaces. The company also offers a GPIB interface for IBM PC/XTs and PC/ATs, as well as a low-cost multifunction board for the same computers. National also demonstrated its LabView 2 instrumentation software system. It is designed for instrument control, data acquisition, data analysis, and signal generation. The system also may be used for designs, simulations, tests, and control.

Photometrics (Tucson, Arizona) presented its IPLab software designed for the Macintosh II. Using pulldown menus, the package offers high-resolution visuals in addition to programming in C, Fortran, or Pascal.

Spectrum Square Associates (Ithaca, New York) has developed software designed for resolution enhancement. The spectral curve-fitting routine enhances overlapping and poorly resolved bands. It operates in conjunction with Spectra Calc from Galactic industries.

NMR AND RELATED INSTRUMENTATION

Bruker Instruments introduced a range of new or enhanced instrumentation and software for magnetic resonance. Bruker Spectrospin, the company's Canadian affiliate, announced the Minispec PC 100 series of automated pulsed NMR spectrometers. These instruments are intended to serve as process analyzers for quality control in the polymer, chemical, and food industries and, accordingly, are designed for low cost, ease of use, and speed of operation.

Among the enhancements made to Bruker's existing NMR systems are:

- Expanded capabilities and support for the AC series of automated NMR instruments.
- Automation hardware for the UNIX-based AMX series, including a bar-code-based automation system, a 120-sample autochanger, individual sample heaters, and the quadro-nucleus probe.
- WIN-NMR software for personal computers, including routines for Fourier transforms, integration, split-screen operation, and plotting.
- Enhancement of the company's X-32 UNIX computer and NS32532 central processing unit. Designed for the analysis of large 1-, 2-, and 3-D data sets, the computer now provides faster processing speed for many NMR routines and has an expandable RAM to 128 Mbytes.

GE NMR Instruments (Fremont, California) introduced expanded capabilities for the QE Plus automated analytical NMR spectroscopy system. The system offers a standard 344-Mbyte hard-disk drive for additional storage, a 256-step pulse programmer for increased experimental capabilities, a 256K data table for improved digital resolution, standard 512K memory capacity, and a 3.5-in. microfloppy disk that reportedly provides 40% more storage capacity in a convenient format.

Hitachi Instruments displayed three NMR systems. The R-1100/R-1200 instruments are permanent-magnet systems for continuous-wave (CW) and rapid-scan (RS) correlation analysis, respectively. The systems are designed for users at all experience levels in application areas such as chemicals, polymers, fermentation, biochemistry, and petroleum. The company's R-1500 FT-NMR system is a 60-MHz permanent-magnet system for routine analyses in fields such as organic chemistry and medicine.

JEOL/USA (Peabody, Massachusetts) introduced the CPF-400 FT-NMR spectrometer, which combines several research-level instrumental features with those of a routine instrument. Software for the system includes a menu-driven package that operates on multiple levels simultaneously: a user-defined automated command set and research operator control. The package can also accommodate user-written data analysis and report software. More than 150 experiments are included, and a graphically displayed pulse-sequence editor is capable of generating pulse sequences up to 1000 steps for the creation of new routines. The CPF-400 is compatible with Digital's DECnet/Ethernet and is expandable for multiterminal operation. The system can be configured for full research functions.

Varian enhanced several of its Gemini FT-NMR systems. The Gemini has been given broadband capabilities, enabling the spectrometer to handle multinuclear experiments. The instrument's memory capacity has also been expanded, with 1.5-Mbyte RAM now standard. A large data table, programmable filters, a PTS-160 frequency synthesizer, and a linear-pulse amplifier have also been added, in addition to an updated keyboard, monitor, and cabinet. According to the company, these additions allow the Gemini to achieve a greater spectral width during experiments, enabling the user to examine a broader chemical shift in which peaks may be separated. Varian reports that this addition also contributes to greater resolution and line shape, enabling more peak definition. The system is available with 200- or 300-MHz magnets and can perform 1-D and 2-D experiments such as APT, DEPT, COSY, and NOE difference.

EPR AND ESR INSTRUMENTS

Bruker announced the ECS-106 electron paramagnetic resonance (EPR) spectrometer series. The series is designed to cost less than, but perform comparably to, standard-sized EPR spectrometers. Though compact, the ECS-106 instrument can accept standard accessories from the company's larger ECS-300 EPR series. The computer-controlled system performs automated functions such as tuning and matching of the microwave bridge and cavity.

JEOL/USA introduced a new line of electron-spin-resonance (ESR) spectrometers. New to the U.S. market, the RE series of ESR instruments are designed for both analytical and research laboratories and are reportedly suited for some of the recent experimentation being conducted in biochemistry. The instruments operate in CW mode and are available with a variety of computer and sampling accessories.

Accessories and Software

Isotec (Miamisburg, Ohio) introduced a multiple NMR tube cleaner that can clean up to five tubes simultaneously, which saves time, space, and solvent. The device works on an inverted NMR tube with its cap positioned on the closed end.

New Methods Research, Inc. (East Syracuse, New York) and BioDesign (Pasadena, California) have established a product development and marketing relationship intended to coordinate BioDesign's Biograf, Polygraf, and NMR-graf molecular simulation software with NMRi's SpecStation computers and data processing and analysis software. Of particular interest to NMR spectroscopists is NMRgraf, an integrated, standalone package that uses NOE and J-coupling data from NMR spectroscopy in conjunction with molecular mechanics and dynamics capabilities of the Biograf and Polygraf simulation programs. NMRi's new NMR2/Model program works with NMRgraf for more accurate 3-D structures of biopolymers and other large molecules.

MASS SPECTROMETRY PRODUCTS

Spurred in part by growth in the environmental and biotechnology markets, vendors of MS equipment brought a wide variety of products to PittCon. With the exception of the products for plasma-source MS, this section addresses all types of MS instrumentation, whether designed for elemental or molecular analysis.

Finnigan MAT (San Jose, California) introduced the MAT 95, a high-resolution magnetic-sector system designed for applications in trace-level environmental analysis, particularly in the determination of trace-level dioxins. Also suited to high-mass determinations, the instrument incorporates a RISC-architecture DEC-station 2100 with an ULTRIX-32 operating system, DEC windows applications, and capabilities for networking to ancillary equipment and data stations. Automation is achieved in part through the company's Instrument Control Language (ICL), which permits the instrument to make data-dependent decisions in real time during data acquisition. The system is amenable to inlet systems and ionization techniques that include gas chromatography, solids probes, desorption chemical ionization (CI) probes, fast-atom bombardment (FAB) and continuous-flow FAB-thermospray LC-MS, and field desorption and field ionization. The instrument can be expanded to a tandem configuration.

FT-MS Products

Bruker launched its CMS-47X FT-MS system in the U.S. market. A highlight of the CMS-47X is its external differentially pumped medium-pressure ion source, which performs EI/CI, Inset desorption and ionization, FAB ionization, field desorption, and GC coupling. The instrument includes a temperature-programmable direct insertion probe and heatable gas/liquid inlets. According to the company, by enabling the operator to use Inset desorption and other ionization techniques, the external design takes advantage of the pulsed nature of FT-MS experiments.

Extrel Corporation (Pittsburgh, Pennsylvania) announced the first products from its FT-MS division, acquired from Nicolet. Based in Madison, Wisconsin, the Extrel FT-MS division produced the 2001 series of upgradable high-resolution FT-MS instruments. A number of ionization techniques are possible with the 2001 series, but the manufacturer is touting laser desorption as perhaps the most significant and powerful method. Extrel's laser-probe mass spectrometer combines a 2001 series FT-MS instrument with the company's Inset-probe accessory to provide spatially resolved laser ablation mass spectra on a variety of samples and materials. A Nd:YAG laser/optical bench accessory for the Inset-probe system was also introduced at PittCon.

GC-MS Systems

Bruker showed its Mobile Environmental Monitor (MEM) for the first time at PittCon. Designed for field analytical work, this quadrupole-based electron impact GC-MS system is designed for use on a four-wheel-drive vehicle/on-site environmental analysis. To permit battery-powered operation, the instrument rquires only 24 V of dc power. The instrument has been designed so that inexperienced technicians can operate it, and it can be fitted with a variety of sampling accessories such as a capillary GC for direct analysis with minimal sample preparation and a direct air/surface sampler for identification of organics in water, air, or solids.

Delsi/Nermag (Houston, Texas) displayed the Automass 150 automated benchtop GC-MS data system. This system combines a Delsi DN 200 gas chromatograph with capillary capability, which interfaces directly with the ion source of a quadrupole MS spectrometer through a 100-mm-long glass-lined tube. The spectrometer has a 1000-amu mass range and includes four cylindrical rods equipped with plug-in prefilters. The ion source has interchangeable ion volumes for El or Cl and plug-in focusing lenses. The company's Lucy software controls operations such as tuning, mass-assignment calibration, data reduction, chromatographic trace, library search, and quantitation.

Extrel entered the GC-MS arena with the THE-400 high-transmission environmental system for analysis of volatile and semivolatile organics in water matrices, in accordance with the EPA's Methods 524 and 525 for drinking water and raw source water. In addition to water analysis, the system is suitable for all EPA methods for which GC-MS is an option, and the company claims low detection levels for a variety of samples encountered in EPA contract laboratories. The system includes a 0.75-in. quadrupole and ion source and a detection system with an electrostatic analyzer and ion extraction system. Software for EPA methods is included.

The Saturn GC-MS system from Varian can be enhanced with a new tuning procedure introduced by the company at PittCon. Designed for environmental GC–MS analysis, the procedure meets all tuning criteria for EPA Methods 500, 600, and 8000. The tuning procedure for volatile and semivolatile compounds is rather tedious, but, according to Varian, the new procedure for the Saturn is a favorable alternative because the instrument has no source lenses to adjust, as do traditional quadrupole-based instruments.

New on the commercial analytical-instrument scene is Viking Instruments (Reston, Virginia), which introduced a miniature mobile GC-MS system for environmental field analysis. Portable instruments seem to be up Viking's alley, as the company provided miniature GC-MS instrumentation on NASA's Viking missions to Mars. Now licensed by the space agency to commercialize this technology among Earthlings, Viking introduced the SpectraTrak 600 and 700 GC-MS instruments. The model 600 is shock-mounted in a weatherproof Mil-Spec case designed for in-vehicle mounting. The SpectraTrak 700 is similar but designed for automated real-time monitoring at stationary field or industrial locations. It is installed in a weatherproof NEMA cabinet with an external computer-control station. Both instruments feature the company's spectraScan/OS operating system for automation, and Viking will supply the instruments with a fully equipped mobile lab. The company also offers a communications system to link users in the field with remote PC data bases and Viking customer service, applications, sales, and engineering personnel. Viking also displayed a prototype of the forthcoming SpectraTrak 500, which is powered by internal batteries.

LC-MS Systems

Extrel also introduced two ThermaBeam particle-beam interfaces for use on VG 7070 and JEOL AX, HX, and SX mass spectrometers, permitting LC and SFC to be performed on these magnetic-sector systems. Perkin–Elmer Sciex, the joint venture well known for its plasma-source mass spectrometers, announced that it will be marketing Sciex's API (atmospheric pressure ionization) III high-performance biomolecular mass analyzer system for LC-MS. This instrument was introduced at the 1989 Pittsburgh Conference in Atlanta.

Vestec (Houston, Texas) introduced a range of LC-MS products. An enhanced thermospray source and a 2000-amu option for the model 201 LC-MS system are available to increase sensitivity for many ionic compounds. The 2000-amu option uses 10-kV post-acceleration with a conversion dynode and the company's negative-ion detection process. The company's model 1700 multipurpose interface in a two-bay portable cart was also introduced in New York. The interface operates with magnetic-sector and quadrupole MS systems and can accommodate a complete HPLC system, thermospray or electrospray ionization sources, and the company's new Separator package. The Separator interfaces LC to electron-impact ionization, making LC-EI routine. The system comprises the company's Universal Interface with a vaporization/nebulizer chamber, a diffuse cell, flow controllers, and electronics, a pumping system and a two-stage momentum separator. Vestec also introduced an electrospray interface for the model 201 dedicated LC-MS system and other quadrupole MS systems. The electrospray interface combines micro-LC and capillary electrophoresis separation methods with quadrupole MS, generating multiply charged species and enabling high-molecular-weight determinations exceeding 20,000 amu.

Quadrupole Systems

Delsi/Nermag introduced Resolver-ES single-stage dedicated electrospray MS system for determinations of ultrahigh-molecular-weight compounds. Based on a quadrupole mass spectrometer and an electrospray interface supplied by Analytica, Inc. (Brantfort, Connecticut), the system has a mass range of 2000 Da and reportedly provides high sensitivity and ease of use without requiring nebulization. The system is designed to circumvent problems associated with delivering large molecules into the gas phase at the low pressures that mass spectral analysis requires.

A major introduction from Finnigan MAT was the TSQ 700 triple-stage and SSQ 700 single-stage quadrupole mass spectrometers, designed to meet analytical needs in the pharmaecutical, biotechnology, and environmental industries. The TSQ 700 is designed for both MS and MS-MS applications. It includes a nonlinear octapole collision cell, a 20-kV dynode detector, and an X-Windows-based version of the ICIS data system that runs on a DECstation 2100 workstation with the ULTRIX-32 operating system. The single-stage version shares the triple-stage version's inlet techniques and post-acquisition applications software. The TSQ's performance is based on the nonlinear octapole collision cell, which increases reproducibility and simplifies tuning in MS-MS work. With the addition of ICIS, the instrument now operates on the UNIX system, and the instrument control language (described previously) optimizes experimental conditions. Finnigan also introduced an electrospray ionization system for routine determination of molecular weight and structural information on compounds at picomole and femtomole levels.

Hewlett–Packard (Palo Alto, California) introduced the HP MS Engine, a high-performance quadrupole MS system for research, methods development, and analytical service laboratories. HP reports that, depending on the application, the sensitivity of this system is 10 times higher than that of previous HP research quadrupole systems. The MS system includes an improved ion optical system, the result of a complete redesign using computer modeling to enhance source performance and detection. The source, lenses, analyzer, and detector are mounted on a single-casting optical bench, and the components automatically index into alignment. The full system includes a gas chromatograph, liquid chromatograph, and peripherals and is controlled by the company's multitasking MS ChemStation. The user interface is based on the X-windows system, and all basic operations are under this control. Standard in the MS Engine is a dual-programmable EI/CI source, which permits quick switching between the two modes. Options include a 2000-amu mass range, redesigned particle beam and thermospray LC-MS interfaces, FAB, desorption CI, direct-insertion probe, and capillary- or packed-column GC-MS operation.

VG Instruments (Danvers, Massachusetts) introduced a new GloQuad glow-discharge quadrupole mass spectrometer. The instrument provides full elemental coverage in <60 s with below parts-per-billion detection levels during direct solids analysis. The instrument combines a dc plasma ion source with a rapid-scanning, high-performance quadrupole mass analyzer.

Time-of-Flight Systems

Bruker introduced a time-of-flight (TOF) system featuring a gridless reflection and a new ion-source design. The TOF-1 mass spectrometer's reflection provides both energy and spatial focusing, achieving high mass resolution (10.000 at m/z 106) and efficient ion transmission. Conventional EI and laser ionization sources are available for volatile samples, and thermally labile samples can be ionized with either Inset desorption ionization or the new multistage ion source in which large molecules are gently desorbed, cooled in a supersonic jet, and selectively ionized with resonant laser radiation.

Vestec introduced the model 2000 UV laser desorption TOF mass spectrometer. The integrated system comprises a linear TOF mass spectrometer with a 2-m flight path and two-stage, 20-kV ion acceleration system. Centroid calculation and mass calibration are performed with a DEC microVAX workstation.

Data Stations and Software Accessories

Extrel introduced two specialized data systems for its Questor series of MS process analyzers. The Questor lamp analyzer and the Questor semiconductor package analyzer add floppy- and hard-disk options for data and analysis methods storage in a single-chassis package. Extrel also introduced a data system for its EXT-1000 glow-discharge mass spectrometer, as well as the Series-8 data acquisition and reduction software for the EL-750, EL-1000, and EL-2000 MS data systems.

JEOL/USA introduced the Complement MS workstation for data acquisition and processing. The package is provided on Hewlett–Packard's 9000 series 300 engineering workstations. The 32-bit, mouse-driven station operates in a windowed environment and includes online documentation and context-sensitive help functions.

Vacuum and Gas Products

Leybold Vacuum Products (Export, Pennsylvania) introduced the Turbovac 340 M magnetic-bearing turbomolecular pump with ISO-K and CF flanges.

Matheson Gas Products (Secaucus, New Jersey) introduced several gas-related products for analytical work, including the Ultraline series of steel cylinders for high-purity corrosive gas, the Leak Hunter portable gas-leak detector, and a series of single-cylinder gas cabinets.

CHAPTER IV.B.5

"WHAT'S NEW IN SPECTROSCOPY INSTRUMENTATION?" 1991 PITTCON WRAPUP

Well, here I am writing about another Pittsburgh Conference. After a great week in the Windy City, I'm still excited about all the new products I saw. Of course, the large companies continue to produce outstanding equipment, but the little fellows didn't do too badly either. As usual, I will cover the spectrum from the ultraviolet (UV) to the infrared (IR) and then describe the new instrumentation for nuclear magnetic resonance (NMR) and mass spectrometry (MS). Equipment within each section will be followed by a discussion of software, then accessories, and manufacturers within each class will be listed alphabetically (this may deteriorate as my eyes blur after the third or fourth day of typing ... if so, forgive me!).

ULTRAVIOLET/VISIBLE

Acton (Acton, Massachusetts) announced a new version of the SpectraPro-500 monochromator/spectrograph. New features of the 0.5 m, multiport, triple-grating instrument include an optional motorized diverter mirror and the availability of interchangeable grating turrets.

American Holographic (Littleton, Massachusetts), well known for its miniature spectrometers, announced a line of dual- and single-beam diode-array instruments for specific, optimized applications. Units are available for medical diagnostic applications, color measurements, and a range of analytical measurements, including modules for environmental scanning spectrometry to near-infrared (near-IR) nonscanning spectrometry.

Beckman Instruments (Fullerton, California) displayed its new programmable DU 7500 diode-array spectrophotometer. The instrument can work with samples as small as 5 pL. Highlighted applications include the analysis of enzymes, proteins, and nucleic acids. Data are "crunched" with FSQ full spectrum quantitation using vector quant mathematics (that is, Fourier transforms with p-matrix math).

GBC Scientific Equipment Ply. Ltd. (Dandenong, Victoria, Australia), which has recently opened an office in Arlington Heights, Illinois, released the 914, 916, and 918 family of UV/vis spectrophotometers at PittCon. Each computer-controlled, double-beam instrument features automatic optimization and reportedly achieves distortion-free spectral scanning as fast as 7000 nm/min. Software is included for the development and storage of methods, and scans, variables, and results can be displayed in three dimensions or stored on disk. A complete range of accessories are available for the instruments.

Hach Company (Loveland, Colorado) introduced a new direct-reading colorimeter that can measure more than 75 common water-quality parameters. The model DR/700 uses interchangeable filter-calibration for tests between 420 and 810 nm. Each unit contains a wavelength filter and factory-programmed calibrations, providing direct readout in concentration units, percent transmittance, or absorbance. The unit operates on four AA-size batteries and has numerous accessories.

Hitachi Instruments (Danbury, Connecticut) introduced the model U-3410 spectrometer. The computer-controlled instrument spans the UV/vis–near-IR range and is especially useful for the analysis of large samples.

Hunter Associates Laboratory (Reston, Virginia) has added a large-area view sensor to its ColorQuest spectrocolorimeter line. The instruments can view a 4-in. area, are flexible (using $45°/0°$ geometry), and have several sources of illumination. The company's new Qual-Probe is designed for online food color measurements.

An interesting CamSpec UV/vis spectrophotometer with an integrated liquid-crystal keyboard and display is available from InterCon Inc. (Champaign, Illinois). The model M330 features a 4-nm bandpass, RS232 output, programs for kinetics and colorimetry, and numerous features found in larger instruments. The company also introduced the Omega 20, a new UV/vis–near-IR spectrophotometer.

Perkin–Elmer (Norwalk, Connecticut) displayed the new Lambda 19 series of UV/vis and UV/vis–near-IR spectrophotometers, which use large collimating mirrors and holographically ruled gratings for maximum throughput. The optical system allows spectral recording from 175 to 3200 nm at band widths from 0.05 to 5 nm in the UV/vis and from 0.2 to 20 nm in the near-IR. Root-mean-square noise (measured at 500 nm) is said to be 5×10^5 AU with a stray light of $<8 \times 10^{-5}\%$. The company's UV-computerized software controls the instruments, and a variety of applications-oriented software packages and accessories are available.

Shimadzu Scientific Instruments (Columbia, Maryland) displayed its model UV-1201 series personally configurable UV/vis spectrophotometers, which now feature a new software package using Microsoft Windows. The compact units have high-resolution optics, a large sample compartment, and a liquid-crystal display screen. Features such as spectral scanning, quantitation, kinetic, or time-course modes are all accessible from insertable firmware cards.

SLM-AMINCO (Urbana, Illinois) introduced the space-saving model 3000 array UV/vis spectrophotometer. Controlled by an IBM-compatible computer, the new system uses the speed of a diode-array detector. An eight-stage sample holder is standard, and temperature-controlled and autosampling devices extend the instrument to other applications. Numerous software options are also available.

Carl Zeiss, Inc. (Thornwood, New York), introduced the models MPM400 and MPM800 microscope photometers, which cover the 240–2100-nm spectral range. The software performs spectrophotometric analysis (using transmission/absorbance, reflectance, and fluorescence spectra), statistical and kinetic analysis, and one-dimensional (1-D) or 2-D photometric mapping.

Accessories

Labsphere (North Sutton, New Hampshire) brought a range of diffuse reflectance and transmittance instrument accessories to the conference. The RSA series, designed to be used with commercially manufactured spectrometers (including Perkin–Elmer, Hitachi, Hewlett–Packard, SLM/AMINCO, and Beckman), are equipped with the company's Spectralon integrating spheres, which are made of a highly diffuse reflectance material designed to extend spectral range, improve signal-to-noise ratio (S/N), and generate reproducible spectral reflectance.

Software

Hitachi introduced a new software package for use with the company's U2000 Data Manager. The Ratio Measurement Utility software, especially useful for determining DNA/RNA protein ratios, provides programming capabilities for as many as six wavelengths with custom reporting and processing, including absorbance, concentration, ratio, inverse ratio, and difference calculations at two wavelengths using a third for background correction.

Time-saving software packages and accessories for the Spectronic line of spectrophotometers were announced by Milton-Roy (Rochester, New York). Micro-Quant software, designed for running linear and nonlinear standard curves, endpoints, or kinetics tests, is now available for the models 301, 401, 501, 601, 1001+, and 1201. Other software packages will extend the capabilities of the various models.

FLUORESCENCE

ISS (Champaign, Illinois) presented its multifrequency phase and modulation fluorometer at PittCon. The K2 fluorometer's optical design and automatic instrument control allow steady-state and dynamic fluorescence studies with frequency capabilities from I Hz to the gigahertz region. Several of the company's software packages for specific applications are available with the computer-controlled instrument.

SLM-AMINCO introduced the AMINCO Bowman series 2 luminescence spectrometer, which offers both pulsed and continuous light sources, making it suitable for fluorescence and phosphorescence work. The instrument's monochromator offers rapid slewing speeds for multiwavelength applications such as intracellular calcium studies using fura-2. Its software features a graphic–user interface, multitasking capabilities, and numerous application routines.

Software

The measurement of intracellular concentrations of Ca^{2+} and other ions with Perkin–Elmer's LS-50 fluorescence spectrometer can now be automated with its new Intracellular Biochemistry software. Any UV/vis wavelength can be used with the software, and analysts ratios for probes such as indo-1. Method calibration and autofluorescence correction are automated, and intensity ratios are plotted during data collection and calibration for process monitoring.

Accessories

Corion Corporation (Holliston, Massachusetts) displayed its range of optical filters and coatings at PittCon, including its improved line of emission and excitation filters for fluorescence spectroscopy. Filter characteristics include sharp transition slope, deep blocking over the detector range, wide bandwidth selection, and high energy throughput.

NSG Precision Cells (Farmingdale, New York) introduced a new fluorimeter scattering cell that can be supplied in optical glass, UV quartz, and IR quartz and ES quartz glass. The 10-mm-pathlength Type 13 cell is designed for applications such as dye laser cell experiments.

NEAR-INFRARED

Bran + Luebbe (Elmsford, New York) announced the introduction of its new InfraPrime in-process analyzer. Based on an acousto-optic tunable filter, the instrument was designed for rapid analysis of food, chemicals, pharmaceuticals, and applications. Also introduced into this country was the company's FT-based InfraProver. Birefringent crystal-wedge optics are used to split the ordinary from the extraordinary rays of the light source, eliminating the need for moving mirrors and reducing sensitivity to vibration. The instrument makes use of a fiber-optic probe to measure samples and principal component regression (PCR) for spectral analysis.

EG&G Princeton Applied Research Corporation (Princeton, New Jersey) introduced three near-IR spectrophotometers. The first, the model 1452NIR, is a multichannel model based on a 256-channel diode array, covering the spectrum from 800 to 1700 nm. The indium gallium arsenide (InGaAs) detector is thermoelectrically cooled, and the system is designed to be used with a fiber-optic probe. A second, acousto-optic–based instrument operates between 1000 and 1700 nm. It is packaged as a portable instrument and optimized for fiber optics. The third is a high-resolution spectrophotometer, capable of scanning from 750 to 2600 nm using a new wide-range detector. It is controlled by an IBM PC and is also optimized for fiber-optic connectors.

Guided Wave (El Dorado Hills, California) announced its new extended-range, near-IR fiber-optic spectrophotometer, which now covers 1000–2200 nm. The model 300P40 offers multiplexing for as many as eight points. Several new fiber-optic probes that are more insensitive to temperature and pressure changes in online systems were also shown. The company's new software, the Unscrambler II, uses partial least squares (PLS) and PCR analysis.

Katrina, Inc. (Gaithersburg, Maryland), a newcomer to the near-IR marketplace in 1991, introduced its Protronics process-control equipment for applications in the food, chemical, pharmaceutical, and paper industries. Based on near-IR–emitting diodes and bandpass filters, the models 112 and 412 work in the 900–1050-nm range (the near, near-IR, or the far-visible regions). The diodes are cycled on and off in $<1/1000$ s and have an estimated life of 30 years. Using higher overtones, the instrument "sees" essentially all that standard near-IR instruments see. The units are contained in NEMA 4X enclosures designed for industrial environments.

LT Industries, Inc. (Rockville, Maryland), announced that its Quantum 1200I Plus process analyzer is now available in a new standalone Class 1, Division 1, Group C-G enclosure. The analyzer meets NEMA 4X requirements, can be field mounted without environmental conditioning, and can be multiplexed through fiber bundles longer than 250 ft. The 1200I is the building block for several dedicated systems now available from the company that incorporate optimized software, fiber-optic cables, and sampling attachments. Systems include the CBA analyzer for beer-blending processes, an octane analyzer, a hydroxyl number analyzer, and a batch-monitoring system. The company also announced its new low-cost fiber-optic bundles and a multiplexer capable of switching 2.5 channels/s.

NIRSystems, Inc. (Silver Spring, Maryland), displayed its new Rapid-ID Analyzer, a material inspection system designed with the pharmaceutical industry in mind. The portable near-IR instrument can identify raw materials anywhere in a plant setting independent of laboratory analysis and uses a hand-held, fiber-optic probe for measurements. The IQ^2 software can identify, qualify, and quantify pharmaceutical samples and compare them to starter or user-generated spectral libraries of reference materials using spectral matching, PCR analysis, or both for identification and quality checking. An operator may specify any quantitative algorithm for accessing substituent levels (for example, moisture and particle size).

MIDRANGE INFRARED

Automatik Machinery Corporation (Charlotte, North Carolina) announced the introduction of a polymer analyzer designed for dedicated spectroscopic applications in quantifying additive, degree of polymerization, and copolymer blending in the production of molten polymers. By selecting specific IR filters, the IROS-SP can analyze for individual components within the melt stream.

Bio-Rad, Digilab Division (Cambridge, Massachusetts) introduced a number of new instruments in Chicago. Its new 696 interferometer features dynamic alignment, step scan, kinetic scan, and 22-bit A/D performance. The 60 Michelson interferometer is designed to operate into the near-UV, and the scan rate varies from 800 Hz (steps/s) to 0.25 Hz. Signal modulation schemes include both amplitude and phase modulation. A second introduction from Bio-Rad, the InfraScan FTIR spectrometer, was built for routine QA/QC testing, raw-material screening, assays, and teaching applications. The keypad-controlled instrument features a large stainless steel compartment that can accept a variety of sampling accessories. Bio-Rad's new FFS 7 PC FT-IR spectrometer is designed for R&D problem solving and features the company's FTS 7 optical bench, a PC-compatible data system, and Galactic Industries' Spectra Calc software. The unit can be expanded with hardware, software, and accessories and is available in a near-IR version (complete with PCR/PLS and discriminant analysis software).

Other systems from Bio-Rad incorporating the company's FTS 7 FT-IR spectrometer include an automated used-oil analysis system designed for analyzing lubricating oils in diesel, gasoline, and natural gas engines. The system comprises a sealed and desiccated version of the FTS 7 spectrometer, an autosampler, and a hydrocarbon alert sensor. Also available is an FT-IR gas analyzer that uses the FTS 7 externally interfaced to a folded-path gas cell housed in a compact side bench.

Bomem, Hartmann & Braun (St.-Jean Baptiste, Quebec, Canada) had a number of interesting new devices on display. Its MB2E FT-IR spectral radiometer has several available operating ranges: 0.75–3, 0.75–22, 2-28, or 2-50 μm (with resolution selectable from 1 to 128 cm^{-1}). The remote measuring device is versatile and may be used for numerous applications, including operation from aircraft and other moving platforms. The model 9100 multicomponent FT-IR gas analyzer is designed for flue gas monitoring. A heated probe, tubing, and absorption cell keep the gas at 180°C to avoid loss due to condensation, adsorption, or secondary effects. Bomem also introduced a new flexible multicomponent FT-IR analyzer for ambient air monitoring. This device may be fitted with cells ranging from 6 to 100 m. Complex mixtures may be analyzed in real time.

General Analysis Corporation (South Norwalk, Connecticut) introduced three LAN systems for gas monitoring. The LAN I, II, and III come with one, two, and three detectors, respectively. The wavelength range is from 1.0 to 20 μm using narrow-bandpass interference filters. The company's LABLAN oil-aromatic hydrocarbon monitor measures the oil concentration in water or soil and displays the aromatic carbon content. The infrared analyzer incorporates a three-wavelength, low-f-number optical system with linearizing circuits that permit direct display in the parts-per-million range.

Laser Precision Analytical (Irvine, California) introduced the diamond-20 compact FT-IR spectrometer. Using Microsoft Windows, this low-cost instrument is said to be comparable with more expensive models. The company also introduced the Analect Advanced Process Development FT-IR system, which may be used to monitor process chemistries from small bench-top reactions to pilot-plant reactions. It measures kinetics and various other parameters without disturbing the reaction.

Mattson Instruments (Madison, Wisconsin) showed up with a number of new offerings. The Galaxy 3000, 5000, and 7000 series FT-IR spectrometers can be used on standard MS-DOS and Macintosh computers and are compatible with hyphenated sampling techniques for operation in the near-, mid-, and far-IR range. The 3000 series, optimized for QC and teaching laboratories, is available in near- and mid-IR models with 2 cm^{-1} resolution. The 5000 series, available in the near-, mid-, and far-IR, features 0.75 cm^{-1} resolution and 15 scan speeds. The Galaxy 7000 series comes with 0.4-cm^{-1} resolution (optional to 0.25 cm^{-1}) and 17 scan speeds from 0.09 to 25 cm/s.

Mattson's Research Series I FT-IR spectrometer covers the range from 25.000 to 400 cm^{-1} for high-level FT-IR applications. Resolution is step selectable to 0.25 cm^{-1} with 17 scan speeds. It features a high-intensity fluid-cooled source and continuously adjustable, computer-controlled iris aperture for optimization of conditions. All models have state-of-the-art electronics and data-handling capabilities.

Midac (Costa Mesa, California) introduced the Outfielder portable emission spectrometer, calling it an advanced spectral FT-IR radiometer. Equipped with Midac's standard interferometer and cryogenically cooled MCT detector, this small instrument is designed for hostile field measurements. Midac also introduced the AnalyzAir gas analysis package. The system, designed for trace gas analysis, includes a 0.5 cm^{-1} resolution Midac spectrometer equipped with a 10-m fixed pathlength gas cell and interfacing optics.

Nicolet Instrument Corporation (Madison, Wisconsin) introduced several new products. Among them were the upgraded model 510 FT-IR spectrometer and the MCT version of its model 205 FT-IR spectrometer. The 510 has an improved frictionless flexpivot interferometer, with resolution improved from 1.5 to 0.8 cm^{-1}. It will work with virtually any computer. The 205 now includes a liquid nitrogen detector option.

Air monitoring is the purpose of Nicolet's new Multipoint Air Sampling System, which operates with the 8220 FT-IR gas analyzer. It comes equipped with numerous automated operating systems and diagnostic programs for unattended work. Nicolet collaborated with Siemans AG to produce the REGA 700 online exhaust analyzer, which is designed to be used for engine design work, fuel developments, catalyst studies, and electronic systems research.

Spectra-Tech's Applied Systems Division (Annapolis, Maryland), developed to transfer FT-IR technology to industry, demonstrated its ReactIR series of FT-IR-based reaction monitoring systems. These provide continuous chemical composition data and quantitative information without disturbing reaction equilibrium. They will tolerate reaction temperatures from -80 to $200°C$ and pressures as high as 1000 psi. The Applied Systems Division also introduced two rugged, online, industrial-type IR systems. These Continuous Concentration Monitors (CCMs) were designed to continuously monitor one or two components. Supplied in NEMA containers, the systems are designed for factory use.

A scanning IR microprobe from SpectraTech ASD provides the ability to correlate molecular structure with physical morphology. The IRμs is designed for use in a wide range of analytical as well as production-monitoring and QA/QC environments.

Sprouse Scientific Systems, Inc. (Paoli, Pennsylvania) displayed its new IPC-100 series of FT-IR industrial analyzers. The 100S is an online system for analyzing multicomponent stack emissions from industrial plants. The 100A is used for ambient atmospheric monitoring, while the 100P is used for analyzing molten polymer additives.

Accessories

The LT-Transform system from ARIES (Concord, Massachusetts) is designed to interface high-performance liquid chromatography (HPLC) with FT-IR spectroscopy. The system comprises two parts—a collector and an FT-IR interface—and is compatible with most commercial HPLC and FT-IR systems.

Axiom Analytical, Inc. (Laguna Beach, California) displayed its SPR-500 series Sparging-IR wastewater analysis system. Designed to work with a standard FT-IR system, the unit performs trace analysis of organic contaminants in wastewater.

Balston, Inc. (Lexington, Massachusetts) introduced its Type 75 Series Air Purifiers for inexpensive purging of FT-IR systems. Said to be less expensive than nitrogen purges, the wall-mountable units offer 24-h protection from moisture.

Bio-Rad offered the Tracer gas chromatography (GC)-IR accessory, which reportedly achieves 100-ppb sensitivity. The UMA 150 FT-IR microscope accessory is designed for low-cost FT-IR spectrometers. Also available is an array of FT-IR sampling accessories, including autosamplers; integrating spheres; fiber optics; and variable-angle reflectance, diffuse reflectance, specular reflectance, attenuated total reflectance, grazing angle, and polarized transmission accessories.

Bruker Instruments (Billerica, Massachusetts) introduced an accessory that can scan 10 × 10-cm TLC plates directly. The plate is introduced, and scattered radiation from the plate's surface is collected with large mirrors and focused on a detector.

Galileo Electro-Optics (Sturbridge, Massachusetts) announced the introduction of optical-fiber materials for accessing the 2–11-μm region. The company's Heavy Metal Fluoride Glass Fibers were developed for use in the near-IR from 1 to 4.5 μm, and the Chalcogenide Glass Fibers were developed for use in the mid-IR from 3 to 11 μm. When coupled to FT-IR spectrometers and sensors, the fiber optics will allow compositional analysis and industrial process monitoring in the IR region. Gallileo also demonstrated the OCS-7 optical channel selector, which allows multiple sensing points in a process line using a single spectrometer, and its Moving Web Sensor for capturing IR spectra of plastic coatings, polymer films, and coated fabrics.

Hatrick Scientific (Ossining, New York) introduced a tool for analyzing optically thick, hard-surface solids such as fibers, paints, and large volumes of liquids or pastes that would be difficult to analyze through conventional accessories. The Ultra-Sample Analyzer (USSA) can be configured for internal or external reflection studies.

Hellma (Jamaica, New York) announced a new line of sampling cells for the IR region. QX cells feature windows made of Suprasil 300, which extends the transmission range in the IR; in addition, the cells reportedly show excellent response in the UV/vis and can also be used for fluorescence measurements.

Two new ATR accessories from Nicolet were designed for its 500, 700, and 800 FTIR spectrometer systems. The Vertical Angle ATR allows surface analysis using continuous incidence angles from 30° to 60°. The Research-Grade Vertical Variable Angle ATR allows 50% throughput and sampling with angles of incidence from 25° to 75°.

Specac (Fairfield, Connecticut) exhibited several new products at PittCon, including the Fiberprobe remote sampling systems for process control, reaction monitoring, and materials evaluation. The transmission, ATR, or diffuse specular reflectance systems can be configured for use from the mid-IR to the UV region. The company also introduced the Minidiff diffuse reflectance accessory, a high-throughput, zero-alignment accessory for teaching and QA/QC purposes; the T-40 autopress, a 40-ton XRF and IR sample preparation press; and the First Aid Kit, which incorporates accessories for the most common sampling techniques (ATR, diffuse, and specular reflectance).

Spectra-Tech (Stamford, Connecticut) introduced three new sampling accessories for IR spectroscopy. The first, the Grazing Angle Objective, provides incident radiation at grazing angles for the analysis of organic and inorganic ultrathin film coatings on metallic substrates when used with the company's popular IR-Plan microscopes. A second accessory, the ATR Objective, is also available for use with its microscopes. Finally, the company's 4X Beam Condenser is now combined with its Diamond Cell for the analysis of small, difficult samples (for example, minerals, hard polymers/plastics, crystals, rubber, and fibers).

Software

In the software arena, Bio-Rad announced the availability of EasyCHECK, a software program for QC designed to run on the FTS 7 spectrometers; the software will also run on any Bio-Rad FT-IR system that uses DDS software. The program qualitatively compares a sample spectrum against a user-generated reference spectrum.

Bruker introduced its icon-based OPUS optical spectroscopy software for data collection and processing. Available for its FT-IR and FT-Raman units that have integral PCs, the multitasking software includes numerous options, including quantitative analysis and spectral library searching.

Mattson also introduced two software packages for FT-IR spectroscopy. Its U-FIRST software can be operated from function keys, a mouse, or a keyboard and includes features such as single-wavelength Beer's law quantitative analysis and correlation charts to aid in peak identification. The FIRST Time Evolved Analysis (TEA) software for GC/FT-IR provides full control in IR spectral acquisition and evaluation of the resultant spectral data. Optional TEA automation packages can control the gas chromatograph and autosampler.

Nicolet's new workstation, the 680D, is equipped with a 3.5-in. 4-Mbyte floppy, a 177-Mbyte DMA SCSI hard drive, and 3 Mbytes of RAM. The company also updated its software: The PC/IR version 3.1 allows control and interpretation of PC/FT-IR and TGA/FT-IR experiments, and many other options have been added.

RAMAN

Bio-Rad introduced an FT-Raman instrument with an optional germanium detector. Its optics allow for 180° and 90° collection of spectra with all filters needed for fluorescence-free Stokes and anti-Stokes Raman data. The standard Rayleigh rejection filter allows Stokes data collection over the 90–3500-cm^{-1} range; an InGaAs detector is also included.

The OMA Vision CCD system from EG&G PARC allows the user to choose UV/vis detection, *thermoelectric* or cryogenic cooling, or some combination thereof. The detector reportedly features low noise, high gain, and high quantum efficiency.

The Jobin–Yvon Optical Systems of Instruments SA (Edison, New Jersey) displayed its new T64000 Ultimate Triple Raman Spectrograph. Based on a 0.64-m focal length monochromator, it operates a single, double, or triple spectrometer/spectrograph. Its aberration-correcting gratings reportedly show superior performance in resolution, stray-light rejection, and imaging capabilities.

Spex (Edison, New Jersey) introduced an enhanced version of the Spectrum One liquid nitrogen-cooled CCD system. Two versions are available: 1 1-L model with 24-h hold time and a 2.8-L model with a hold time greater than three days. A wide-mouth fill system and mounting flanges are compatible with the company's photodiode-array focusing adapter.

GENERAL EQUIPMENT AND COMPONENTS

Burle Industries Tube Products Division (Lancaster, Pennsylvania) announced improved performance for its line of side-window PMT components. Reportedly, the components feature better gain, dark current, hysteresis, and spectral response.

Edmund Scientific (Barrington, New Jersey) itnroduced several new products at PittCon: an autocollimator for measuring small angles and calibrating and aligning optical instruments; a laser-scattering demonstration tank designed for the laboratory and classroom; optical bench plates; and an assortment of lenses, optics, and optical components for various applications.

EG&G Reticon (Sunnyvale, California) announced the addition of the Value-D family of linescan image sensors to its line of solid-state image sensors. The devices operate at speeds as high as 10 MHz and feature the company's charge-coupled photodiode (CCPD) technology. In contrast to charge-coupled device (CCD) technology, these devices reportedly have excellent photo response in the blue and UV portions of the spectrum and extended response in the near-IR. The sensors are available with 256, 512, 1024, and 2048 photodiodes located on 13-μm centers with an aperture of 13 μm.

Epitaxx (Princeton, New Jersey) showed a series of detectors capable of working in the near-IR. The most interesting was the Gr series extended-wavelength InGaAs room-temperature photodetector covering the 800–2600-nm range. The detectors are available with active device diameters ranging from 75 μm to 1 mm at cutoff wavelengths of 1.9, 2.2, and 2.6 nm. The photodiode can be operated either with reverse or zero bias for speed or thermal noise-limited performance. Cooling is available for this and several other InGaAs and Si/InGaAs arrays the company offers.

In addition to numerous new electron tubes and related components, Hamamatsu (Bridgewater, New Jersey) showed its line of metal oxide semiconductor (MOS) linear image sensors. The MOS series includes 12 serial-current-output types and 12 serial-voltage output types ranging from 128 to 1024 elements, and pixel heights of 2.5 mm to as small as 0.5 mm. The line also includes random-access readout types in sizes from 512 to 1024 elements.

Oriel Corporation (Stratford, Connecticut) announced new software for its InstaSpec III diode-array system that makes it suitable to capture fast laser pulses. The instrument uses low dark current diode arrays, a 15-bit A/D converter, and a fiber-optic data link. Laser pulses are captured with the company's Autotrigger feature, and a simple programming language is provided for automated control and data manipulation.

In addition to its extensive line of hollow cathode tubes for atomic spectroscopy, Imaging and Sensing Technology Corporation (ISTC, Horseheads,

New York) announced two new deuterium lamp power supplies. The model 10232, designed to lengthen lamp life, can be customized to meet specific applications. The model 10262 allows operation of a lamp independent of an instrument.

Holographic diffraction gratings are now offered by Physical Optics Corporation (Torrance, California). The plane, concave, and echelle gratings range in size from 400 to 6000 lines/mm; custom sizes can also be manufactured. The company has also developed a line of holographic optical diffusers that incorporate a new method of light-beam shaping that uses volume holography to scatter collimated light into a controlled pattern with little variation in light intensity.

Spex's DataScan is a controller/photometer package introduced by the company that drives two monochromators, controls shutters, and provides data acquisition and storage. The package comprises an on-board microcontroller, memory, and a 12-bit A/D converter. It can be interfaced to a PC or to the company's hand-held controller through RS-232C or optional IEEE 488.

COMPUTER HARDWARE AND SOFTWARE

Chemical Concepts (Weinheim, Federal Republic of Germany) introduced an advanced software package that features extended capabilities for structure determination of molecules not previously known to the database. SpecInfo comprises NMR, IR, and MS spectra and integrated software for spectral interpretation, calculation, and structural determination. Users' data may also be added to the company's database.

Galactic Industries (Salem, New Hampshire) unveiled Quant Classic, a software application that supports both single and multicomponent analyses of spectroscopic data by performing conventional peak height and area calculations. The company also announced the addition of QTools to its library of QC software. The software needs minimal programming and includes support routines for over 20 instrument-collect drivers.

LabData (Pittsburgh, Pennsylvania) displayed LabData 100, a lower-cost version of the company's model 200. The model 100 is a single-board, two-channel data acquisition and analysis package designed for use with the company's chromatography, AA, UV/vis, thermal applications, and dispersive IR software packages. The company also offers a LIMS and a diode-array spectroscopy package.

Labtronics (Guelph, Ontario, Canada) unveiled a software package designed to automate and enhance older IR, UV, and photometric spectrometers. Direct*DataTalk can prompt the instrument for results, acquire results, and filter out unwanted data.

Molecular Design (San Leandro, California), formerly Biodesign, introduced a chemical drawing package designed to allow scientists to create a range of scientific presentations. ISIS/Draw provides users flexibility because it can run on various major computing platforms in a windows environment and comprises drawing tools with chemical intelligence.

Softshell International (Grand Junction, Colorado) introduced an updated drawing program, ChemIntosh II, which is designed to create publication-quality graphics for chemical reports and papers. The ChemIntosh II updates the ChemIntosh with more than 45 new features. The software program is compatible with other Macintosh drawing applications and word processors.

NUCLEAR MAGNETIC RESONANCE

American Microwave Technology (Brea, California) announced the availability of its series 3000 solid-state pulse power amplifiers for nuclear magnetic resonance (NMR) and magnetic resonance imaging (MRI) applications. Units in the series cover 6–200 MHz (300 and 1000 W) and 200–500 MHz (50, 150, and 300 W). A dual-mode integrating rf power protection system allows high-power pulses to be intermixed with reduced-power continuous-wave levels.

Bruker (Billerica, Massachusetts) announced a number of new NMR instrument accessories and upgrades in Chicago. The gradient-enhanced spectroscopy (GES) accessory is an upgrade to existing hardware for the company's AC, AM, or MAX spectrometers. Bruker's new multichannel interface (MCI) for the AMX series of spectrometers provides as many as three independent rf channels with equal capabilities, allowing for a range of homo- and heteronuclear triple-resonance experiments.

Bruker also added a 400-MHz NMR spectrometer to its AC series of instruments. The AC 400 includes comprehensive automation software for 1- and 2-D experiments and hardware for multinuclear applications, inverse experiments, and solvent-suppression techniques. A new probe head from Bruker makes possible the analysis of trace compounds in biochemical and electrochemical studies using liquid chromatography (LC)-NMR. Yet another enhancement to the AMX spectrometers is the introduction of a new software package that runs on industry-standard computers for processing 1- and 2-D NMR data. UXNMR/P, designed for a networked laboratory environment, is implemented on Sun SparcStations, SGI IRIS, and IBM RISC/6000 computers. Lastly, Bruker announced that WIN-NMR, a PC software package for processing NMR data, is now available on Macintosh computers.

GE NMR Instruments (Milwaukee, Wisconsin) has added spectral library capability to the QE Plus NMR spectroscopy system. Users can create their own ^{1}H and ^{13}C spectral libraries for compound identification by inputting compound name, CAS registry number, comments, solvent, temperature, shift reference, and chemical structure. Data for the proton library are stored as reduced data sets; ^{13}C data are stored as stick spectra.

GE also presented its Omega NMR system, which features refined pulse shaping and decoupler modulation for flexibility in advanced 2- and 3-D applications. The multitasking Omega system, based on Sun Microsystems' SPARC

Engine, now incorporates linear modulators on each transmitter channel and new, user-programmable RAM-based decoupler modulation. Computer-synthesized imagery (CSI) for a variety of biomedical, industrial, oil core, and agricultural applications is now available from GE NMR. The Omega CSI system delivers user-friendly NMR spectroscopy through easy-to-use menus, and an integrated UNIX operating system simplifies data acquisition and processing in a multiuser environment.

The Sadler Division of Bio-Rad Laboratories (Philadelphia, Pennsylvania) introduced CSearch, a tool for assigning chemical shifts and estimating NMR spectra from input structure proposals. The software program runs on VMS or UNIX in an X-Windows environment and is supported by a database of 50,000 carbon entries; it also provides a user database for proprietary materials.

Varian (Palo Alto, California) announced in Chicago the availability of a new NMR software package. Solids Analysis Software, available for use with the company's Unity NMR spectrometers, allows both the simulation and fitting of solid-state NMR spectra. The software is composed of several standalone programs that may be run from UNIX or within VNMR, Varian's NMR software.

Varian also introduced a family of 200- and 300-MHz probes, which allow users of the company's Unity and Gemini FT-NMR spectrometers to observe four nuclei (^{1}H, ^{19}F, ^{13}C, and ^{31}P) within a sample without retuning.

MASS SPECTROMETRY

The focus of growth in the mass spectrometry (MS) marketplace continues to be driven by two major forces: the ever-higher requirements for sensitivity and specificity in the biomedical lab in analyzing molecules that are large and complex; and the requirements for conducting routine analyses of target compounds, such as those found in a drug- or environmental-testing laboratory.

Kratos Analytical (Ramsey, New Jersey) debuted the concept 32 modular mass spectrometer system, which allows for varying configurations for different applications as well as the ability to modify the system on site with accessories and upgrades. Users can choose from two master operating consoles, and the spectrometer can be configured with as many as four analyzers and a variety of detectors. According to the company, the system is the first mass spectrometer to be controlled by a windows-driven data system.

Leybold Infocon (East Syracuse, New York) introduced the Auditor I and II process mass spectrometers, which include tutorial software programs for training and improving staff skills. Auditor I, a 650-amu quadrapole-based mass spectrometer, is designed for use in a process-plant environment. Auditor II, a lower-cost instrument, is designed for laboratory environments in which protection is not required. Both instruments offer interactive controls, mass-selectable software on a PC/AT-based data system, the ability to quickly change the ion source and electron multiplier, and turbo-molecular vacuum pumping.

Fisons/VG Instruments (Danvers, Massachusetts) claims that its new

AutoSpec Q system is the first hybrid, high-resolution (>60,000) magnetic-sector mass spectrometer capable of electrospray ionization (ESI), expanding its potential for biotechnology applications. The instrument combines MS-MS technology with multiple-inlet and ionization techniques.

Hyphenated Systems

Headlining Extrel's (Pittsburgh, Pennsylvania) exhibit at the conference was the company's Benchmark LC-MS detection system. The compact, benchtop mass detector includes the company's patented ThermaBeam particle-beam LC-MS interface; Extrel's Thermospray is available as an ionization technique. Benchmark also offers a set of GC options and allows for switching between GC and LC operation.

Finnigan MAT (San Jose, California), which has been acquired by Thenno Instrument Systems since last year's Pittsburgh Conference in New York, launched its INCOS XL GC-LC-MS-DS (difference spectroscopy) system. The benchtop quadrapole system allows a variety of inlet and ionization techniques. A workstation handles data processing, and system operation is integrated with automated software control of the gas chromatograph (the standard configuration includes a Varian 3400 gas chromatograph) and autosampler parameters. For greater productivity, Finnigan also offers the INCOS XLe, which comes with a second terminal, software packages, and a NIST library of spectra.

Fisons/VG Instruments unveiled the TRIO-1000 LC-GC-MS, a benchtop quadrapole mass spectrometer. The instrument combines research-grade LC-MS and ultrasensitive GC-MS for applications in agrochemicals, bioanalysis, and the environmental arena. LINC, the company's particle-beam interface, produces library-searchable electron ionization (EI) and chemical ionization (CI) spectra from LC-MS analysis. The gas chromatograph and the LINC interface can be connected to the ion source simultaneously for rapid switching between methods.

VG also introduced updated versions of the TRIO-1S GC-MS quadrupole system and the TRIO-2000 ESI multipurpose mass spectrometer. The company has increased the sensitivity of the TRIO-1S to picogram detection limits in scanning acquisition and has made available negative and positive CI as field-upgradable options for existing instruments. The TRIO-2000 now features an electrospray LC-MS interface.

Hewlett–Packard (Palo Alto, California) presented several new and upgraded instruments at PittCon. The second-generation HP 5971A quadrupole GC-MS system features new capabilities, including software and PC enhancements that increase analysis speed and improve reproducibility. The new A-series RTE GC-LC-MS from HP provides high-throughput, target-compound analysis in environmental and drug testing laboratories. The system includes new Revision F software with automatic QA/QC functions, the LAN-based HP 5971A mass-selective detector (MSS), an HP A900 Micro 29 minicomputer, computer peripherals, and HP networking. As many as three MSs, which can be located as far as 185 m from the computer, can be incorporated with and controlled by the system via LAN.

HP also announced that its research-grade quadrupole mass spectrometer, the MS Engine, is now available with an interface to the HP 5965B FT-IR detector for GC-FT-IRMS analysis, which can provide simultaneous MS and IR structural information.

Hitachi (Danbury, Connecticut) introduced an LC-MS system that features atmospheric-pressure chemical ionization (API) for high sensitivity and ESI, which allow the analysis of high-molecular-weight compounds using a mass analyzer of nominal range. The API interface of the model M1000 LC/CMS accomplishes ionization with a corona discharge, which eliminates the need for special traps (for example, liquid nitrogen), and is compatible with reversed and normal LC mobile phases.

A low-cost magnetic-sector mass spectrometer for dedicated LC-MS applications was presented by JOEL (Peabody, Massachusetts). The LC/AX spectrometer features a frit-fast atom bombardment (FAB) LC-MS interface combined with a pressure-regulated, variable-flow splitter, which is reportedly compatible with any type of HPLC system. The instrument has a 3-kV accelerating potential and a maximum resolving power of 2500.

Perkin–Elmer's (Norwalk, Connecticut) entry into the low-cost GC-MS market is represented by the Q-Mass 910 benchtop mass spectrometer. Optimized for use with PE's gas chromatographs, the instrument features innovative hardware and easy-to-learn software for routine and automated GC-MS analysis. The heart of the spectrometer is a quadrupole mass filter that yields library-searchable EI mass spectra. The company also claims the fastest pumpdown in the industry: An analyst can reportedly break vacuum, change any system component, pump down, and be back on line in 5 min.

Perkin–Elmer Sciex (Thronhill, Ontario, Canada) announced the newest member of its atmospheric pressure ionization (APt) LCMS family of products. The API I is a single-quadrupole mass-spectrometer version of the company's well-known API III LC-MS-MS system. The smaller and less expensive API I has the same mass range and accepts all of PE Sciex's liquid introduction systems (the Heated Nebulizer, the IonSpray, and the new Articulated IonSpray) but is designed for rapid, high-throughput, targeted analyses, such as QC functions.

After its acquisition by Baker Hughes last year, Tracor Instruments is now known as Tremetrics (Austin, Texas). At PittCon, the company introduced the model 850 quadrupole mass spectrometer, which interfaces with the 9000 series gas chromatograph and an IBM-compatible PC controller. The instrument has a mass range from 4 to 650 amu—appropriate for EPA analyses—and generates EI spectra that can be searched against the 49,000 NIST/EPA/MSDC library spectra stored in the controller.

In addition to the IR, NMR, and MS spectral libraries and software outlined previously, Chemical Concepts also introduced version 6 of MassLib, a software package for mass spectral interpretation, documentation, and archiving.

New features include an index option, which allows storage and retrieval of retention times and chromatographic indexes; neutral loss similarity search for structural elucidation; fragment coding and searching of structures; and multivariate statistics.

Delsi–Nermag Instruments (Houston, Texas) now offers a PC-based data station for the company's Resolver MS and MS-MS systems. The Lucy workstation is available as a 386 or a 486i with a 25-MHz processor, 416 Mbytes of RAM, and a 135–664-Mbyte hard disk. It is delivered with the 50,000-spectra NIST library or Wiley's database of 130,000 spectra in PBM format with chemical structure display.

Extrel introduced the IONstation, an updated data system for the company's series 400 mass spectrometers. The system uses Sun Microsystems–based technology that provides high-resolution graphics, internal memory up to 96 Mbytes, and processor speeds as fast as 28.5 MIPS. A Postscript laser printer is included.

Fisons/VG Instruments introduced Transform electrospray software for interpreting electrospray data. Complex series of peaks are transformed to a single peak on a molecular scale for each protein in the sample. A profile of proteins in the sample mixture is produced, giving the relative intensity and molecular mass of each component.

Hewlett–Packard introduced ChemLAN software and hardware packages that provide networking capability to HP's MS-DOS series ChemStations and 1000-based RTE Aseries mass spectrometers. ChemLAN provides data-file conversion from MS-DOS to HP 1000 RTE-A file formats, data analysis on remote computers, remote printing, file transfer, and logging of network activity.

CHAPTER IV.B.6

INTERNET SITES FOR INFRARED AND NEAR-INFRARED SPECTROMETRY—PART I: ONLINE INSTRUCTION AND DIRECT COMMUNICATION

Elizabeth G. Kraemer and Robert A. Lodder

The chapter authors are Rob Lodder and his graduate student Elizabeth G. Kraemer of the University of Kentucky's College of Pharmacy (Lexington, KY 40536-0082). Lodder and Kraemer have assembled a monumental list of Web sites and newgroups for the IR community, as well as some commentary on the future impact of electronic media on scientific education and publishing. Citing examples from their institution and others, the authors discuss important changes the Internet has already brought about and share some of their vision of its future influence. As if to prove one of their points about the limitations of "paper" journals, space in this issue did not permit the publication of the full list of Internet resources. This month, we'll focus on educational and communication-oriented sites specifically dealing with IR- and near-IR analysis—laboratories and departments, spectroscopists' personal home pages, courses and tutorials, discussion lists and newsgroups, and related sites. In the next chapter we will list locations of promotional and reference information—home pages of key journals, conferences, societies, instrument companies, and publishers; online abstracts and articles; and software, spectral databases, and images. In the meantime, if you wish to access the entire article (complete with hot links), check out Lodder and company's electronic journal Wave of the Future at http://kerouac.pharm.uky.edu/ASRG/wave/wavehp.html.

In the 1990s, the use of the Internet has expanded dramatically in all areas of science. The World Wide Web (WWW or "the Web") on the Internet is a network through which information can be accessed from millions of computers around the world. The growth of the WWW has been tremendous, increasing from 20,000 to an estimated 4 million Web sites in just 1–2 years. Countless articles have been published about this virtual explosion in network communication and its effects on different disciplines. This chapter provides a survey of Internet resources of use to infrared (IR) and near-IR spectroscopists. This chapter is available online for the convenience of readers at the *Wave of the Future* home page at *http://kerouac.pharm.uky.edu/ASRG/ wave/wavehp.html.* The list of Internet and WWW resources is also available at *http://kerouac.pharm.uky.edu/ASRG/cnirs/ir_ spec.htm.* These links are provided online to enable readers to explore the links with a minimum of typing.

TRENDS IN INTERNET CONTENT

Recently, the Internet and the WWW have seen a new trend—various student groups and schools have placed diverse class projects online. This is a trend that should continue for several reasons. When students put term projects on the WWW, they give something of academic value back to the community while they are still in school. Specifically, it allows students to help other students, as well as the public, learn new things in a new and convenient way. Moreover, students can often learn things from other students that they cannot seem to learn from their teachers. Both the public and other students discover these new WWW offerings using search engines that provide an online virtual libary. Furthermore, WWW term projects give students a product- and goal-oriented mindset, instantaneous feedback via e-mail from users of the project, and lessons in team building, quality control, and cooperation. For example, multimedia pages on HPLC are provided on the WWW as a project of University of Kentucky graduate students in "PHR 510 Modern Methods of Pharmaceutical Analysis" (*http://kerouac.pharm.uky.edu/ASRG/HPLC/hplcmytry.html*). Readers of the student project are invited to submit a grade online for the work of the students as well as provide any additional comments on content, presentation, graphics, additional links, and so forth. If readers choose to provide their name and e-mail address, students can respond to the comments and questions. The pharmaceutical analysis class fulfills the following learning objectives:

1. Students learn to perform in groups effectively, manage the quality of a project, and achieve their selected goals.

2. Students publish a peer-reviewed research paper on the WWW using modern analytical methods to solve a problem of interest to industry or academia.

Online review papers also become a ready reference for others on campus who use Netscape to learn which instrumentation is available for research use, and how it should be operated.

3. Students become familiar with state-of-the-art analytical instruments and techniques.

This Web-based course in analysis utilizes quality-management principles in a problem-based approach to learning. Corporations use these management techniques to make certain they meet all of the needs of their customers. Major companies (the future employers of the students) are now advocating quality management for the academic world to make certain that universities satisfy all of their customers (including the students and the future employers). Through the group WWW projects, the students learn how difficult it can be to motivate any person to be productive in a project that person may not be interested in. In the corporate environment, these persons are reassigned to other duties, resign to find more appropriate employment, or are terminated. In this way, the corporate entity maintains the unity of purpose necessary to satisfy the customer. Giving students a WWW term project as a product goal ultimately creates a similar unity of purpose. Quality management is geared to meeting customer requirements, which is necessary if universities are to survive and thrive in the modern environment. Quality management makes students better problem solvers and more able to work effectively in groups. WWW term projects are beneficial at the undergraduate level as well. Undergraduate students in "PHR 395 Introduction to Home Diagnostics Kits" created the Home Diagnostic Kits home page (*http://kerouac.pharm.uky.edu/HomeTest/KitsHP.html*) that provides new information annually about test kits available in retail pharmacies (blood glucose tests for diabetic patients, pregnancy tests, and so forth).

The University of Minnesota College of Biological Sciences operates a number of chemistry classes for their registered students on the WWW (*http://dragon.labmed.umn.edu/ ~lynda/eva/internet-class/*). Dr. Larry Wackett offers "Ecological Biochemistry and Advanced Biochemistry I: Protein Structure and Function." Dr. Lynda Ellis has placed "Computational Analysis of Biological Sequences" in the WWW, and Dr. Karin Musier-Forsyth provides "Chemistry of Nucleic Acids" online. Students in these classes learn to make and post their own WWW pages. For example, in one class students examine one incomplete biodegradation pathway and document what is needed to complete it. To do this, the students use the University of Minnesota Biocatalysis and Biodegradation Database (UM-BBD) that is available on-line.

At some institutions, term projects of multiple classes are brought together to create a scientific e-zine (3), such as the Notre Dame Science Quarterly on the WWW (*http://www.nd.edu:80/%7escienceq/*). This journal is published in print four times during the academic year (December, March, April, and May) by the undergraduate students of the College of Science at the University of

Notre Dame. The WWW site is a new expansion of the Quarterly, which is celebrating its 35th anniversary. Readers who stop by the Internet site are encouraged to comment to the e-zine home page and are encouraged to link Notre Dame to their own relevant sites. (Notre Dame asks that webmasters e-mail them when links are made so that they can return the courtesy.)

At the University of Kentucky, our previously mentioned e-zine *Wave of the Future* provides online peer review of manuscripts as part of a vision of a new scientific publication system, comprising a large number of computer servers linked logically by search engines, with each server storing years of quality articles, graphics, and data from a handful of researchers for easy retrieval. Many complaints are often heard about publishing in the scientific literature. Authors claim that it takes too long to get an article published (posting of articles can be instantaneous on the WWW), not enough page space is available (space on the WWW is limited only by the computing hardware, which grows almost exponentially), page charges are too high (publication is free on existing machines on the WWW), and anonymous peer review does not allow readers or authors to gauge the quality of the review process (on the WWW, all reviews can be signed and published along with the article just as practiced in certain fields, such as in statistics literature). With this online peer review format, *Wave of the Future* hopes to address these problems on an experimental basis.

Comments are invited via a forms page on all of the manuscripts provided in *Wave of the Future*. All comments are posted along with the reviewer's name and e-mail address in the page following the manuscript. Authors are encouraged to respond to comments in the same way, making all the information available in a package so that readers can judge the merits of all sides of the issues presented. At some point in the future, e-zines like *Wave of the Future* may be used as a feeder for paper journals. Editors may select papers from the e-zines for print dissemination based on peer review comments and number of accesses as an indication of reader interest, thus increasing the quality of the papers they spend money on publishing as hard copy. Papers could actually be perfected on the Web as a result of the online comments received. In the future (probably years from now), scientific WWW publication seems likely to increase. Journals may eventually cut the number of pages they print to zero, go to all Web publication (or whatever online equivalent replaces the Web), and move their advertisements to the WWW servers. Until then, filling the journals with papers that review well and attract a lot of reader interest can only increase their readership and advertising revenue.

HOW THE LIST OF SITES WAS ASSEMBLED

The list of IR and near-IR sites was assembled by online searches during the spring of 1996. The search engines used were Webcrawler (*http://webcrawler. com/*), Metacrawler (*http://metacrawler.cs.washington.edu/*), and Alta Vista (*http://www.altavista.digital.com/*). The authors apologize in advance for omitting the many IR and near-IR sites not indexed by their search engines. If your site was omitted and you would like to have it added to the online version, please notify the authors by e-mail at *Kraemer@pop.uky.edu* or *Lodder@pop.uky.edu*.

REFERENCES

1. T. C. O'Haver, *Spectroscopy* **11**(1), 12–13 (1996).
2. K. L. Busch, *Spectroscopy* **11**(6), 32–34 (1996).
3. E-Zine (noun): Short form of Electronic Magazine. A publication published partly or completely through electronic means, (such as a BBS, the Internet, or Usenet) that can range from a simple, one-page text document to an elaborate presentation with graphics and sound. (From *http://www.dakota.net/pwinn/dict/ezine.shtml.*)

LABORATORIES AND DEPARTMENTS

The Infrared Space Observatory
   http://isowww.estec.esa.nl/
University of Rochester's Near Infrared Astronomy Group
   http://sherman.pas.rochester.edu/
The InfraRed Army, Downs Laboratory of Physics, Caltech
   http://www.cco.caltech.edu/ ~ira.html
Vibrational Spectroscopy Laboratory
   http://optic2.chem.su.oz.au/
Vibrational Spectroscopy
   http://www.ruhr-uni-biochum.de/www-public/goetzhbv/ftir.htm
JPL Physics & Astrophysics
   http://www.jpl.nasa.gov/mip/physics.html
Institute for Instrumental Analysis (IFIA)
   http://www-ifia.fzk.de/
Open University Astrophysics Group
   http://assp01.open.ac.uk/ ~ajnorton/astro/activities.html
James A. de Haseth Research Group
   http://dehsrv.chem.uga.edu/
Laboratory for Spectrochemistry at Indiana University
   http://rustico.chem.indiana.edu/
Optical and Infrared Astronomy Division (Harvard University)
   http://cfa-www.harvard.edu/cfa/oir/oir.html/

Analytical Spectroscopy home page
  http://kerouac.pharm.uky.edu/ASRG/Grpcont.html
Computational Chemistry Center Erlangen
  http://schiele.organik.uni-erlangen.de/ak/tgup.html
Center for Process Analytical Chemistry
  http://www.cpac.washington.edu/index.cgi
Standard Test Configuration and Data Products
  http://odo.elan.af.mil/pages/facilities/rsom/irsys/irsttcdp.html
Infrared Analysis Measurements and Modeling Program
  http://infrared.nswc.navy.mil/projects/irammp/irammp.html
Australian National University Research School of Chemistry Laser and Optical
  Spectroscopy
  http://rsc.anu.edu.au/%7Ekrausz/laser.html
The Infrared Science and Technology Group
  http://www.ecs.soton.ac.uk/labs/irst.htm
Instrumentation and Sensing Laboratory at the USDA
  http://www.ars-grin.gov:80/ars/Beltsville/barc/nri/isl/index.htm
McGill IR Group
  http://www.agrenv.mcgill.ca/foodsci/irg/irhomepg.htm
The Wilson Group
  http://www-wilson.ucsd.edu/
  QuickTime movies show basics of spectroscopy at
  http://www-wilson.ucsd.edu/education/spectroscopy/spectroscopy.html
Spectroscopy at the Institute for Biodiagnostics
  http://www.ibd.nrc.ca/ ~mansfield/spectroscopy.html
The Center for Quantized Electronics Structures (QUEST)
  http://www.quest.ucsb.edu/
University of Duisburg, Physical Chemistry home page
  http://www.theochem.uni-duisburg.de/PC/
Photoacoustics at the ETD Program of the U.S. Department of Energy's Ames
  Laboratory
  http://www.etd.ameslab.gov:80/etd/technologies/projects/pas/index.html
Infrared Astronomy at the University of Calgary
  http://iras2.iras.ucalgary.ca/
Department of Chemical Engineering: Gas Transport and Polymer Spectroscopy
  Lab (at Virginia Tech's College of Engineering)
  http://www.eng.vt.edu:80/eng/che/gastlab.html
Faculty of Pharmaceutical Sciences—IR information
  http://www.interchg.ubc.ca/tilcock/mainmenu.html
The Membrane Technology Group
  http://utct1029.ct.utwente.nl/documents/membrane.html
  The Membrane Technology Group also has a page on characterization of
  mass transport through polymer membranes by IR spectroscopy at
  http://utct1029.ct.utwente.nl/documents/infrared.html.

Snap Shots of Research in the School of Biochemistry
  http://iptunix.bcm.bham.ac.uk/staff/research.html
ASU Infrared Subnode home page
  http://esther.la.asu.edu/asu_ test/TES_ Editor/VIRTUAL_ FACILITY/orig_ homepg.html
ASU Mars Thermal Emission Spectrometer home page
  http://esther.la.asu.edu/asu_ tes/
Infrared and Raman Analysis
  http://www.chem.run.nl/amsir/www/amsir.html

PERSONAL HOME PAGES OF IR SPECTROSCOPISTS

James B. Callis, University of Washington
  http://www.chem.washington.edu/fac-callis.html
James K. Drennen III, Duquesne University
  http://www.duq.edu/pharmacy/FACULTY/DRENNEN.HTML
Bruce R. Kowalski, University of Washington
  http://www.chem.washington,edu/fac-kowalski.html
Thomas P. Greene, University of Hawaii Institute for Astronomy
  http://www.ifa.hawaii.edu/%7Egreene/
Peter R. Griffiths, University of Idaho
  http://esther.la.asu.edu/sas/officers/pgriffiths.html
  http://www.chem.uidaho.edu/faculty/griff/index.html
Gary Hieftje, Indiana University
  http://rustico.chem.indiana.edu/hieftje/group/cv.html
Robert A. Lodder, University of Kentucky
  http://kerouac.pharm.uky.edu/ASRG/People/lodder/Rob-Vita.html
Jim Mansfield, Institute for Biodiagnostics, National Research Council of Canada
  http://www.ibd.nrc.ca/ ~mansfield/
W.F. McClure, Weaver Laboratories
  http://www.bae.ncsu.edu/bae/people/faculty/mcclure/index.html
David Moss, Institute of Instrumental Analysis
  http://www-ifia.fzk.de/Moss/David.htm
James M. Palmer, University of Arizona
  http://www.opt-sci.Arizona.EDU:80/summaries/James_ Palmer/ jmphompg.html
Mark Winter. The University of Sheffield
  http://www.shef.ac.uk/%7Echem/staff/mjw/mark-winter.html

COURSES AND TUTORIALS

Courses

International Course on Infrared Spectroscopy (Utrecht University)
   http://www.chem.ruu.nl/amsir/www/course.html
SAS Short Course home page
   http://dehsrv.chem.uga.edu/sas/courses/
SAS FT-IR Spectrometry Short Course
   http://dehsrv.chem.uga.edu/sas/courses/ftir.html
SAS Biological Infrared Spectroscopy Short Course
   http://dehsrv.chem.uga.edu/sas/courses/birs.html
Interpretation of Infrared Spectra (Utrecht University)
   http://www.chem.ruu.nl/amsir/www/interpretation.html
Organic Chemistry 351, Infrared Spectroscopy (University of Wisconsin)
   http://www.ems.uwplatt.edu/sci/chem/fac/sundin/351/351h-ir.htm

Tutorials

Analytical Spectroscopy (Virginia Polytechnic Institute and State University)
   http://www.chem.vt.edu/chem-ed/analytical/ac-spectroscopy.html
Quick Procedures for Infrared Analysis (California State University, Stanislaus)
   http://wwwchem.csustan.edu/quickir.htm
Infrared Light (Produced by Newton's Apple, a production of TICA Twin Cities
   Public Television. Educational materials developed with the National Science
   Teachers Association.)
   http://www.mnonline.org/ktca/newtons/11/infrared.html
Infrared Analysis (Petroleum Technologies)
   http://www.oil-lab.com/Infrared.html
Spectroscopy Problems Using Chemical Mime Types (Dr. R.J. Lancashire,
   Mona Campus of the University of the West Indies, Jamaica)
   http://wwwchem.uwimona.edu.jm:1104/spectra/index.html
List of Surface Science Techniques (University of Surrey)
   http://www.surrey.ac.uk/MSE/ESCA/ESCA/tech/list.html
Spectroscopy (Wilson Group, University of California, San Diego)
   http://www-wilson,ucsd.edu/education/spectroscopy/spectroscopy.html
ASD: Near-Infrared Analysis (NIRA) Tutorial (Analytical Spectral Devices)
   http://www.csn.net/asd/apps/nira_tut.html
Infrared Spectroscopy (Brian Tissue, Virginia Polytechnic Institute and State
   University)
   http://poe.acc.virginia.edu/ ~roa2s/spectroscopy-ir.html
Fourier Transform Infrared Spectroscopy (University of Wisconsin)
   http://www.chem.wisc.edu/ ~hamers/FTIR.html
Basic Infrared Spectroscopy (University of Illinois, Chicago)
   http://gopher.chem.uic.edu/organic/tutorial/IR.html

Fourier Transform Infrared (Michigan Molecular Institute [MMI])
  http://www.miep.org/mmi/standard/anser/ftir.html
Fourier Transform Infrared Spectroscopy (Surface Science Laboratories)
  http://www.surface-science.com/ftir.html
NIST Guide—Far-Infrared Spectroscopy
  http://www.nist.gov/item/NIST_ Far-Infrared_ Spectroscopy.html

DISCUSSION LISTS AND NEWSGROUPS

HIRIS-L: High Resolution Infrared Spectroscopy
  http://www.tile.net/tile/listserv/hirisl.html
sci.techniques.spectroscopy
  news://sci.techniques.spectroscopy

ONLINE LISTS OF RELATED SITES

FAQ and miscellanea from sci.techniques.spectroscopy
  http://lolita.colorado.edu/faq.html
Umeå University's Analytical Chemistry Springboard
  http://www.anachem.umu.se/jumpstation.htm
Spectroscopy, Radiometry, Photonics, Quality and Validation Resource Center
  (SciOptics Corp.)
  http://www.gate.net/ ~quanta/
Spectroscopy Links on the WWW
  http://www.ibd.nrc.ca/ ~mansfield/spec_links.html
Chemistry Resources on the Internet
  http://alchemy.yonsei.ac.kr/sites/chem-resources.html
Yahoo—Science/Chemistry/Spectroscopy
  http://www.yahoo.com/Science/Chemistry/Spectroscopy/

CHAPTER IV.B.7

INTERNET SITES FOR INFRARED AND NEAR-INFRARED SPECTROMETRY—PART II: PROMOTIONAL SITES AND ONLINE PUBLICATIONS

Elizabeth G. Kraemer and Robert A. Lodder

As we began to explore in the preceding chapter, the infrared spectroscopist with Internet access can select from a vast menu of free online information. Rob Lodder and Elizabeth Kraemer (University of Kentucky School of Pharmacy, Lexington, KY 40536-0082) have compiled a detailed listing of Internet and World Wide Web (WWW) resources specifically dealing with the world of infrared (and near-IR) analysis. Here the authors wrap up this compendium with a look at sites we can consider "promotional," in that they provide information on goods or services that generally cost money—products from commercial vendors, conferences, society memberships, or journal subscriptions. The authors also have assembled a useful list of online articles and abstracts for the IR community.

The authors have included all commercial vendors whose web sites were accessible through the following search engines: Webscrawler (http://webcrawler.com/), Metacrawler (http://metacrawler.cs.washington.edu/), and Alta Vista (http://www. altavista.digital.com/). Because this obviously did not generate a thorough, up-to-date list of all commerical IR sites known to the editors, where possible they have taken the liberty of adding other vendors' WWW addresses.

The full list of IR Internet sites is available on-line through Lodder's on-line "e-zine," Wave of the Future (http://kerouac.pharm.uky.edu/ASRG/wave/wavehp.html).

INTERNET SITES DEALING WITH IR AND NEAR-IR SPECTROSCOPY AND RELATED TOPICS

COMMERCIAL INSTRUMENT MANUFACTURERS AND OTHER COMPANIES

Infrared Incorporated
  http://www.infrared.com/
Cincinnati Electronics
  http://www.tiac.net/users/nets/cincinn.htm
Optical Hybrids
  http://www.3doptics.com/
Fiveash Data Management
  http://www.intaccess.com/fdm/
Remspec
  http://www.remspec.com/
Used Lab Instruments
  http://www.opennet.de/labexchange/english/
Servo Corporation of America
  http://www.servo.com/index.htm
Sterling Analytical Laboratory
  http://www.scseng.com/lab.html
Andruss-Peskin Corporation
  http://www.andruss-peskin.com/search.html
Galactic Industries
  http://www.galactic.com
Bruker (USA or Germany)
  USA: *http://www.bruker.com/*
  Germany: *http://www.bruker.de/*
SE-IR
  http://www.seir.com/
Surface Science Laboratories
  http://www.surface-science.com/
Perkin–Elmer
  http://firewall.perkin-elmer.com:80/
  http://www.perkin-elmer.com:80/index.htm
Bio-Rad Laboratories, Sadtler Division
  http://www.sadtler.com/
CIC Photonics
  http://www.cicp.com/
CVI Laser
  http://www.nmia.com/~cbs/cvi.html
Axiom Analytical
  http://www.goaxiom.com/axiom/

Nicolet Instrument Corporation
  http://www.nicolet.com/
Analytical Spectral Devices
  http://www.csn.net/asd/index.html
Wilmad Glass
  http://www.wilmad.com/index.html
Acton Research
  http://www.acton-research.com
Air Instruments & Measurements
  http://www.aimanalysis.com
Buck Scientific
  http://ourworld.compuserve.com/homepages/Buck_ Scientific
Galileo Electro-Optics
  http://www.galileocorp.com
LECO
  http://www.lecousa.com
Oriel
  http://www.oriel.com

JOURNALS, SOCIETIES, AND MEETINGS

Journals

Spectroscopy
  http://www.techexpo.com/toc/spectros.html
CRC Press
  http://www.crcpress.com/
Wave of the Future
  http://kerouac.pharm.uky.edu/ASRG/WAVE/wavehp.html
Food Testing and Analysis
  http://www.worldsys.com/labinfo/journal/fla/fla.htm
Applied Spectroscopy
  http://esther.la.asu.edu/sas/journal.html
HyperSpectrum
  http://www.techexpo.com/WWW/opto-knowledge/hyperspectrum/
Journal of Near-Infrared Spectroscopy
  http://www.impub.co.uk/nirp/jnirs.html

Societies

FACSS
  http://FACSS.org/info.html
Council for Near-Infrared Spectroscopy
  http://kerouac.pharm.uky.edu/asrg/cnirs/cnirs.html

American Chemical Society
  http://www.acs.org/
The International Society for Optical Engineering
  http://www.spie.org/
Society for Applied Spectroscopy
  http://esther.la.asu.edu/sas/
American Association of Cereal Chemists
  http://www.scisoc.org/aacc/info.html

Meetings

Second International Symposium on Advanced Infrared Spectroscopy: Development and Applications of Dynamic and Muli-Dimensional Infrared Spectroscopy (AIRS II)
  http://www.chem.duke.edu/special/airs/
International Conferences on Fourier Transform Spectroscopy
  http://dehsrv.chem.uga.edu:80/icofts/
International Symposium on Molecular Spectroscopy
  http://molspect.mps.ohio-state.edu/symposium/
Eastern Analytical Symposium (EAS)
  http://www.eas.org/%7Eeasweb/
The Pittsburgh Conference (PittCon)
  http://www.pittcon.org/
7th International Conference on Time-Resolved Vibrational Spectroscopy
  http://rchs.1.chemie.uni-regensburg.de/gopher/
  *pub/.maillist/MOL-DYN/TRVS-7_*

ABSTRACTS AND ARTICLES

Abstracts

Far-Infrared Magneto-Spectroscopy of Deep Mesa Etched Quantum Wires
  http://hq.aps.org/BAPSMAR95/abs/SB1703.html
NSFCAM—A New Infrared Array Camera for the NASA Infrared Telescope Facility
  http://blackhole.aas.org/meetings/aas183/abs/S11701.html
Infrared Analysis of Wastes: Novel Laboratory and On-Line Measurements by Photoacoustic and Transient Infrared Spectroscopies
  http://cmst.ameslab.gov/Cmst-Cp_reports/Jun95/CH131001_mixed_waste.html
Examination of Some Misconceptions About Near-Infrared Analysis
  http://esther.la.asu.edu/sas/journal/ASv49n1/ASv49n1_ sp1.html
Determination of the Equilibrium Constant and Resolution of the HOD Spectrum by Alternating Least-Squares and Infrared Analysis
  http://esther.la.asu.edu/sas/journal/ASv49n10/ASv49n10_ sp5.html

Nanosecond Time-Resolved Infrared Spectroscopy with a Dispersive Scanning Spectrometer
http://esther.la.asu.edu/sas/journal/ASv48n6/ASv48n6_ sp7.html
External Infrared Reflection Absorption Spectrometry of Monolayer Films at the Air–Water Interface
http://www.annurev.org/series/physchem/vol46/pc46ab16.htm
Analysis of Carbon-Black-Filled Rubber Materials by External Reflection FT-IR Spectrometry
http://esther.la.asu.edu/sas/journal/ASv48n7/ASv48n7_ sp14.html
Field Validation Testing of Fourier Transform–Infrared Spectrometry Method for Measurement of Formaldehyde, Phenol, and Methanol
http://airwm.awma.org/ta3201.html
Matrix-Isolation Infrared Spectroscopy of Hydrogen-Bonded Complexes of Triethyl Phosphate with H_2O, D_2O, and Methanol
http://esther.la.asu.edu/sas/journal/ASv48n7/ASv48n7_ sp4.html
Fourier Transform–Infrared Spectroscopy Studies of the Secondary Structure in mt-PA6 and mt-PA6 Treated with L-Arginine, L-Asparagine, and (+)-citrulline in Aqueous Solution
http://esther.la.asu.edu/sas/journal/ASv48n9/Asv48n9_ sp13.html
Infrared Analysis of Refrigerant Mixtures
http://www.ari.org/rt/p54000.html
Monitoring Hydrofluoric Acid Activity by Vapor Phase Infrared Spectroscopy
http://esther.la.asu.edu/sas/journal/ASv48n7/ASv48n7_ sp5.html
Screening Pap Smears with Near-Infrared Spectroscopy
http://esther.la.asu.edu/sas/journal/ASv49n4/ASv49n4_ sp5.html
Dynamic Two-Dimensional Infrared Spectroscopy. Part I: Melt-Crystallized Nylon 11
http://esther.la.asu.edu/sas/journal/ASv49n8/ASv49n8_ sp2.html
Near- versus Mid-Infrared Spectroscopy: Relationships Between Spectral Changes Induced by Water and Relative Information Content of the Two Spectral Regions in Regards to High-Moisture Samples
http://esther.la.asu.edu/sas/journal/ASv49n3/ASv49n3_ sp5.html
Double-Chambered Flow Cell for In Situ Infrared Spectroscopy Studies of Chemical Reactions in Ziegler–Natta Catalyst Systems
http://esther.la.asu.edu/sas/journal/ASv48n8/ASv48n8_ sp12.html
Near-Infrared Spectroscopy of Luminous Infrared Galaxies
http://blackhole.aas.org/meetings/aas183/abs/S7708.html
Infrared Spectroscopy of Formate on Oxygen-Predosed Cu(100): Broadband Reflectance and Low-Frequency Vibrations
http://www.tufts.edu/ ~rtobin/abs.jvst95.html
Near-Infrared Spectroscopy of Seyfert Galaxies
http://blackhole.aas.org/meetings/aas185/as/S2702.html
Near-Infrared Spectroscopy of Seyfert 1 Nucleus NGC 4151
http://www.aas.org/meetings/aas185/abs/S10802.html

Articles and Reports

Precise Chemical Analyses of Planetary Surfaces
  http://cass.jsc.nasa.gov/hottopics/psiw/psiw.3.html
Annual Report to the American Astronomical Society
  http://charon.nmsu.edu/baas95.html
WIRE—Wide-Field Infrared Explorer
  http://sunland.gsfc.nasa.gov/smex/wire/wire_top.html
Infrared Imaging and Photometry
  http://decoy.phast.umass.edu/hillenbrand/imaging/node4.html
Cassini Saturn Orbiter
  http://rbarney.gsfc.nasa.gov/html/configuration.
  management/QMW.ICD.html
Near-Infrared Imaging and Spectroscopy in Stroke Research: Lipoprotein Distribution and Disease
  http://kerouac.pharm.uky.edu/ASRG/Wave/Lipo/lipo.htm
Driven to Depth: Biological and Medical Applications of Near-Infrared Spectrometry
  http://kerouac.pharm.uky.edu/ASRG/Papers/Depth/depth.html
Near-Infrared Optical Measurements to Probe the Effects of Intense Far-Infrared Radiation on Quantum-Confined Carriers
  http://lemonzest.qi.ucsb.edu/cerne/johnexp.htm
Depth-Resolved Near-Infrared Spectroscopy
  http://kerouac.pharm.uky.edu/ASRG/WAVE/jim/jim.html
Use of Near-Infrared Spectroscopy in Cereal Products
  http://www.worldsys.com/labinfo/journal/fta/v1n2/infrared.htm
Redox-Linked Conformational Changes in Proteins Detected by a Combination of IR Spectroscopy and Protein Electrochemistry. Evaluation of the Technique with Cytochrome *c*.
  http://www-ifia.fzk.de/Moss/Eur-J-Biochem.htm
Determination of Cholesterol Using a Novel Magnetohydrodynamic Acoustic-Resonance Near-IR (MARNIR) Spectrometer
  http://kerouac.pharm.uky.edu/ASRG/MAReNIR/MARtle.html
The Stellar Populations of Deeply Embedded Young Clusters: Near-Infrared Spectral Classification
  http://donald.phast.umass.edu/latex/meyerH/meyer.heidel.html
Improved Identification of Pharmaceutical Tablets by Near-IR and Near-IR/Acoustic Resonance Spectrometry with Bootstrap Principal Components
  http://kerouac.pharm.uky.edu/ASRG/Papers/nirars/nirars.html
Sloan: Long-Slit Infrared Spectroscopy of Mars
  http://www-space.arc.nasa.gov:80/ ~sloan/mars.html
Chemometrics Fundamental Review 1994
  http://gopher.udel.edu/chemo/chemo94.htm
The Use of Shift Reagents in Near-Infrared Spectrometry
  http://rustico.chem.indiana.edu/hieftie/SAMPRK.html

Passive-Remote Fourier Transform Infrared Spectroscopy for Detecting Organic Emission
http://www.anl.gov/LabDB/Current/Ext/H598-text.001.html

ONLINE SERVICES, SOFTWARE, AND BOOK PUBLISHERS

Analytical Infrared Spectroscopy (Elsevier)
http://www.elsevier.nl:80/catalogue/SAA/205/06020/06024/501094/501094.html
Spectacle
http://www.labcontrol.com/spectac.htm
Science Hypermedia
http://www.scimedia.com/index.htm

SPECTRA, SPECTRAL DATABASES, AND IMAGES

AEDC/EPA Spectral Database
http://info.arnold.af.mil/epa/welcome.htm
Ice Analogs Database
http://www.strw.leidenuniv.nl/ ~shutte/database.html
HMT in Cometary Ices
http://www-space.arc.nasa.gov/ ~max/HMTvsRes.html
ASU Infrared Subnode
http://esther.la.asu.edu/asu_ test/TES_ Editor/INFRARED_ SUBNODE/infrared_ subnodeMENU.html
Satellite Data (Images and MPEGs from Purdue University)
http://thunder.atms.purdue.edu/gopher-data/satellite/
Satellite Images from Ohio State University
http://asp1.sbs.ohio-state.edu/satimages.html

INDEX